In-Situ Characterization of Soils

In-Situ Characterization of Soils

Editors
K.R. Saxena
V.M. Sharma

A.A. BALKEMA PUBLISHERS LISSE / ABINGDON / EXTON (PA) / TOKYO

Published by : A.A. Balkema, a member of Swets & Zeitlinger Publishers
www.balkema.nl and www.szp.swets.nl

ISBN 90 5809 244 5

Printed in India

Preface

Professor W. Skempton, while dealing with the history of Geotechnical Engineering at the San Francisco Conference of the International Society of Soil Mechanics and Foundation Engineering, described 1910-1927, as the first phase of modern soil mechanics. In the subsequent phase, according to R.B. Peck, number of far reaching effective changes took place, but the 'observation method', a very powerful tool, became discredited by misuse. According to him it was 'invoked by name but not by deed'. Adopting a course of action and 'observing the consequences' did not mean 'observational method' as applied in soil mechanics. In this procedure, very often overlooked was the element of subsurface investigation. The bore logs are the net results of certain *facts,* obtained at sampling points and interpretation of these required relevant combination of *other facts.* Judgement based on consideration of facts is much more complex as it has to see beyond the facts on display and dredge the relevant facts from elsewhere termed as *experience.* Terzaghi emphasized that "Laboratory investigations should not receive too much attention unless the validity of the conclusions has been demonstrated by adequate field observations on full sized structure".

For a long time soil investigations, primarily depended upon collection of soil samples and for this various types of samplers were developed. This brought into focus the length, diameter and thickness of the samplers and terms like *area ratio, recovery ratio* etc. were coined to define the satisfactory undisturbed soil samples. It was found impractical to recover ideal undisturbed soil samples due to various limitations like type of soils, thickness of samplers and transit problems of the samples to the laboratory. In-situ tests thus were found imperative to assess the *nearly* true characteristics of soils. This was a great step forward in developing economical and safe foundations of structures and provided methods for their predictable performance. Significant developments have taken place in the methods of in-situ testing of soils, which in turn have helped in development and verification of theories for design purposes.

It is erroneous to consider that utilizing sophisticated calculations in designing a structure can compensate or replace the laborious subsurface investigation, in-situ testing, complex loadings, geometrics and soil conditions. Fieldwork, involving open pits, deep bores, besides permitted in-situ testing to evaluate subsurface features, give confidence about safety and desired performance of structures. In this background the publication has been attempted by the editors to collect the present practices used for in-situ characterization of soils in various countries by different geotechnical engineers and concerned organizations. The gamut of in-situ characterization of soils is mostly related to geology of sites and investigators' experiences and the choice of methods adapted for investigation as warranted by site conditions. Hence the contributions made to this publication, are technically interesting due to its coverage of various geological formations and it is expected that these would lead to newer techniques in methods of in-situ characterization of soils. The publication does not cover the entire field of in-situ characterization methods of soils, but is thought provoking to keep on the dynamic aspect of the subject.

The papers cover a wide range of topics, both the respect to practices and experience of individuals with the similar tests and the regional deposits. It is rightly emphasized in number of papers that there cannot by a general procedure of soil investigation. Soil conditions, the problem or purpose of testing control the procedures and primarily, the availability of appropriate equipment with the investigator. Above all this, site geology plays a major role, which sometimes warrants fabrication of special type of equipment for in-situ testing. The paper from Swedish Geotechnical Institute has dealt the subject exhaustively, suggesting that a 'reliable model of soil conditions' can be obtained by the combination of field and laboratory tests with the knowledge of local geology. Dr. Sharma has described the various methods of in-situ characterization of soils, the advantages and shortcomings of these and the factors influencing the results. Comparing same parameters obtained by different tests, quoting the example of plate load and pressuremeter test results, he advocates choice of tests to yield parameters to be compatible with design techniques to be employed. Lehane further clarifies that the evaluation of insitu tests, within a framework so as to assist geotechnical design, does not provide a direct measure of the parameter required. Through examples he demonstrates that the interpretation, of both in-situ and laboratory tests, require a rational soil mechanics framework for the extrapolation to full scale performance of structures.

Ting has been more specific in recommending soil models for the various deposits to evaluate soil parameters, corresponding to the geotechnical structure load and support. It is also considered that a theoretical framework is unavoidable in the analysis and in-situ characterization of soils for their use in design and construction of geotechnical structures. The significant conclusion in the same paper is that weathered formations require an empirical approach for the load supported by natural stratum for Mohr-Coulomb parameters. Swamy and Saxena, through case studies, subscribe to the same conclusion.

Sanglerat in his contribution has shown that AMAP'sols CPT rig is useful and economical in static-dynamic soil investigation, when compared to other traditional methods. Justo and others have shown the use *of viscoelastic rheological* models to measure the creep of granular materials. But the fundamental parameters, as mentioned by them, elasticity modulus, creep ratio and creep time, though can be estimated at preliminary stage from the laboratory tests, require correction by in-situ testing for settlement measurement during and after construction.

Vertical drains are normally applicable for stabilizing a deep layer of soft soil, like organic and inorganic silts and clays, having low permeability. This topic very efficiently dealt by Indraratna and others describe the factors influencing the technique and the significance of draining properties and their measurement. Installation and monitoring of vertical drains is illustrated by typical example and the requirement of monitoring equipment, before and after construction. It is also emphasized that to predict the behaviour of structure, an appropriate constitute model is necessary. Prakash and Puri describe the significance of measurement of Dynamic properties of soils and the factor affecting these. Field techniques though very advantageous, they measure the dynamic properties at small strain only. There is need of methods to measure 'G' at large strain, particularly in strata, like sand and gravel, where sampling and laboratory testing is difficult.

There has been significant development in in-situ testing of soils, as is seen from the papers presented in the publication, as it is realized that the samples collected do not give the true sub-surface conditions. It is also noticed that during the last three or four decades, certain age-old techniques, like SPT or Plate load tests, have great appreciation, perhaps due to large number of case histories are supporting these techniques. Hence there is the necessity of collecting large number of case histories, particularly where new techniques of sub-soil investigations are used. This would require the systematic, accurate forecasting techniques developed on the basis of actual performance records. It is also emphasized that besides the engineers, the technicians operating equipment and recording data, require personnel especially trained for the job.

It is expected that the publication will prove a useful reference to the geotechnical engineers involved in in-situ characterization of soils for sub-soil investigation, observational methods and design of sub-structures.

27th July 2002
New Delhi

K.R. Saxena
V.M. Sharma

Contents

The Contributors

A.S. Ramanuja Swamy took his B.Tech in Civil Engineering from J.N.T.U. Hyderabad in 1975 and obtained M.Tech in Geotechnical Engineering with distinction from Regional Engineering College, Warangal. He started his career at the A.P. Engineering Research Laboratories, Hyderabad and had the unique opportunity of working in Hydraulics, Engineering Material and Geotechnical Laboratories. Working in the Geotechnical Laboratory since 1974 and presently heading one of its sections as Research Officer, he has made significant contributions to the site investigations of earth dams, instrumentation and post construction problems of embankment dams. He shared the C.B.I. & P. National 'Cash Your Ideas' award in 1983 for the development of Electronic Probe for measurement of settlement in earth dams and has won many other awards and certificates of merit for best paper contributions to various national conferences. He is the life member of Indian Geotechnical Society (IGS), Indian Society of Construction Materials and Soils and the Clay Mineral Society of India. He was the member, representing IGS on the International Committee of 'Partially Saturated Soils' TC-6 of ISSMFE during 1991-1996 and also organized the national seminar on the subject. He has made valuable contributions to the State Disciplinary and Vigilance proceedings on investigations involving earth dams. He has authored/co-authored more than 25 papers contributed to National and International conferences and Symposia.

Barry M. Lehane has been involved in design and research in geotechnical engineering since 1984. He has worked for six years with Ove Arup & Partners (UK & Hong Kong) as a Geotechnical design consultant, four years as a research assistant at Imperial College, London (UK) and eight years as a lecturer and researcher at Trinty College, Dublin (Ireland). In addition to a general interest in the design of all types of geotechnical structures, his specific research areas have involved the interpretation of measured field behaviour, laboratory & field-testing and numerical analysis & software development. He has published many papers in International Journals, dealing with foundation design and soil behaviour. His contact address is Arup Geotechnics, 13 Fitzroy Street, London W1P 6BQ, U.K.

Buddhima Indraratna graduated in Civil Engineering from Imperial College of Science and Technology, University of London in 1982. Subsequently he completed Master degree in Soil Mechanics from Imperial College and Ph.D. in Geotechnical Engineering from University of Alberta, Canada. He worked with the Geomechanics Research Centre in Sudbury, before taking an academic position at Asian Institute of Technology, Bangkok in 1988.

In 1991, he joined University of Wollongong, where presently he is Professor of Civil Engineering and also acts as the Director of Studies, Postgraduate Research in the Faculty of Engineering. Professor Indraratna has published over 180 articles in referred journals and conferences. He is the author of two books and was Keynote Speaker and Guest Lecturer at number of International conferences. In addition to soft clay engineering his fields of interest are embankment engineering, rail-track geotechnology, hydromechanics of jointed rock and environmental geomechanics. His contact address is: Professor of Civil Engineering, University of Wollongong, Northfield's Avenue Wollongong, and NSW2522, Australia. E-mail : <engineering@uow.edu.au>

E. Justo is a Mechanical Engineer and is author of scientific papers and his main fields of interest are finite elements and boundary elements.

Guy Sunglerat graduated from the 'Ponst-et-Chaussees' and in 1949 he worked on difficult foundation problems at French Equatorial Africa and Belgian Congo. He worked as Director of the Katanga branch of SECO. In 1951 he took up research and analysis on Dutch penetrometer (C.P.T.) and studying the completed structures. In 1956 at France as a representative of Securitas-SOCOTEC in Haute Savoie and at Lyon, he worked as advisor of the Appalette Court. During this period he studied both the conventional methods of investigation and penetrometer. He traveled widely and collected information about the application of static and dynamic penetrometers and methods of analyses of data. He became Professor of Soil Mechanics at the Ecole Centrale Lyonnaise in 1966 and continued his research on penetrometers till 1988. He is author of Books on Penetrometers, and 'Practical Problems of Soil Mechanics' translated in various languages. Author of numerous papers he delivered lectures in 29 countries. He is expert witness *<pres La Cour d Appel de Lyon, agree par la Cour de Cassation>* His e-mail address is: Dactylo.print@wanadoo.fr

J.L. Justo, born in Spain, got his degree in Civil Engineering in 1963 and Master of Science degree in Soil Mechanics from University of London in 1966. He took Ph.D. from University of Madrid in 1969. He worked as Professor of Ground Engineering at University of Seville and was Vice-rector of Research from 1961 to 1975. As Head of the Department of Continuum Mechanics, University of Seville, he worked from 1987 to 1997. He is the member of number of societies, concerned with Earthquake Engineering. He is the author of 9 books, 34 book chapters, 65 scientific papers and 50 communications to scientific conferences. He has been consulting engineer in earth-rock dams, subways, motorways, high-rise towers, etc. His contact address is: Head of Department, E.T.S. Arquitectura, Sevilla, Spain.

K.R. Saxena, graduated in Civil Engineering from Osmania University, India and in 1965 and 1969, took D.I.C. and M.Sc. in soil mechanics and foundation engineering from London University. Starting his career at the Soil Mechanics Laboratory of A.P. Engineering Research Laboratories in 1952, he rose to the position of Director of the Laboratories. Through out his career at the research laboratories, he was associated with Investigations, design and monitoring of structures, like earth dams, major and multistoried Buildings, bridges and highways. From 1983 to 1990 he also served as national Director at the International organizations, like ICOLD, ICID, ISRM, ITA etc. and was the Secretary of the International Committee on Expansive Soils of ISSMFE and organized the 6th International Conference on Expansive Soils. He was President of Indian Geotechnical Society and member of International Committee on *Tailings Dams.* Author of number of technical papers he co-authored/co-edited books, *Geotechnical Engineering, Value Engineering in Project Management, Instrumentation of Embankment Dams, In-situ Characterization of Rocks etc,* published by Oxford & IBH Publishing Co. Pvt. Ltd., India and A.A. Balkema Publishers, Netherlands. His e-mail address is *kr_saxena@yahoo.com*

Leif Viberg working at the Swedish Geotechnical Institute since 1964 and currently a senior researcher obtained his degree of Licentiate in geotechnical engineering in 1972. During the academic year 1974-75 he held a position as visiting scholar at Purdue University, USA. His main activities have been the development of geotechnical maps for city and infrastructure planning. He has used database and GIS technique in his development work. He has been the secretary of Swedish Geotechnical Society and has served as a member on International committees on slope stability.

P. Durand, born in Peru, took his degree in civil engineering in 1988 and Ph.D. in 2000 from University of Seville. He is author of several scientific papers, including the one published in International Jr. for Numerical and Analytical Methods in Geomechanics.

Rolf Larsson started his career in 1964 as research engineer at the Geotechnical Department of Chalmers University of Technology. He joined the Swedish Geotechnical Institute in 1973 as a researcher. His research activities are related to development laboratory equipment, field investigations, settlement and stability. In 1993 he joined the Chalmers University of Technology as part time Adjunct Professor in the rheological properties of soil materials. Author of more than 100 paper and technical reports, he served on Swedish and International committees on standardization of soil testing.

Shamsher Prakash is presently, Professor Emeritus of Civil Engineering, University of Missouri, Rolla (MU) 65409. He is the President of Shamsher Prakash Foundations, Anand Kutir, 1111 Dnane Ave. Rolla (MO) 65401. He is the President, Roorkee University Alumini Association.

His major research area has been non-linear Solutions in Geotechnical Engineering and Predictions and Performance of Structures. He was President of Indian Geotechnical Society (1971-1975) and Indian Society of Earthquake Technology (1985-1987). He has published more than 200 paper and 15 books covering the subject of Foundation Engineering and Soil Dynamics. His e-mail address is <prakash@umr.edu>

Ulf Bergdahl started his career at the Swedish Geotechnical Institute in 1962. Through 1969-1973 he worked as geotechnical consultant with Scandiaconsult and returned to the Institute in 1973. From 1976 to 1988 he was Head of the Department of Site Investigation and Field Measurements and 1988 to 1994 was Head of the Department of Earthworks. Till 1997 held a position of Senior Researcher and consulting Chief Engineer. His main research areas have been site investigations, pile foundations, slope stability and shallow foundations. Consultancy covered areas, roads on soft grounds, foundations of tall buildings and bridges and tunnels. He was deeply involved in developing standards, codes and handbooks and served as member of national and international committees on geotechnical engineering. He is the author or co-author of more than 100 papers and reports related to geotechnical engineering. His contact address is Swedish Geotechnical Institute (SGI), SE-581 93 Linkoping, Sweden. E-mail: <sgi@swedgeo.se>

V.M. Sharma, B.Tech (Hons) and Masters degree in Civil Engineering took his Ph.D. in Rock Mechanics from I.I.T. Delhi. Starting his career at Central Water Commission, Government of India, he rose to the position of Director of Central Soil and Materials Research Station. In various capacities he has been closely associated with investigation, design and construction of Major Dams, Tunnels and Underground Structures. Presently working as Chief Consultant, Advanced Technology & Engineering Services, (A Division of AIMIL Ltd.), New Delhi, Dr. Sharma is involved in Numerical Modeling, Reinforced Earth Technology and Instrumentation of structures. He was the president of Indian Geotechnical Society and Organizing Committee Member (Technical) of 13th ICSMFE, New Delhi. Author more than 100 technical papers, he is co-editor of number of books, the latest being *Distinct Element Modeling in Geomechanics, Instrumentation of Embankment Dams and In-situ Characterization of Rocks,* published by Oxford & IBH Publishing Co. Pvt. Ltd. New Delhi and Balkema, The Netherlands. His e-mail address is <vmsharma@aimil.com>

Vijay K. Puri is currently the Professor in Department of Civil and Environment Engineering at South Illinois University, Carbondale, IL 62901, USA. He obtained his M.S. degree in Civil Engineering from University of Roorkee, India and PhD in Civil Engineering from University of Missouri, Rolla, USA. His major areas of expertise are, Geotechnical Engineering, Soil Dynamics, Soil Improvement and utilization of coal preparation and coal combustion by products. He has co-authored a graduate level textbook, entitled 'Foundation for Mechanics, Analysis and Design', published by John Wiley, New York.

Ting, Wen Hui, in 1958 graduated from the University of Malaysia, took Master Degree in Soils and Highway Engineering in 1961 from Ohio State University and Ph.D. in Soft Clays from Cambridge University, U.K. in 1968. After graduation he had joined PWD, Malaysia, Looked widening of the Johore Causeway, PWD Soils and Materials Laboratory and was involved in Subang International Airport Project. From 1973-78 worked as Lecturer in Soil Mechanics and Foundation Engineering at University of Malaysia. From 1978 worked as Consulting Engineer in field of Highways, Airports and various Civil Engineering works. He was past President of the South-east Asia Geotechnical Society, Past President of Institution of Engineers, Malaysia and its Geotechnical Division. He was a Theme Lecturer on Foundations at 9th Asian Regional Conference of ISSMFE, Bangkok, 1991. In April 2000, he was elevated as Fellow of the Academy of Sciences of Malaysia. His contact address is: 18 Jalan SS 20/10 Damansara Kim, 47400 Petaling Jaya, Selangor Darul Ehsan, Malaysia. E-mail: tingwh@pop.jaring.my

1

In-Situ Characterization of Soils

*V.M. Sharma**

INTRODUCTION

It happens quite frequently that one comes across a very intense and heated discussion regarding the merits of in-situ versus laboratory tests. According to Jamiolkowski and Presti (1994), this discussion is of little practical significance, because laboratory and in-situ testing are complementary rather than contrasting methodologies. In-situ characterization of soils coupled with laboratory testing is probably the best policy. The normal tendency, however, is to avoid in-situ testing due to the time and cost involved. There are some situations though, when in-situ testing becomes essential, for example, for:

1. Soil profiling and characterization
2. Evaluation of in-situ horizontal stresses
3. Assessment of horizontal permeability
4. Evaluation of stiffness of cohesionless soils
5. Calculation of settlement and bearing capacity directly from in-situ test results specially for cohesionless soils.

The techniques used for soil profiling and characterization are either based on geophysical methods, or on drilling, sampling and testing with Penetrometer.

Pressuremeter tests are conducted for the evaluation of ground stiffness, bearing capacity and horizontal stresses.

In-situ permeability tests are conducted for assessing the permeability of ground, before and after the ground treatment, if any.

A brief discussion on the recent developments in the techniques mentioned above follows:

GEOPHYSICAL METHODS

Several geophysical methods are in use, but the most commonly used are seismic methods and electrical resistivity method.

*Chief Consultant, ATES (A division of AIMIL Ltd.), Naimex House,
A-8 Mohan Co-operative Industrial Estate, Mathura Road, New Delhi—110 044, India.

SEISMIC METHOD

A mechanical shock is created, at or near the surface, and a vibration passes outwards in all directions to be detected at varying distances from the source by geophones (on land) or by hydrophones (in water). The initial pulse of limited duration is created usually by firing a shot of high explosive in a hole tamped with soil or water but, under certain surface conditions and shallow investigations, it may also be created by striking a steel plate embedded in the ground with a hammer or by allowing a heavy weight to fall on it. In water a variety of acoustic sources are available in addition to explosives. Besides transient pulses, continuous waves may be used by coupling mechanical or electromechanical vibrators to the ground.

Seismic exploration depends fundamentally upon the velocity of propagation of elastic waves and their capabilities of being reflected and refracted at boundaries. Two waves are capable of being transmitted through the body of the material:

Compressional or *P*-wave : $V_p = \sqrt{\dfrac{3B + 4G}{3\gamma}}$

Shear or *S*-wave: $V_s = \sqrt{\dfrac{G}{\gamma}}$

where

B = bulk modulus of elasticity
G = rigidity modulus of elasticity
γ = density.

Surface waves can also be set up in addition to the above body waves; of these the most important is the Rayleigh (ground-roll) at the earth's surface carrying a large fraction of the elastic energy released by the source.

In the refraction method of investigation, time-distance data are acquired by firing shots at a line of geophones (geophone spread) and plotted as graphs (Fig. 1).

Refraction techniques suffer from the following drawbacks:

1. Velocities must increase downwards
2. Spread lengths long if velocity contrasts not high
3. Velocities and dip of boundary must be such that sum of angles of refraction is less than 90°.
4. Depths determined at key points and boundary assumed planar between.
5. Ability to draw straight lines through data affected very much by surface topography and variability in velocity.

The reflection method requires a contrast between velocity or density

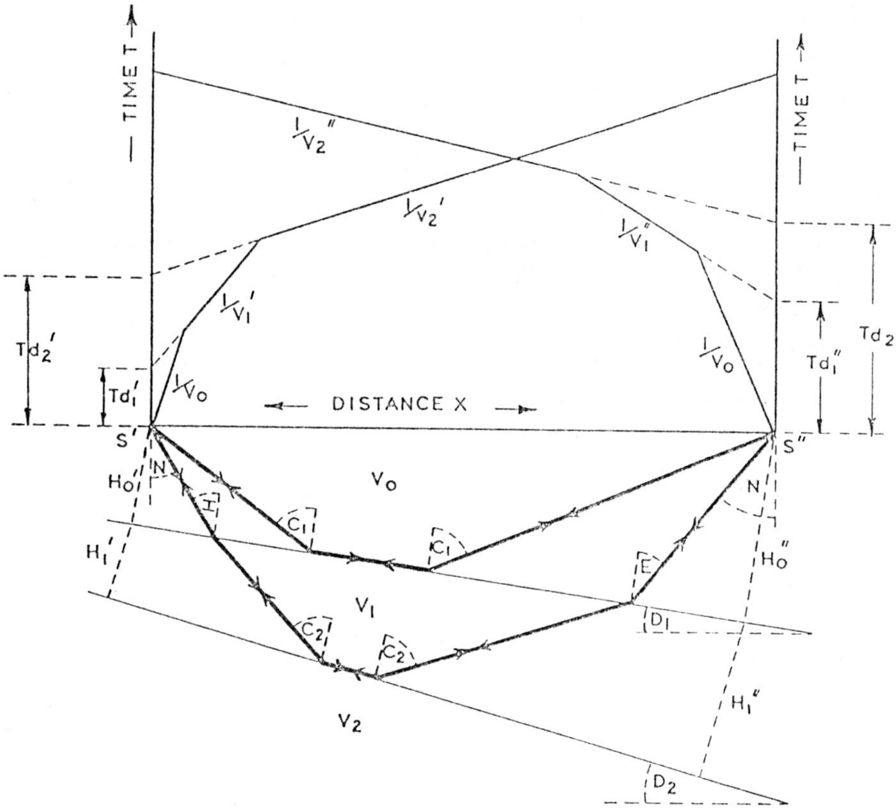

Fig. 1 *Seismic refraction method of geophysical investigations.*

for it to be effective. As a reflection cannot be a first arrival, on land it suffers from interference with earlier arrivals (P-wave, S-wave, ground roll) and cannot be used in shallow investigations. No such interference occurs at sea and hence its usefulness in over-water surveying. The system is very similar to echosounding where source and receiver are close together, the collected time data being interpreted directly in terms of depth provided the velocity conditions are known.

ELECTRICAL RESISTIVITY METHOD

The passage of an electrical current through the various formations beneath the Earth's surface produces certain effects—particularly a variability in potential distribution—which allow deductions to be made concerning those formations. The current may be created in several ways. It may be:

1. A natural earth current of uncertain magnetic origin
2. A natural earth current of electrochemical or electrofiltration origin:

the former is associated with base metal sulphides, the latter with leakage in earth dams.

3. Artificially produced.

Within the artificial current methods, the most important in site investigations are those which depend on measuring any variability in surface potential distribution created by resistivity change in the direct (or low frequency) current field.

Resistivity of a material is defined as the ohmic resistance of a 1 m^3 with the current flowing parallel to one edge. As mineral grains are mostly insulators (apart from natural metals and clay minerals with adsorbed water), resistivity of a rock or soil depends on:

1. Quantity of water present
2. Salinity of water present.

The electrical formation factor F of a soil is given by:

$$F = \frac{\text{Resistivity of the saturated soil}}{\text{Resistivity of the pore fluid}} = \frac{\delta\, s}{\delta\, w}.$$

This is related to the porosity and permeability of the material.

Measurements of resistivity variations are made by passing a current through the ground between two electrodes (A, B) and determining the consequent surface potential distribution with another two electrodes (M, N). Two basic techniques are possible depending upon whether the resistivity variations occur in a vertical or in a horizontal direction.

For depth determinations the current base is expended about a central point measuring apparent resistivity from the relationship.

$$\rho a = k\,\frac{\Delta V}{I}$$

where k is a geometric factor depending on the relative position of the four electrodes. For an in-line situation

$$k = \frac{\pi}{4MN}\left(AB^2 - MN^2\right)$$

Thus Wenner system: $\rho a = 2\pi\ MN\dfrac{\Delta V}{I}$ (where AB = 3 MN)

And Schlumberger system: $\rho a = \dfrac{\pi}{I}\left(\dfrac{AB}{2}\right)^2 \dfrac{\Delta V}{MN}$ (where AB $\geq$ 5 MN)

The measurement is an apparent resistivity for, as the current base expands, the current penetrates deeper: if an imaginary boundary exists in a homogeneous medium at a depth equal to the current base, only one-third of the total current would penetrate beyond it. Thus by plotting

$\rho\, a$ against some function of the current base (in the Schlumberger system it is AB/2; in the Wenner system AB/3), a unique curve is obtained of the resistivity/depth relationships. This curve may be compared to theoretically-derived curves, based upon resistivity-depth models, so that a deduction concerning the field situation may be made. The technique suffers from the fact that the interpretation is not unambiguous: various theoretical models can exist to match the field data. Strict borehole control is necessary.

For lateral variations in resistivity—steeply dipping boundaries between two contrasting resistivities such as faults and dykes—the measurements must be made with mobile electrodes across the strike of the structure. Two techniques are possible:

1. The current base is kept large and fixed; the potential electrodes are moved along lines parallel to the current base within a rectangle centred on the mid-point of the current base.
2. All four electrodes are moved together with a fixed electrode separation. The current base is chosen so that surface fluctuations in resistivity are minimized. The technique requires strict borehole control and at least two resistivity depth probes per traverse or, alternatively, each traverse is surveyed twice, once with a small current base and once with a relatively large electrode separation. In mapping variations in thickness of the overburden

$$H = 0.5\, AB\, \frac{\rho 1}{\rho\, a} \quad \text{(Schlumberger system)}$$

$$H = 0.46\, AB\, \frac{\rho 1}{\rho\, a} \quad \text{(Wenner system)}$$

where
AB = large current base
$\rho 1$ = resistivity of surface medium
ρa = apparent resistivity measured with large AB at any point.

BOREHOLE GEOPHYSICS

The principal applications of borehole geophysical logging are correlation, identification and discovery. In recent years, by virtue of their continued sophistication in the mineral exploration work, the methods have become more attractive to the geotechnical engineer. The subject falls naturally into two parts.

First is 'down-hole logging', where measurements are made within a single borehole at the time and data acquired is restricted to the zone near the borehole wall. Second is 'cross-hole logging' where measurements are made between two, three or more boreholes at one time to make assessment of the volume of rock between the holes, between two, three or more boreholes at one time to make assessments of the volume of soil between the holes.

Down-hole logging

Down-hole logging suites are built up around three major types of measurements—sonic velocity, gamma radiation and electrical.

Sonic Logging

It consists of a transmitter (an electromechanical device which periodically creates a pressure wave in the fluid filled borehole) and a receiver spaced a fixed distance apart (usually between 0.5 m and 2 m in our field). The 'pressure wave' travels out from the transmitter to the borehole walls, where some portion of the energy is spent in creation of two other forms of elastic wave. These are referred to as the 'shear wave' and 'boundary wave'.

These waves travel in all directions from the vicinity of the transmitter. One of these directions is down the borehole wall towards the receiver, where they are 'picked up' again by the receiver.

The major problem with sonic logging is that it requires both an uncased and fluid filled hole. Developments, however, are presently in hand to remove the necessity of the latter.

Gamma Radiation Logging

Soil masses are naturally radioactive due to the traces of radioactive salts which they contain. It is, therefore, possible to detect variations in their composition by counting the gamma radiation emitted throughout the borehole depth. Casings within the borehole reduce the activity level but do not mask the changes in emission levels. The count can be taken in the wet or dry hole but it is desirable to log under a consistent set of conditions throughout an individual borehole when possible.

The Gamma Ray Log is an 'uncalibrated log' in that only charges in count level are recorded. As radioactivity depends on the age (and hence the state of decay) of the source minerals and their concentration, two soils of differing type and age could give the same 'signatures'. Most of the soils, however, have distinctive radiation levels.

An excellent lithological correlation can be found with the Gamma Ray Log. Good correlation is found between density as determined from the Gamma-Gamma Log and that measured from small laboratory samples.

Electric Logging

Electric logging techniques can be divided broadly into two groups:

1. Spontaneous potential logging (S.P.)
2. Resistivity logs of various types

Most electric logging operations involve a combination of both groups and, although interpretation procedures are not fully rational, the wealth of practical experience available makes the logging a very powerful tool.

Other Logging Methods

A Caliper Log (which measures hole diameter) is run with all sonic and radiation logging packages, as hole diameter is necessary for part of the calculation and interpretation of results. If in-situ testing of another kind is to be carried out, it can identify areas where the borehole walls are of acceptable profile. In deep exploration, a temperature log may also be recorded but this has little application in geotechnology.

Cross-hole Shooting

Cross-hole shooting is the term applied to measurement of sonic velocity through the ground between boreholes. While in theory there is no reason why electrical measurements should not be made in the same manner, in practice it is rarely done because resistivities can more easily be determined from surface measurement.

The majority of work done in this field has used explosive sources, the practice being to start shooting at the base of the hole and gradually work up the hole as it collapses after each shot.

Cross-hole shooting is generally considered for two types of problems:

1. The location of geological or man-made structures which are not two-dimensional to any extent in plan such as tunnels, caverns, faults and dykes.
2. The determination of sonic velocities for calculation of dynamic elastic parameters.

In practice, it is most important to develop a proper geometry for the test. There are two major difficulties in practical application of the technique. Firstly boreholes are rarely vertical; an acceptable tolerance is 1:100 for holes up to 30 m. Borehole survey techniques are such that a hole of this depth can probably be logged for x and y departures to about $\pm$ 75 mm. Consequently, the distance between a pair of holes could be $\pm$ 150 mm. This value dramatically increases as holes become deeper. Secondly, geological formations are rarely homogeneous and isotropic. They are usually layered and naturally each layer will have different sonic velocities. However, seismic waves will travel from A to B along the fastest route possible.

ECONOMIC JUSTIFICATION FOR USING GEOPHYSICS

Apart from the technical justification for the use of a geophysical method, its use must also be economically sound. Economic justification for the use of geophysics is mainly to decrease the financial risk of subsequent exploration. While the resistivity technique is not as uniformly precise in

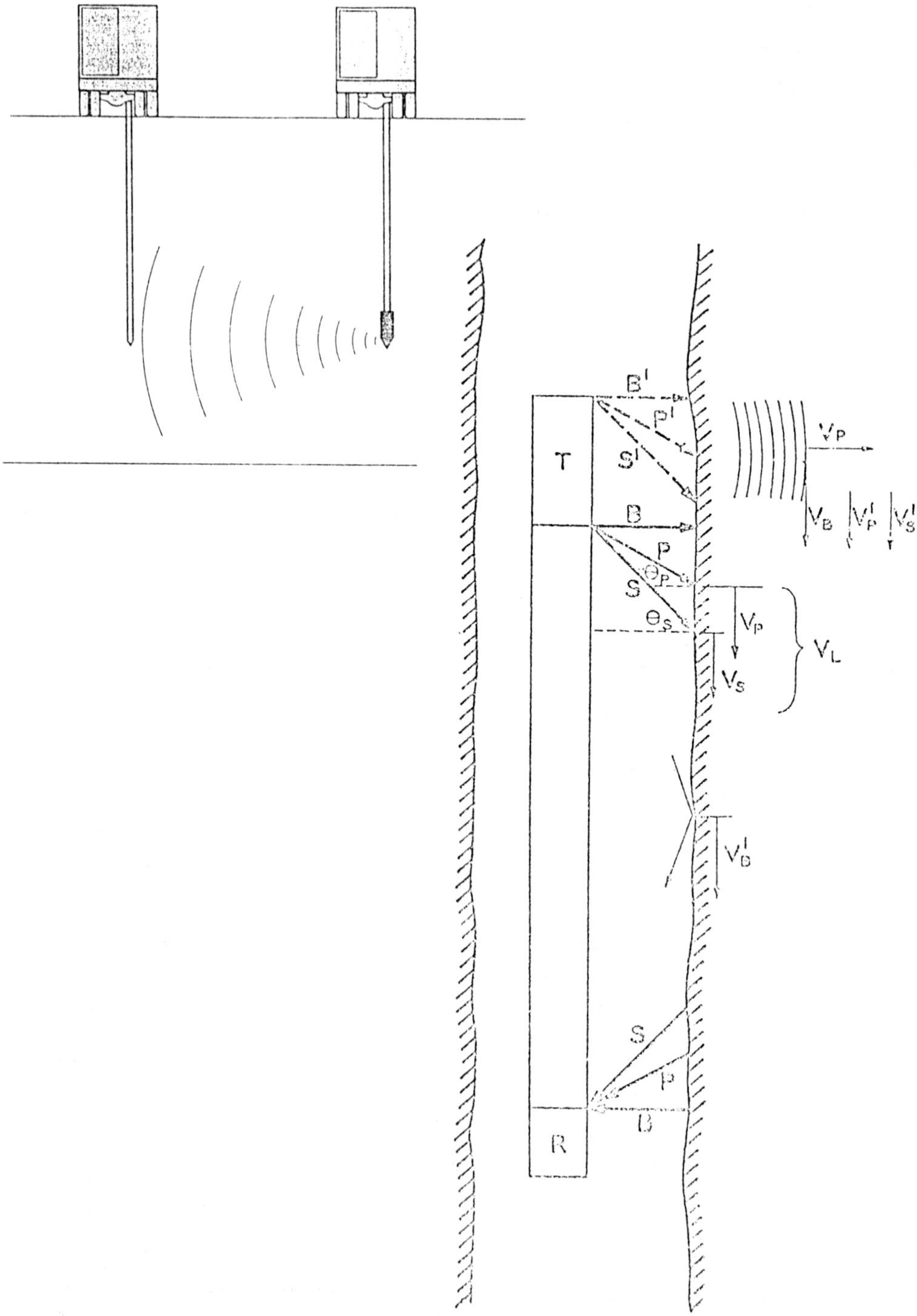

Fig. 2 *Cross-hole and down-hole methods of geophysical investigations.*

depth determinations as the seismic refraction system, its general simplicity and speed of operations often gives it a low-cost preference for many field situations. On the other hand, better definition is often achieved by using more than one geophysical technique: to locate a dyke intruded into a till-covered area, it would be sensible to use both seismic and resistivity techniques.

However, in assessing costs, the higher cost of drilling as compared to geophysical work has to be balanced against the fact that borehole information is more acceptable to the engineer.

SEISMIC RESPONSE OF DIFFERENT SOILS

For the cyclic characterisation of seismic regions, the soil properties, particularly the plasticity index and those determined by the cyclic strain-controlled tests, are used. Mladen Vucetic (1994) states that the seismic response of different natural soil deposits must reflect different dynamic properties of soil material. To explain the effects of the type of soil on the seismic response in a systematic manner, it is convenient to link the cyclic behaviour to the standard Unified Soil Classification used in geotechnical engineering practice. Such link, can be conveniently established via PI, and can be summarized in the following three points:

1. **High plasticity soils** have relatively small damping and a large range of linearly elastic behaviour, which are both prerequisites for a resonance. Therefore, when seismic waves propagate through a deposit composed of such soil, a resonance of the deposit may take place. Such resonance can greatly amplify the incoming seismic motion and cause considerable damage. Soil failure due to cyclic degradation of stiffness and strength in the highly plastic clay deposits is unlikely, because the degradation is small.
2. **Low plasticity fully saturated soils and soils with no plasticity**, like fully saturated sands, exhibit a highly non-linear behaviour and considerable damping even at small shear strains, and during continuous cyclic loading their stiffness and strength degrade significantly. An amplification of motion through such soils caused by the resonance of the deposit is therefore unlikely. However, due to the significant softening of the soil, the predominant period of the ground surface motion can considerably increase (Vucetic and Zorapapel, 1990; Zorapapel and Vucetic, 1990), and the structures with similar larger predominant period may be affected. The most critical aspect of the response of such soils is large drop in soil strength due to cyclic degradation, which may cause the failure of the foundation soil and collapse of supported structures. The extreme examples of such failure are the liquefaction-included failures.

3. The seismic response of **medium plasticity soils, i.e., the medium plasticity clays**, is somewhere between the two extreme cases described above. Such soils can moderately amplify the incoming seismic motion, and their bearing capacity can be moderately decreased by a continuous cyclic loading.

DRILLING AND SAMPLING

Geophysical investigations cover a large area in a short time. The information obtained is not point specific as in the case of drilling. Still, the geophysical methods are only indirect and therefore need some reliable data for calibration. Therefore, drilling is unavoidable.

Several techniques are available for drilling borehole and collecting samples of soil. For making a profile of the soil strata, either samples can be collected with the help of thin walled tubes or a continuous sampler may be used. A brief discussion follows:

Boring Methods

There are many methods of boring exploratory holes. These may be grouped in different ways:

1. Those which produce usable samples and those that do not, and
2. Those that are accessible and those that are not.

In the investigation of foundations or materials, the principle purpose of exploration is to secure samples for inspection and testing. Therefore, these procedures which do not produce samples should be used only where the nature of materials to be penetrated is already known and where sufficient samples for testing have already been secured.

In soft or loose soil formations soil may flow from the sides to the bottom of the hole. Often, filling the hole with water will suffice to hold the materials in place. In more severe cases wall stabilizer or drilling mud must be used. This is a specially prepared fluid which has sufficient density due to the addition of finely divided solid materials to prevent the sloughing of the sides of drill holes.

Boring

Wash Boring

Wash boring is extensively used in reconnaissance exploration and quite often in advancing and cleaning boreholes between samples taken in detailed investigations.

The principal advantages of the wash boring method are that it can be used in borings of small and large diameters, that the equipment is inexpensive, light, and can be made portable, and that the method does not cause sealing of the bottom of the hole.

Boring with Augers

Augers can be of different sizes. Small helical augers are hand operated and can be used for boring holes of 100 mm to 300 mm in diameter. These can be used for exploration down to about 6 metres. Large helical or warm type augers are used for holes 75 mm to 400 mm in diameter. Barrel and disk augers up to 1200 mm are used for accessible holes. One can physically go down in these holes for collection of samples and visual inspection.

Casing is required for auger borings in unstable soil and borings extending below the ground-water table. Auger borings should be kept dry since water in the hole increases the danger of losing the soil in the auger since the soil-laden auger acts as a piston and tends to force water above it out of the hole. In hard, dry, clayey soils water may be added to moisten the soil and speed up the augering. Very dry sands or silts may be made slightly cohesive by adding water, thus making them easier to recover on the auger.

Rotary Drilling

Rotary drilling is probably the most expensive and most common technique used for the detailed exploration of soil and rock. The rotary drill is efficient, highly versatile, and has the ability to penetrate great depths (>300 m) in any type of soil or rock because of the adaptability of the method.

In rotary core drilling, the rig consists of a motor connected through a transmission to a chuck which rotates a drill rod with a bit at the end. The drill rod is rotated through the chuck and pressure is applied on the bit by hydraulic pistons or screw jacks acting on the chuck. The bit cuts, chips, and grinds away material which is then lifted from the hole by circulating drilling fluid. The drilling fluid is commonly pumped down the drill rods, through the bit and back up between the drill rods and the wall of the hole or the inside of the casing. In some cases where fluid return is good, the fluid is recycled back down the hole after passing through a "desander" which removes the cuttings.

Drilling machines and accessories should be chosen on the basis of the expected formations to be drilled. Rigs should preferably be equipped with hydraulically raised, single unit masts. These masts can be raised in a matter of seconds unlike the three tripod legs which require hours to erect. Moving time is also greatly reduced as the mast is an integral part of the rig whereas the tripod legs must be moved separately.

Experienced drillers are required to ensure that satisfactory samples are recovered with the least delays. For good coring in either soil or rock an experienced driller keeps careful watch on the speed of rotation of the bit, the bit pressure, bit feed, and the pump pressure and discharge.

Overburden

Overburden drilling is much more difficult and requires highly experienced personnel. The driller's responsibility is to operate the drill efficiently, ensure that the helper mixes the drilling fluid in the right proportions and carry out any sampling and testing.

In the case of rotary drilling in overburden the hole is usually started and advanced between samples by triconing.

Penetration rates in overburden vary much more than for soils and depend on the types of materials drilled and the drilling equipment. Drilling is fastest in fine soils and slowest in very coarse gravel deposits containing cobbles and boulders. In coarse deposits the best rates are attained with slow rotation and substantial bit pressure.

The most commonly used drilling bit is the tricone, although diamond coring bits, carbide bits and hardened steel bits can be used in overburden to obtain continuous cores.

Sampling

Samples can be secured from holes supported by drilling mud or casing either with the Denison or drive samplers. To minimize sample disturbance, piston-type samplers are preferable for very soft or very loose soils.

Special Sampling

In the majority of site investigations, it is necessary to obtain samples of the soil. Where recovered samples are to be subjected to laboratory testing for the determination of the engineering parameters or used for stratigraphic or structural investigations, it follows that they should be both undisturbed and of the appropriate dimensions. In what may loosely be termed as traditional types of site investigations, undisturbed sampling has often been by means of driven thin walled tubes. In recent years, however, the significance of soil fabric or structure and sample disturbance on the engineering parameters has been more widely recognized and in some investigations the thin walled tube is unable to provide an appropriate sample. There are several soils which cannot be satisfactorily sampled in a conventional investigation (e.g. very loose silts or sands, weathered or fissile soft rocks, some boulder clays). These can sometimes be dealt with by special sampling techniques. Depending upon the soil type and the information required, the Delft continuous sampler, large diameter piston sampler, Bishop sand sampler, Shrew sampler or block sampling techniques can be used.

Delft Continuous Sampling

The Delft continuous sampler is available in 29 mm and 66 mm diameter sizes and is generally suitable for use in soft or loose fine-grained alluvial soils. Where thick deposits of sand are encountered, the Delft system can prove difficult to operate successfully and consideration should also be given to the use of the Bishop sand sampler. The maximum length of sample which can be taken with the Delft sampler is about 18 m.

The sampler is pushed into the ground by means of a Dutch deep sounding machine; generally a 2 Mg machine is used in the case of the 29 mm sampler and a 17 1/2 Mg machine for the 66 mm unit. As the sampler advances, the stockinette is pulled off the magazine to cover the cylindrical sample of soil. The sampled soil is supported laterally in the sampling head and extension tubes by bentonite supporting fluid. The specific gravity of this fluid can be adjusted so that its density is the same as the soil being sampled. This ensures that the sample suffers a minimum of disturbance after cutting. To prevent contamination of the soil by the supporting fluid and subsequent moisture content changes, the stockinette is made impervious by impregnating it with a waterproofing compound. In the twin liquid variety of sampler, a vulcanizing liquid is added to the magazine chamber during assembly of the equipment and this coats the stockinette as it is drawn from the magazine. The liquid cures to make the stockinette impermeable when it comes into contact with the water based supporting fluid. The twin liquid sampler is only manufactured in 29 mm size. In the single liquid type of sampler the stockinette is waterproofed during its manufacture and consequently the only liquid required on site is the supporting fluid.

When the sampler has been pushed to the required depth, the sampler must be locked before withdrawing the equipment, in order to prevent loss of the sample during the withdrawal.

The sample is recovered by withdrawing the equipment from the ground, removing one tube at a time, together with a metre or so length of sample contained in it. This operation involves loss of the supporting fluid to that metre of sample, which sometimes tends to compress longitudinally due to the stockinette tension set up during sampling. This problem occurs principally with the 29 mm sampler and implies that small length corrections should be made during the logging of the hole. Recovered samples are normally laid out on specially designed trays for splitting, description and selection of laboratory test samples.

The operation of splitting the Delft sample is to first remove the stockinette by cutting along one side of the sample and rolling the sample out. The sample should then be cut into manageable lengths of about 10 to 15 cm and split in half using a broad bladed knife. One half of the sample should be used for laboratory test samples for moisture content, index property testing, etc. and for consistency assessment in description, whilst the other half is left untouched and allowed to dry out. Due to variations in grain size or porosity, samples dry differentially and in the partially dry state, structure is emphasized by colour variations. It is suggested that permanent photographic records should be obtained of at least representative types of material as it is virtually impossible to describe in words the structure of some complex alluvial soils.

Where continuous sampling is a necessity (e.g. in detailed stratigraphic investigations), the Delft method has the inbuilt advantage that it is a truly continuous sampling procedure, whereas even 'continuous' piston or thin walled tube sampling must be considered as only semi-continuous. Even where continuous sampling is not a strict requirement, the speed and financial advantages of using the Delft sampler dictate its consideration for all fine alluvial soil investigations. It should not be regarded as a replacement for conventional borings but rather a complementary technique. It is particularly useful in getting together an early picture of the site and subsequent boring, larger diameter sampling and in-situ tests can then be located to provide samples and data from typical soil types involved.

TESTS IN BOREHOLES

Standard Penetration Test

The Standard Penetration Test (SPT), when carried out carefully and conscientiously, provides a useful and cheap empirical index of the ultimate strength of the ground. Although intended for use in sands, it is used (and hopefully interpreted) in everything from soft clays to weak rocks. The results of tests may significantly be affected by a number of factors relating to test procedure and certain of these are in turn influenced by the ground conditions. For this reason the quality of the results of the majority of tests is at best uncertain and consistent correlation between dynamic penetration tests and other soil properties are rare. Nevertheless, the test remains the widest used of all in-situ tests and remains an important procedure in most site investigations in soils.

The test was first developed about 1927 by the Raymond Concrete Piling Co. in the USA and is still performed in that country in wash bored holes from 60 mm to 100 mm diameter. The procedure for the test and its correlation with relative density and allowable bearing pressures for sands was first published by Terzaghi and Peck (1948).

Features of Test Procedure Influencing Results

HAMMER WEIGHT/DROP EFFECTS

The energy per blow delivered to the top of the rods for the test is 140 1 bf x 30 in = 4200 in 1bf. Several workers have shown that different hammer weights and drops, although giving the same energy, do not give the same N value. Automatic trip hammers should be insisted upon today for all work. Serota (1971) has suggested that the blow count for a trip hammer is substantially less than that obtained using a slip rope hammer. It must be remembered that Terzaghi and Peck's work was based on slip rope hammer data. The method of delivering the blow to the rods has been one of the most important operator variables which have caused difficulties in interpreting results. Unfortunately, there are insufficient documented records to allow a reassessment of Terzaghi and Peck's original work taking this factor into account.

Diameter of borehole and tools

When 'shelling' methods of boring are employed, disturbance of the material beneath the bottom of the hole is inevitable due to suction created during shelling, even when the water pressure in the borehole is balanced. B.S. 1377 requires that the shell diameter be not more than the 90% of the casing diameter and small shells should be used in difficult ground for cleaning out at least above the test level. Every attempt should be made to bore in the smallest diameter possible.

Size, length and weight of rod

The original work was carried out using square boring rod but B.S. 1377 specifies the use of a rod. Gibbs and Holtz (1957) concluded that the effects arising from using A, B or N rod was not large, although the heavier rods gave slightly reduced blow counts. Generally the effects are negligible except for soft and loose materials where present practice may give a larger blow count than original equipment.

Comparison of split spoon and solid cone

Palmer and Stuart (1957) replaced the open ended split sampler with a solid pointed 60° cone for tests in materials coarser than sands. They and subsequent authors have concluded that the results of either penetrometer are of the same order with the cone giving slightly higher results in coarser materials.

Position of water table during test

In boring through and carrying out SPTs in granular material below the water table, it is a basic requirement that the head of water in the hole is maintained at or above standing water level. This avoids upward percolation at the base of the hole resulting in disturbance (usually referred to as piping). Disturbance due to unbalanced water heads can extend to a depth of five times the borehole diameter without piping being apparent. Even when water levels are carefully controlled, the position of the water table can have a marked effect on results. Schultze and Menzenbach (1961) have reported a reduction in N value of 15% for tests performed immediately (1 m) above and below the water level. Terzaghi and Peck (1948) and Gibbs and Holtz (1957) suggest that the difference is more pronounced for tests in loose material. Many engineers feel that tests below the water table in sands and silts are at best unreliable and wherever possible use the Dutch deep sounding in these soils.

Terzaghi and Peck (1948) suggested that allowance should be made for the position of the water table by reducing the N values measured in the field.

Features of Ground Conditions Influencing Results

Overburden pressure

The blow count obtained is a measure of the ultimate bearing capacity of the ground and therefore reflects both the in-situ relative density and the effective stress at the test level. It is, therefore, possible to get the same blow count from two locations with different relative densities and levels of effective stress. Gibbs and Holtz (1957) demonstrated that the test underestimated the relative density at shallow depths and presented correction factors for adjusting the N value. These were given in a more convenient form by Tomlinson (1969) (Fig. 3). Thorburn (1963) has also produced a correction chart which is based on field data recorded using UK investigation practice related to shallow foundations (Fig. 4). The depth of overburden refers to dry ground above the water table. There is not sufficient justification as yet for the indiscriminate use of the Gibbs and Holtz correction but the Thorburn correction, which is more conservative, is widely used today.

Particle size

The penetrometer was originally designed for use in sands where the ratio of diameter of penetrometer to maximum particle size is large and most correlations available to present are based on data from sands. As the particle size increases, the reliability and reproducibility of the test drops off. Meyerhof (1956) suggested that results from soils where particle size and penetrometer size are comparable are unreliable and that lowest N values determined should be used where the particle size is above 12 mm.

Tests in fine sands and silty deposits are likely to give abnormally high *N* values because of their relatively low permeability. Terzaghi and Peck (1948) recommended a correction for these soils when $N > 15$.

$$N_{corr} = 15 + 1/2\,(N-15)$$

This is generally applied today.

Grading

Many workers have indicated the importance of grading but as yet there is insufficient well documented case histories to be able to be definitive. Meigh (1964) stated that with increasing width of grading, the *N* value became

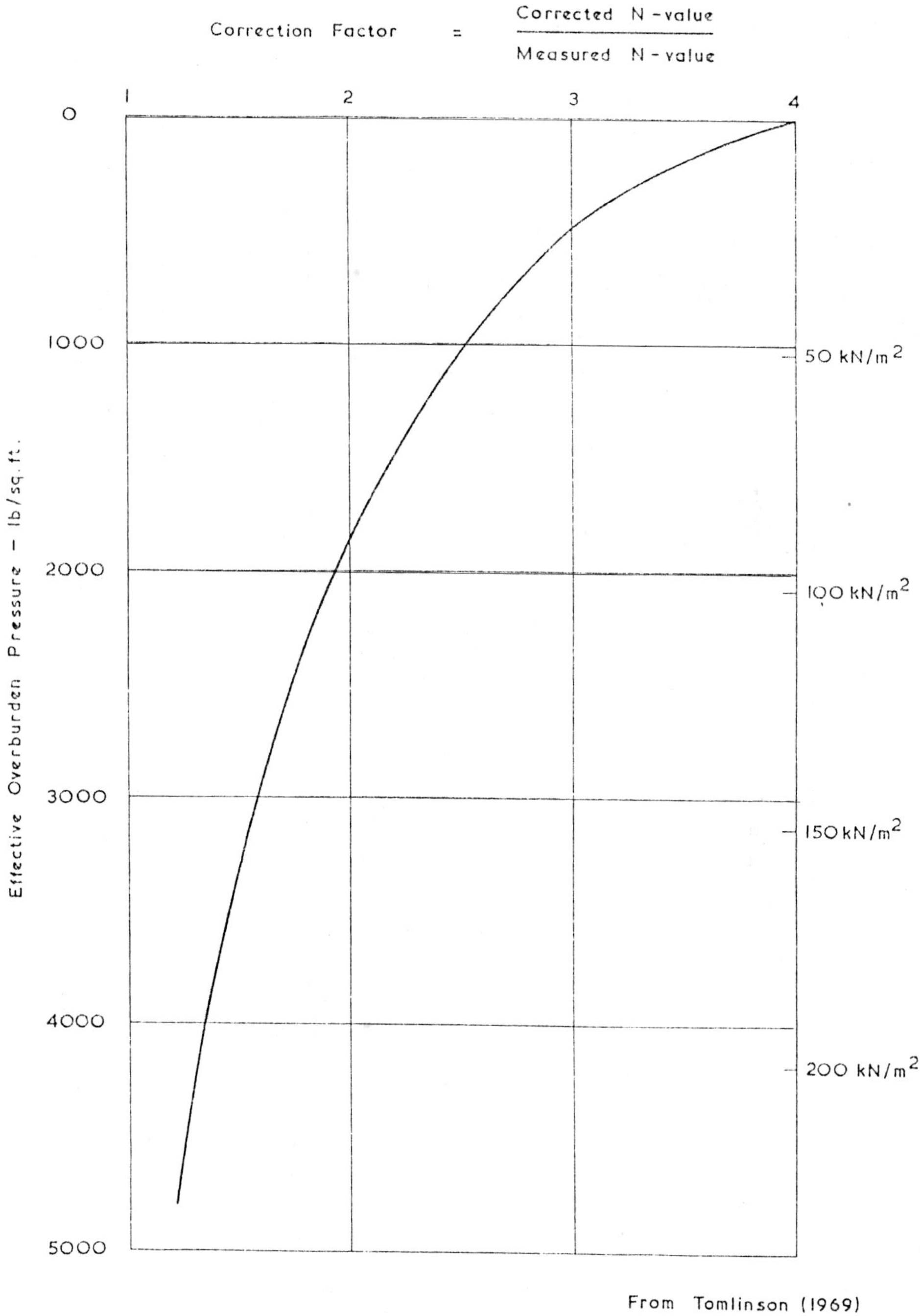

Fig. 3 *Correction factors for standard penetration tests for overburden effects after Gibbs and Holtz.*

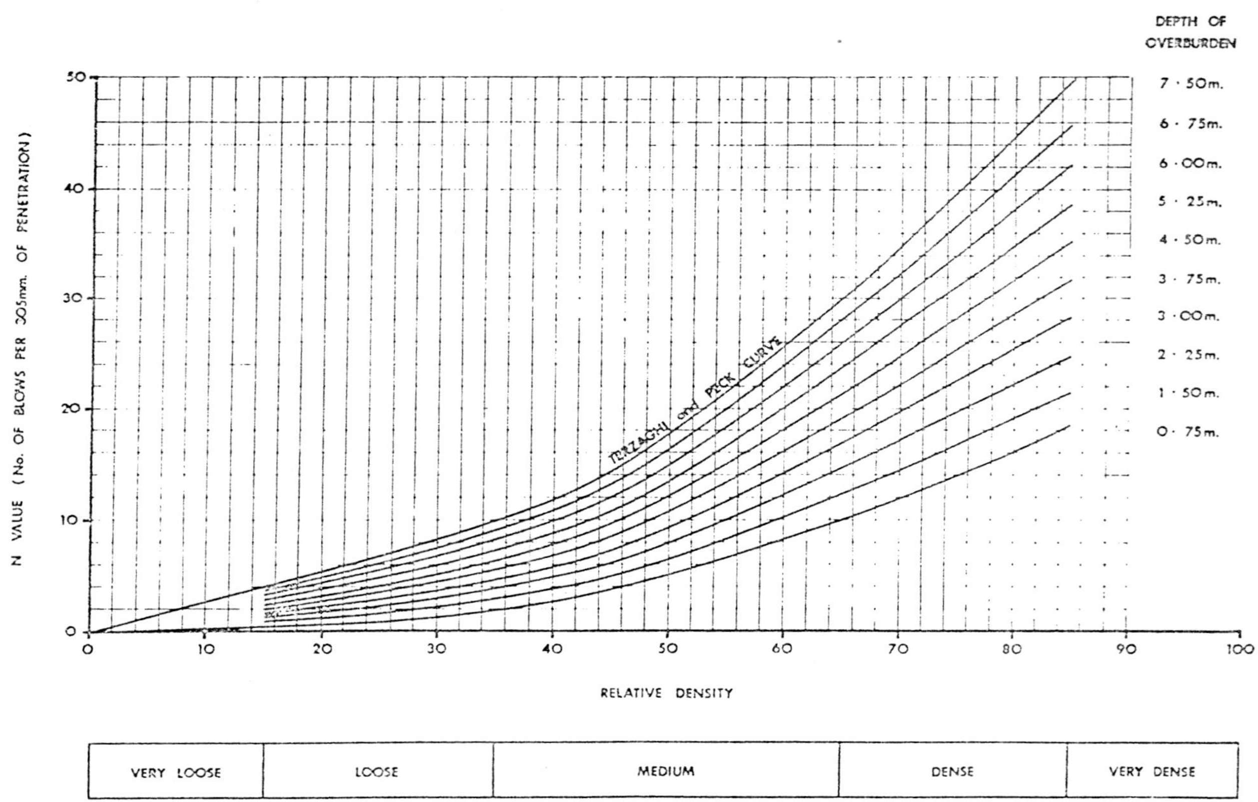

Fig. 4 *Correlation of standard penetration tests for overburden effects after Thornburn.*

more conservative, whereas it has been suggested that uniform fine sands perform as predicted using Meyerhof's method of calculation. Meigh and Hobbs (1973) suggest that grading is the most important aspect of ground conditions affecting the test and set out a correction procedure for taking grading into account.

Schmertmann and Palacios (1979) tried to measure the energy transmitted to the drill rods and rationalize the use of SPT.

Current SPT Energy Calibration

The existing method of SPT energy measurement as specified in ASTM Standard D4633-86 and in the ISSMFE (1988) International Reference Test procedure is based on the force measurement approach developed by Schmertmann and Palacios (1979). The method consists of attaching a load cell near the top of the drill rods and measuring the force time history during hammer impact as shown schematically in Fig. 5. On hammer impact, a compression stress wave is generated which travels down the drill rods at a consistent speed of about 5120 m/s in steel. Upon reaching the bottom of the sampler, the compression wave reflects as a tension wave back up the drill rods. At the top of the drill rods (or anvil), the upward travelling tension wave reflects once again with opposite sign, this time from tension to compression. The reflected compression wave from the anvil then travels down the drill rods a second time, but with much reduced amplitude.

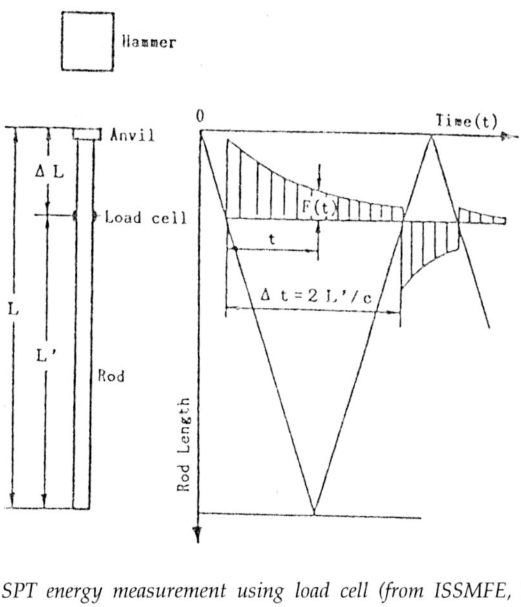

Fig. 5 *SPT energy measurement using load cell (from ISSMFE, 1988).*

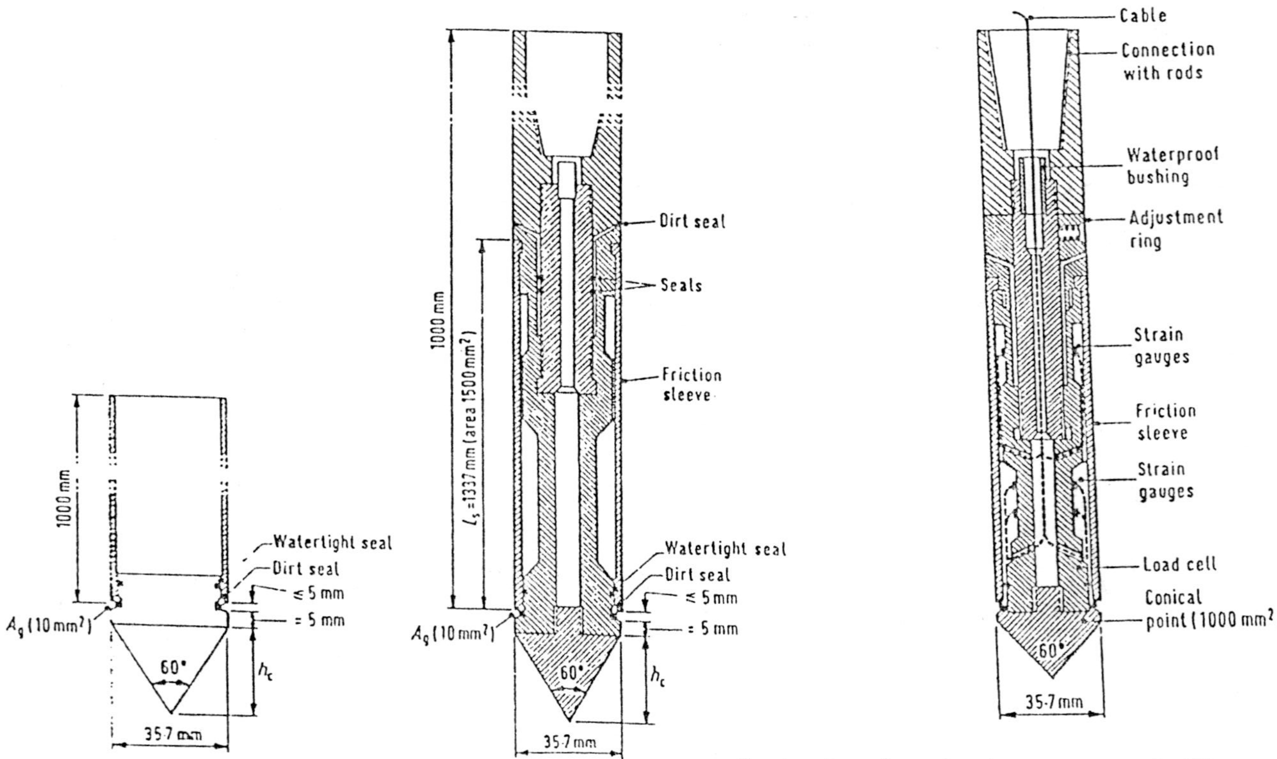

(a) Without friction sleeve

(b) With friction sleeve

Section through an electric penetrometer tip (R)

Fig. 6 *Section through electrical penetrometer tip.*

Recent Developments

Despite problems associated with its repeatability and reliability, the Standard Penetration Test (SPT) continues to be the most widely used in-situ test for foundation design, liquefaction potential assessment and compaction control in sands and silty sands. It is known that the most important factor affecting the SPT blow count (or *N* value) is the amount of hammer energy actually transferred into the drill rods. For liquefaction analysis, for example, the measured *N* values are commonly corrected to a reference energy level of 60% of the theoretical free-fall SPT hammer potential energy. The transferred energy is currently determined by attaching a load cell near the top of the drill rods and measuring the force time history during hammer impact. This method, however, has some shortcomings as indicated by several investigators in recent years.

In gravelly soils where the SPT *N* value is not reliable due to the large particle size relative to the sampler diameter, the Becker Penetration Test (BPT) has found useful applications. The BPT consists of driving a specially designed double-walled, closed-end pipe into the ground with a double-acting diesel pile hammer and recording the driving resistance or blow count per 0.3 m of pipe penetration. The BPT simulates the driving of a displacement pile and is like a large-scale continuous penetration test. Numerous attempts have been carried out in the past to correlate the BPT blow counts to the SPT N values, in order to make use of the large foundation performance database currently available for the SPT. Most of the BPT-SPT correlations, however, have limited applications since they did not take into account the inherent variable output of the diesel hammer used in the Becker system and they ignored the shaft friction acting on the pipe during driving.

An extensive study of the SPT and BPT has been conducted at the University of British Columbia (UBC) with the ultimate objective of obtaining reliable correlations between the two tests. The research project involved performing SPT, BPT and other in-situ tests in a controlled pattern at several sites in British Columbia, Canada. As part of the field testing, dynamic measurements were carried out which included force and acceleration near the top of the SPT drill rods and BPT drill pipes, as well as bounce chamber pressure in the double-acting diesel hammer during the BPT.

Recent developments in energy calibrations of the SPT and BPT show that the energy transferred into the SPT drill rods and BPT drill pipes can be determined by force and acceleration measurements, similar in principle to that used in pile driving. The proposed approach provides a fundamental and unified method of measuring transferred energies in the SPT and BPT, and is also applicable to other dynamic Cone Penetration tests. The measured energy data are then used in a consistent manner to correct the recorded blow counts to a reference energy level in each test. The energy correction allows the SPT and BPT blow counts from different drill rigs or hammers to be compared, and allow more reliable BPT-SPT correlations to be established.

Penetrometer Tests

Standard penetration tests carried out in boreholes suffer from the major disadvantage that the hole-making process usually disturbs the ground to be tested below the bottom of the hole. In most ground conditions where penetration tests are useful in predicting the performance of engineering structures, deep sounding techniques are a viable alternative and in many cases the latter provide both more reliable and more extensive data. Deep penetrometer tests can be divided into three major groups related to the method in which penetration is effected; static, static-dynamic and dynamic tests. In general, sounding equipment is light and portable and tests can be performed very rapidly. Average all-in work rates range from 5 m to 10 m per hour compared with 5 m to 10 m per day for conventional boring and sampling.

Cone Penetration Tests

In the cone penetration test, a cone on the end of a series of rods is pushed into the ground at a constant rate and the resistance to the penetration of the cone is measured either continuously or intermittently. If required, measurements are also made of either the combined resistance to penetration of the cone and outer surface of the rods or the resistance of a surface sleeve.

The test results are used :

1. to determine the soil profile and identify the types of soils present
2. to interpolate ground conditions between control boreholes
3. to evaluate the engineering properties of the soils and to assess bearing capacity and settlement

Normally the practice is not to rely on CPT alone for getting the design parameters unless the geology is fairly uniform and a lot of experience of working in a similar terrain and correlations exist.

The CPT has two main advantages over the usual combination of boring, sampling and standard penetration testing.

1. It provides a continuous although indirect record of ground conditions
2. It avoids the disturbance of the ground associated with boring and sampling particularly that which occurs with the Standard Penetration Test (SPT). Although, there is some disturbance which results from the advancement of the cone, this disturbance is consistent between one test and the other. The use of the Cone Penetration Test is also very cost effective particularly when a number of tests are to be carried out along the alignment of a canal or a roadway.

The Equipment

The CPT apparatus consists of a thrust machine with a reaction system, and a penetrometer including measurement and recording equipment. Machines generally have a thrust ranging from 20 to 200 kN. They can be classified into three categories light, medium and heavy. Hand-held and light weight 'suitcase' apparatus are also available.

Penetrometer Tips

Penetrometer are of two main categories: mechanical and electrical. In mechanical penetrometer, the force to mobilise cone resistance and local side friction is applied through rods and measured at the surface. In Electrical penetrometers, the forces are measured by electrical devices built into the tip and measurements are transmitted to the surface through a cable.

Electrical penetrometers are an improvement over mechanical devices. A section through an electrical penetrometer tip is shown in Figure 6.

Procedure

Generally CPTs are used in conjunction with borings and sometimes trial pits. In the light penetrometer rig the mechanical tip the cone is advanced by hand operation of a chain drive at an approximate rate of 20 mm/s. The load is measured by hydraulic load cells with pressure gauges or with a proving ring and a dial gauge. Tests are normally carried out at the intervals of 200 mms.

In the medium penetrometer rig, penetration is obtained by use of a hydraulic ram or rams. Usually the local side friction is measured in addition to the cone resistance. Penetration rate is kept at 20±5 mm/s. Mechanical tip is used in most cases. Electrical penetrometer tips are generally used with heavy rigs. This is also advanced at a rate of 20 mm/s and continuous readings are taken of cone resistance and of local side friction where a friction sleeve is incorporated.

The cable connecting the load cells within the tip to the read-out instrument is put through the hollow push-rods. Sufficient length of connecting cable is put inside push rods to reach the expected depth of testing.

With the purpose built truck mounted rig and a heavy penetrometer, the rate of testing can be between 150 and 300 m per day. With a trailer mounted medium rig, the rate of progress is nearly 50 to 100 m per day depending upon the terrain and the soils.

The equipment must be calibrated from time to time to make sure that the measurements are accurate.

Use of CPT Results

The CPT results can be used for the following purposes,

1. Soil profile along the depth
2. Identification of soil types
3. Properties of in-situ soils
   (a) Cohesionless soils
      (i) Relative Density
      (ii) Strength
      (iii) Deformability
      (iv) Correlation with SPT
   (b) Cohesive soils
      (i) Undrained shear strength of normally consolidated clays
      (ii) Undrained shear strength of over consolidated clays
      (iii) Deformability
   (c) Design of spread footings
   (d) Design of piles
   (e) Control of compaction and ground improvement
   (f) Liquefaction potential

Piezocone

Piezocone incorporates a porous element in the tip. With this modification it is possible to gain certain advantages over the conventional tip in the following activities:

(i) Interpretation of cone resistance and skin friction in terms of shear strength and deformation characteristics.
(ii) In-situ permeability and consolidation characteristics
(iii) Assessment of stress history and OCR of cohesive soils
(iv) Measurement of static porewater pressure.

It is essential that pore pressure measuring system should be saturated, because even a small quantity of entrained air significantly increases the compressibility of fluid affecting the response time of the system.

Following activities can be carried out with the help of a piezone:

(i) Soil profile
(ii) Identification of soil
(iii) Equilibrium pore pressure
(iv) Correlation of CPT values for pore pressure effects
(v) Undrained shear strength
(vi) Over-consolidation ratio
(vii) Effective stress shear strength
(viii) Coefficient of consolidation
(ix) Permeability

Case Study

One of the few organizations in India which possesses a proper Cone Penetrometer Rig is the Central Soil and Materials Research Station, New Delhi. A case history of foundation design on silty sands was presented by a team from CSMRS at a conference on Problematic Subsoil Conditions (1996).

Conventional tests were conducted for working out the bearing capacity of the foundations, besides using their 20T Truck Mounted Cone Penetration Rig (A.P. Van der Berg, Netherlands). Their findings were that the design based on conventional laboratory determined deformation characteristics may be somewhat misleading since the settlement computed using the laboratory test data appears comparatively less, prompting the engineer to use higher bearing capacity. Cone Penetration Tests using an electrical cone were also conducted at the same site (a multi-storeyed complex) and the foundation design based on properties interpreted from the cone tip resistance profiles suggested a lower foundation bearing capacity. The study concluded that Electrical Cone Penetration Tests gave more reliable and dependable bearing capacity values.

A summary of their findings is given below:

A. Bearing Capacity Based on Consolidation and Triaxial Shear Tests

The effective shear strength parameters obtained in the laboratory were $c = .06$ kg/cm^2, φ 36.1°. Adopting a value of $\varphi = 36°$ and using the procedure given in IS Code, the bearing capacity was worked out for different *D/B* ratios. Their values ranged from 214 to 820 KPa for different *D/B* ratio.

The safe pressure on foundation soil was also calculated based on one-dimensional consolidation test results. The settlement for square footings of different sizes is shown in Figure 7.

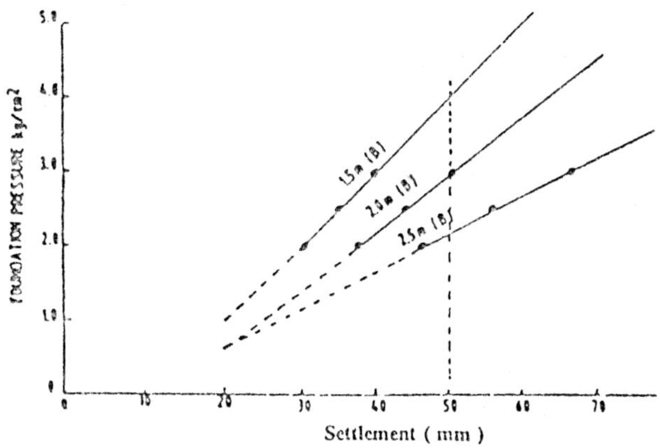

Fig. 7 *Foundation pressure and total settlement for square footing at 2 m below ground level using consolidation test results.*

B. Bearing Capacity based on CPT data

Figures 8 and 9 show the cone data in the form of tip resistance, sleeve friction and friction ratios. The two locations for CPTs were about 25 m apart. The uniformity of the soil deposit in the area can be seen from the two plots.

The correlation given by Robertson and Campanella (1986) was used and a φ value of 34° was determined. Settlement computations were done with the help of the formula given by Schmertmann (1978). Creep settlements were also computed using the same formulae. The results are shown in Figs. 10 and 11.

The following conclusions were drawn by the authors.

Significant differences in the allowable bearing pressures has been observed from the applied pressure versus settlement plots obtained from the consolidation test data and those determined from the Cone Penetration test data. Of the two approaches, CPT results have yielded more conservative and accurate values in view of the fact that in the case of CPT, the deformational properties of all the soil layers are taken into account to the finest possible degree.

Based on the continuous CPT plot, it is possible to identify the loose layers and such layers can be densified by vibrofloatation technique etc.

The penetrometer is driven with a standard hammer blow and the blow count for a given interval of depth is logged. The equipment is light and satisfactory penetration can be effected in most ground conditions. As only the cone is pushed statically, the required reaction for the test is substantially less than for an equivalent pure static sounding. Static-dynamic soundings have been found most useful in estuaries, where a stratigraphy of soft alluvium over gravel overlying over-consolidated clay or soft rock is common.

Dynamic Sounding

Again a wide variety of dynamic penetrometers have been developed with cone diameters ranging from about 15 mm to 110 mm. They fall into three broad categories. The 'single diameter' variety uses sounding tubes of the same diameter as the cone, although an anti-friction ring is sometimes used to reduce 'parasitic' friction on the sounding tubes, which of course increases with depth. 'Lost point' penetrometers consist of a cone of about twice the diameter of the sounding tube, which is loose fitted and is lost when the driving tubes are withdrawn. They are widely used in France, where it is felt that parasitic friction is reduced to negligible proportions. U.K. engineers hold the view that it makes the friction conditions uncertain and probably widely variable from sounding to sounding. Finally, 'flushed' penetrometers are being used particularly in France, where bentonite mud is pumped out just above the cone to form a frictionless skin on the sounding rods.

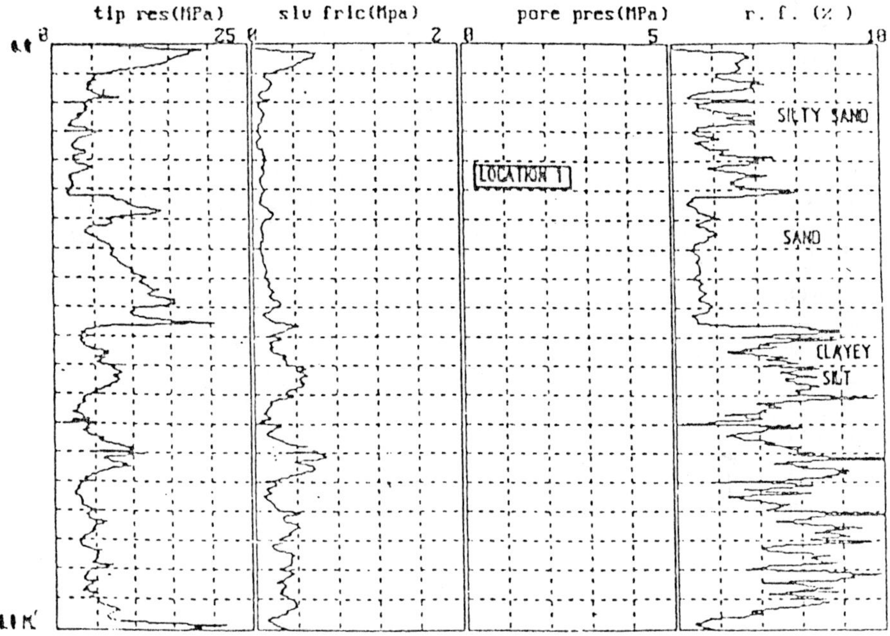

Fig. 8 *Cone penetration tests data from location 1.*

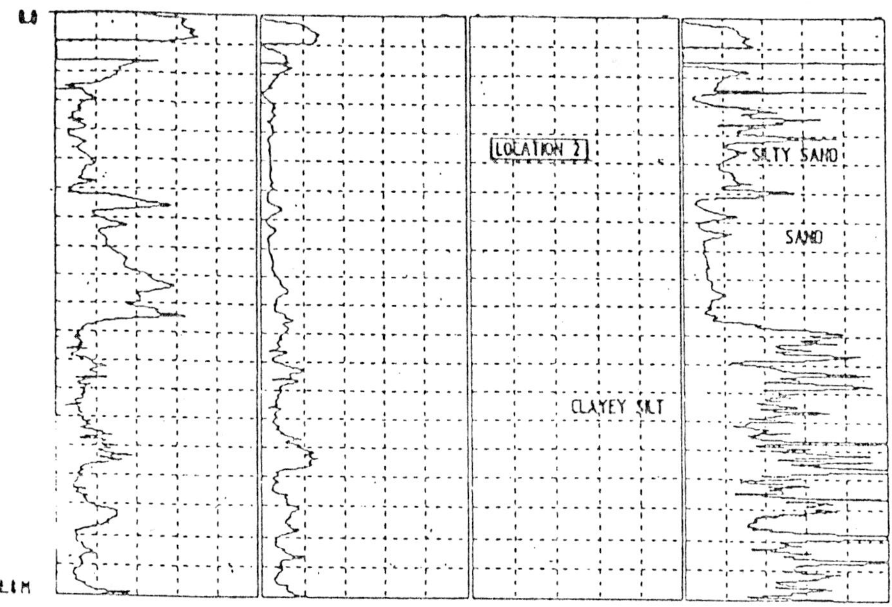

Fig. 9 *Cone penetration tests data from location 2.*

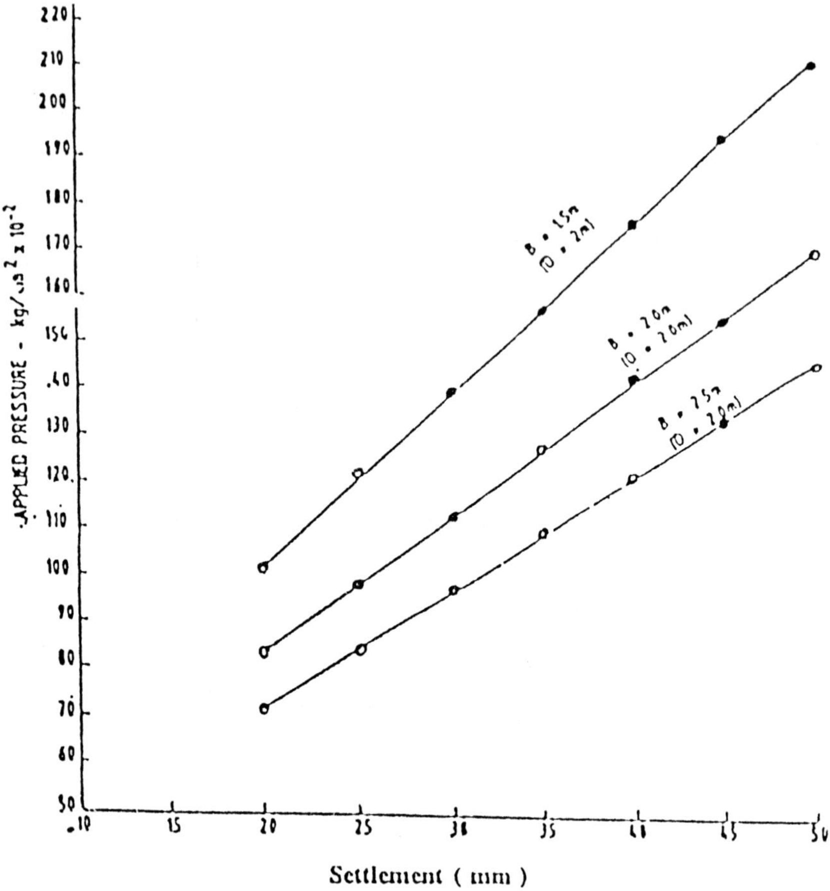

Fig. 10 *Foundation pressure and total settlement based on CPT data from location 2.*

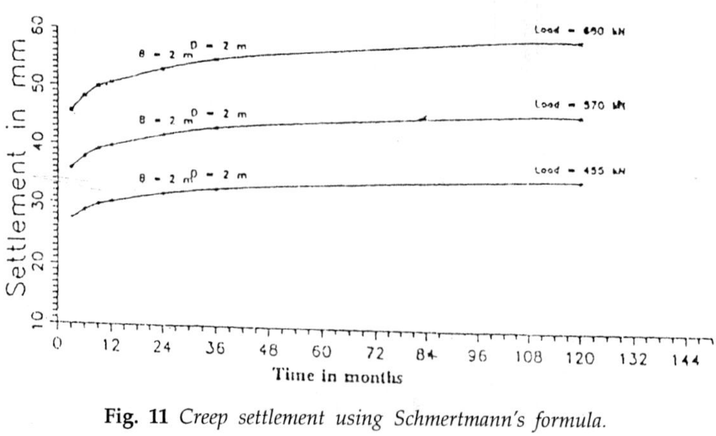

Fig. 11 *Creep settlement using Schmertmann's formula.*

In-situ Vane Test

The vane test was developed to determine the in-situ undrained shear strength or clays within the consistency range of very soft to firm. The major impetus for its development was that shear strength parameters obtained from laboratory tests were found to be lower than those obtained from stability analyses. It was considered that an in-situ method could reduce errors resulting from sample disturbance inherent in the laboratory approach.

Development took place simultaneously in Sweden by Olsson (1928) and in Germany (patent dated 1929). In 1947 the Swedish Geotechnical Institute designed several improved vanes and this work was reported by Cadling and Odenstad (1950).

In essence, the vane now consists of four thin rectangular blades welded on to a small circular shaft to form a cross. Generally, the height to diameter ratio is equal to two. Testing consists of advancing the vane from its protection shoe into undisturbed soil. A torque head is then fitted at the surface to rotate the vane at a constant rate of shear and record the torque causing soil rupture along the cylindrical surface passing through the edges of the vane blades and along the horizontal surfaces at the top and the bottom of the blades. For each test the relationship between torque and angular rotation can be obtained, and knowing the vane dimensions and the maximum torque applied, the undrained shear strength is calculated. Remoulded strength is determined in the same manner after destroying the soil fabric by rapid rotation of the vane, and allowing a short time period for pore pressures to decay. Sensitivity is determined as the ratio of natural to remoulded shear strengths.

In addition to the testing of weaker cohesive soils, several researchers have investigated its use in cohesionless and organic soils but interpretation in these materials is less certain. Used in suitable soil conditions the vane test is a relatively cheap and rapid method of obtaining soil parameters.

Factors Influencing the Results

Installation

Unfortunately the vane cannot be installed in the ground without causing disturbance to the soil around it. Changes of stress are caused by boring and a zone around the bottom of the hole will be disturbed.

Where the vane body is forced into the ground from the surface, a plastic zone is formed around the vane protection shoe with the contact surface often completely remoulded. Measurements have been made which indicate that the immediate pressure at the contact surface could be 8 to 10 times the undrained shear strength.

The intrusion of the vane itself into the clay may be sufficient to break clay bands in sensitive clays and also to induce an increase in pore pressure surrounding the blades, which may not have dissipated when the test is carried out. These effects will lead to strength estimates lower than the in-situ state.

Time Effect

The above disturbing effects and the ability of clay to regain strength with time means that the shear strength measured by the vane will vary according to the time elapsed between installing the vane and the commencement of shearing (Flaate, 1965). A standardized procedure is, therefore, important and a delay of between 3 and 5 minutes between installation and the start of the shearing process is used. The standardized rate of strain during the shearing is 0.1 degree per second.

The engineer must, therefore, carefully consider the rate of loading associated with any project compared to the rate of strain applied in the vane test.

Anisotropy

The direction of failure has been shown, by means of both compression and extension tests, to have a significant effect on the shear strength value. In many problems it is necessary to consider the shape of the failure surface to allow for correction due to an isotropic effects. As has been mentioned earlier, it is possible to estimate the anisotropy by using two different vanes at the same position or the same vane in the horizontal and vertical plane.

Progressive failure

In some soft clays, in particular those of high sensitivity, the stress-strain curves exhibit a sharp peak followed by a rapid and large reduction in strength after failure has occurred. Simultaneous mobilization of peak shear strength can only occur if the strains undergone by the clay are uniform. Therefore, progressive failure may be initiated at the edges of the blades where the clay is most intensely stressed. Again, in a soil with lithological layering the clayey bands may reach a peak more readily than the silty bands and consequently the vane test stress strain curve will be a combination of two strength components which do not reach peak at the same strain. Further research is required to evaluate the significance of this effect.

PRESSUREMETER TESTS

In the pressuremeter system, there is a fluid filled flexible membrane or dilatometer giving an all-round constant pressure, which applies pressure on borehole walls. The resulting deformation of borehole walls is measured with the help of a volumeter.

The most widely used dilatometer is the *Menard Pressuremeter* (Menard 1957) which may be used in materials varying from stiff clays to soft rocks and with some additional equipment in soft clays, sands and gravels. Fig. 12 shows the layout of the test equipment which is made up of a down hole probe and surface volumeter connected by a pressure line. Pressure is applied to the measuring section of the probe via a water column and the volumeter measures volume change of the probe. Guard cells are used both ends of the measuring section of the probe in order to contain the expansion into a purely radial sense. With the increase in pressure, volume change of the measuring cell is monitored at the surface using the volumeter with measurements taken at 15, 30, 60 and 120 seconds after the application of

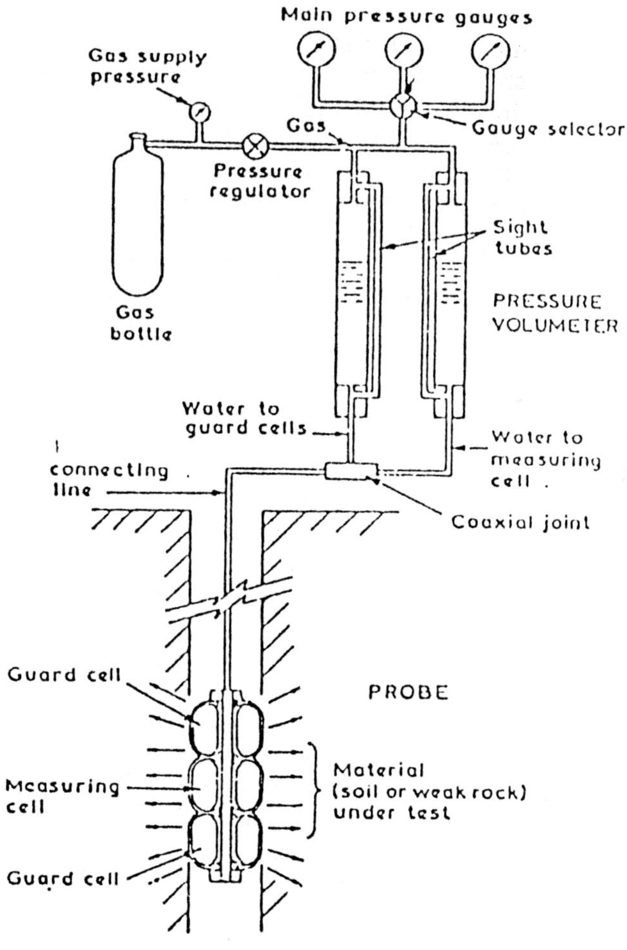

Fig. 12 *Menard pressuremeter.*

the load increment. Figure 13 shows a typical pressure/volume curve in which pressure corrected for various effects and the volume reading 120 seconds after increase in pressure are plotted. A 'creep' curve is also plotted as the difference between the 30 s and the 120 s reading. The test is normally carried out on an NX sized pocket about 1 m long, drilled at the bottom of a larger sized borehole. It is also possible to test within a special slotted casing which may be driven into soft materials and some sands and gravels. As may be seen in Fig. 13 the results provide a loading and generally reloading modulus, *Po* a measure of the lateral earth pressure and *Pf* and *Pl* which indicate the failure characteristics of the material. The moduli are calculated as a diametral change based on a measured volume change and require the assumption of a suitable Poissons ratio for the material being tested.

A simpler form of pressuremeter may be used to estimate levels of horizontal ground stress at depth. This is known as the *Geocell*, which is a flexible membrane installed at the required depth and set at a particular pressure. By monitoring pressure in the geocell over a period, which may be extended, this will rise or reduce until it becomes constant, at which point it indicates the mean surrounding horizontal pressure.

Developed at Cambridge University and fitting generally into the dilatometer category is the *Camkometer*, which is designed to measure Ko and behaviour characteristics of soft and very soft materials (Wroth and Hughes 1972). The instrument is in two sections, the lower part containing

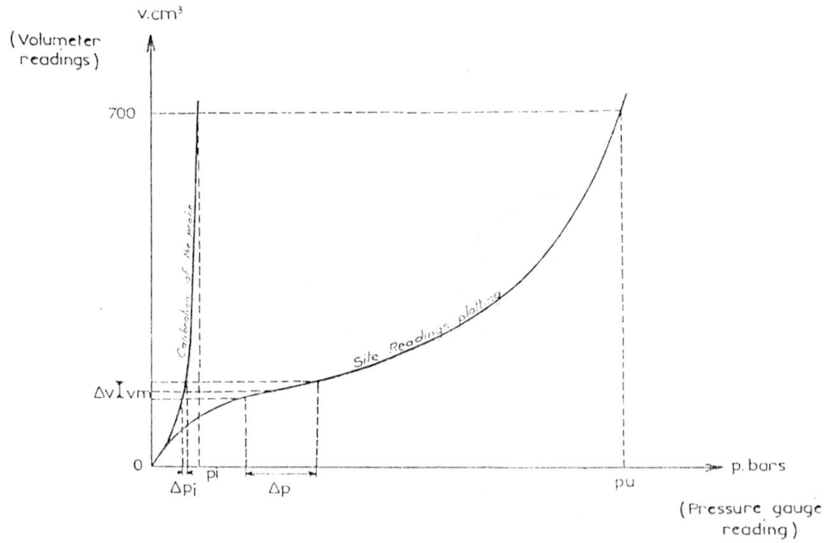

$$E = K\,(vm)\,\frac{\Delta p - \Delta p_i}{\Delta v} \simeq K\,\frac{\Delta p}{\Delta v}$$

Fig. 13 *Typical pressure volume curve—menard pressuremeter.*

load cells for the measurement of total lateral stress and hence, from assumptions of vertical stress, *Ko* is determined. The upper section of the probe has a flexible rubber membrane which may be expanded hydraulically against the soil with strain measured using resistance strain gauges on a mechanical feeler. This is needed to determine stress/strain behaviour of the soil. The complete probe is carefully jacked into the ground while the soil is excavated by the action of an internal cutter rotating within a cutting head. Water circulating within the central tube on which the cutter is mounted washes the soil cuttings up the annulus formed between this tube and the outer cylinder and this procedure ensures that the probe is inserted with a minimum of disturbance.

Pressuremeter testing is traditionally carried out within predrilled vertical boreholes or with the self-boring pressuremeter. Soil moduli are measured in the horizontal direction since expansion of the pressuremeter probe imposes a lateral stress on the surrounding soil inducing plane strain conditions. Anisotropy and stress dependency may mean that the vertical modulus of the soil differs from the horizontal modulus, thus rendering test results inappropriate to the design situation.

Comparisons of plate and pressuremeter test results have been carried out in various instances, for example Powell and Uglow (1985), Windle and Wroth (1977), Wroth et al. (1979), Walker and Jewell (1979) and Marsland et al. (1983). The correlation between pressuremeter and plate moduli varies according to a number of factors such as material type, the technique employed to derive moduli from test results and the stress or strain ranges over which moduli are measured. The aim therefore when deriving material parameters from test results should be to ensure that values are compatible with design techniques to be employed. With this in mind, the ideal situation is where parameters are back calculated from full-scale structural behaviour and compared to those derived from pressuremeter curves. The most appropriate parameters should then be identified. An example of this can be seen where the unload-reload moduli from Cambridge self-boring pressuremeter have been compared to values obtained from analysis of the settlement of a reactor (Wroth et al., 1979).

Clough (1990) notes that unload-reload moduli from pressuremeter curves seem to give the most appropriate values for design and in addition, these values are relatively unaffected by drilling disturbance, which is often a large consideration, especially in sands.

Pencel Pressuremeter

The Pencel pressuremeter is a small diameter pressuremeter initially developed for use in airport and highway design in North America in the 1970s and 80s, principally by Briaud and Shields (1979), Briaud (1979), Briaud et al. (1982) and Briaud et al. (1987). The Division of Roads and

Transport Technology (DRTT) of the CSIR in South Africa also applies this pressuremeter to pavement evaluation design (Sanders, 1991). In particular, the pressuremeter is being used to measure in-situ pavement layer material stiffnesses and being applied to the South African Mechanistic Pavement Design Procedure (Freeme, 1983).

Two pressuremeters have been used at DRTT: 32 and 22.5 mm diameters, with active lengths of 225 and 150 mm respectively, thus maintaining a length diameter ratio of around 7 which is comparable with a typical ratio of between 6 and 6.5 used for general commercial pressuremeters. Ideally the length diameter ratio should be as large as possible in order to simulate ideal radial expansion over as much of the length of the membrane as possible, thus reducing end effects to a minimum.

The smaller pressuremeter was developed principally to test pavement layers down to 150 mm thick where holes are quite easily predrilled in materials with an electric power drill. The holes are drilled deeper than the length of the pressuremeter probe to provide relatively equal confining stresses within the zone of influence of the pressuremeter. A schematic representation of the pressuremeter test is given in Fig. 14.

Before testing, the pressuremeter is calibrated by measuring the membrane resistance to expansion. The expansion of the entire apparatus when pressurized is also measured for calibration purposes. After drilling, the hole is cleared of all drilling spoil, the probe inserted to the required depth and the test conducted by addition increments of approximately 2.5% of the initial probe volume. Unload reload cycles are carried out where required during testing to obtain elastic moduli, the amplitude of which should be restricted to twice the undrained shear strength (C_u) in cohesive materials and for non-cohesive materials, the pressure difference between the passive and active failure conditions (Wroth, 1982):

$$\Delta p = \left(\frac{2 \sin \phi'}{1 + \sin \phi'}\right) * (p - u_0)_{\max}$$

Where Δp represents the allowable pressure difference, ϕ' the angle of shearing friction and $(p\text{-}u_0)_{max}$ the effective radial pressure at the wall of the cavity at the start of the unloading cycle.

On completion of the test the membrane stress calibration is again carried out using the same increments of volume as used during the test.

Following the manufacture of the small pressuremeter the possibility of using the device of horizontal insertion in large diameter boreholes was considered. In this way vertical soil stiffness parameters, as opposed to horizontal parameters obtained via plate bearing and conventional pressuremeter testing, should be obtained. Another perceived advantage was that as the apparatus is small and light, tests would be carried out

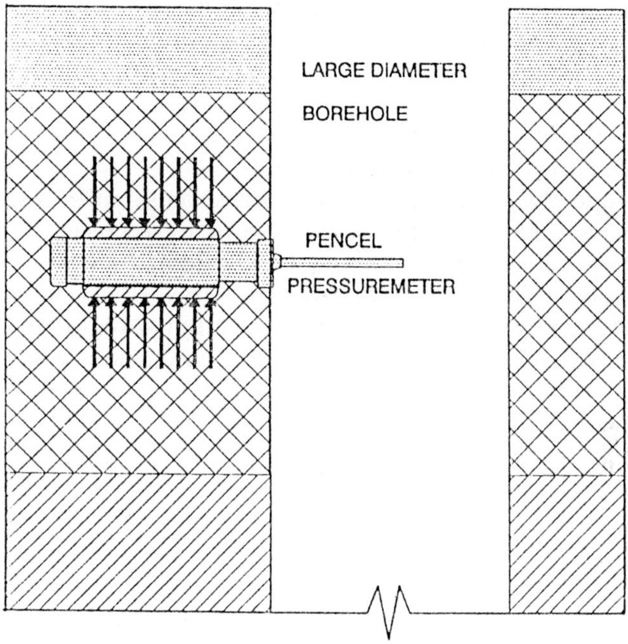

Fig. 14 *PENCEL pressuremeter—horizontal insertion.*

more easily than lateral plate bearing tests for instance. Equipment required to insert the probe is a brace and bit or a power drill to drill the horizontal hole and a tubular reaming tool to smooth the sides of the hole and to remove drilling spoil. The *Pencel* pressuremeter is then inserted and the test conducted.

LARGE DIAMETER HOLES

In several countries, extensive use of large diameter (750 mm) auger holes for soil profiling and sampling is made, especially where geotechnical investigations for larger structures are carried out. Investigations typically involve inspection and sampling of materials for laboratory testing. This is possible over most of the interior of the country where permanent water tables are relatively deep, i.e. typically below 20 m depth. Areas where large diameter boreholes would not be used include soil profiles where seepage is present from shallow perched water tables, as well as in coastal areas where mainly estuarine and beach deposits are associated with shallow permanent water tables. The only other important constraint to using this method is suitability of the soil profile for auguring. For soils and very soft rock, virtually the only significant problems are when large boulders in transported soils or corestones in residual soils are encountered.

Using large diameter boreholes for investigation has the advantage of obtaining a first hand chance of evaluating the soil profile in terms of some of its primary properties, i.e. moisture content, colour, consistency, structure and origin. It also allows "undisturbed" block samples to be cut from the sides of these holes. Although good quality samples are relatively easy to obtain for laboratory testing, the effects of changes in stress and moisture content are difficult to quantify and avoid.

Consequently, laboratory tests conducted on samples obtained from large diameter holes are often augmented by in-situ tests. Typically this may comprise lateral plate jacking tests within the holes or possibly pressuremeter testing in smaller diameter holes. Despite being fairly popular, lateral plate bearing tests are often difficult to carry out satisfactorily due mainly to the awkwardness of handling the apparatus inside a large diameter borehole and preparing suitable bearing surfaces. These surfaces should be parallel to each other and smooth, necessitating plaster of paris of a similar rapid hardening seating agent. This increases the time taken for the test to be carried out and introduces more complexity to the test procedure and analysis.

IN-SITU PERMEABILITY TESTS

In-situ permeability tests range from packer tests in rock through piezometer permeability tests of the rising, falling or constant head type, to tests carried out during boring. The theory applied to borehole and piezometer tests is identical. The borehole test is more appropriate to granular materials, while the piezometer test is suitable for fine grained deposits. An arbitrary boundary might be placed at coefficient of permeability, $k = 1 \times 10^{-6}$ m/s.

The soil around a borehole is subject to stress change during boring and there are pore water pressure changes in response to these stress changes. In granular materials of higher permeability, the pore water pressures stabilize quickly and borehole tests are appropriate. In fine grained materials the pressure recorded by a piezometer must be allowed to come to equilibrium before a test can be performed. Since pore pressures are depressed by boring, filling the piezometer with water assists the equalization process. Because of the swelling/consolidation behaviour of the soil under changing pressure (from the piezometric head), it is possible for the piezometer to 'overshoot' equalization and then return to a steady level. After equalization the sand pocket of the piezometer is surrounded by soil at a constant pore pressure.

If an excess pressure is applied to a piezometer or borehole and this pressure is sustained, the following effects will occur:

1. There will be a change in pressure on the boundary of the sand pack and in response to this, a change in stress and in pore pressure in the surrounding soil.

2. Seepage will occur from the pocket into the soil (assuming a positive head change) with a reduction in effective stress in the soil resulting in swelling of the soil. The flow rate will depend on the permeability and the coefficient of swelling of the soil. Ultimately (where spherical flow occurs) steady flow conditions are reached and in this condition the flow rate is governed by the permeability alone.

It is immediately obvious that the variable head test may be a rather unsatisfactory test because of the constantly varying conditions. Nevertheless, provided that small head differences are used in the finer grained soils, their results can be interpreted satisfactorily. In general, however, the constant head test is to be preferred in all materials, although it is substantially more time consuming and costly.

Borehole Permeability Tests

Various falling, rising or constant head tests in open boreholes or piezometers are carried out. The pumping of water into selected lengths of borehole is used to examine several properties of soils all of which relate to permeability. Single or double packers are used for sealing the boreholes.

Pumping Tests

The other category of tests is called Pumping Tests.

1. *Single Packer Testing:* Carried out as the borehole are advanced in which the lowest portion of the hole, generally 3 or 6 metres, is isolated for testing using a single packer.
2. *Double Packer Testing:* Taken up after drilling is complete and two packers at a fixed distance apart (usually 3 or 6 m) are used to isolate successive lengths of borehole.

The test consists of pumping water under pressure into selected lengths of boreholes isolated by the use of pneumatic or mechanical packers and the test is carried out in one of two ways.

Water is injected into the ground under pressure applied in increments, with the highest pressure in the test zone being generally no greater than the effective overburden pressure. For each pressure increment, water flow is maintained until a constant flow rate is recorded.

From the recorded pressures and rates of flow the permeability may be calculated using the formulae presented in the US Bureau of Reclamation Earth manual (1968). In their simplest form these formulae may be written as:

$$k = C_p \frac{q}{H} \tag{1}$$

where permeability (k) is related to the constant rate of flow into the hole (q) and the corrected excess head of water (H) by a shape factor (C_p). The shape factor depends upon the relative magnitudes of the test length (*L*) and hole diameter (2*r*) and values are tabulated in the Earth Manual for various typical drillhole sizes. The most valid results are obtained when the thickness of the stratum tested is at least 5m. The test results are strictly accurate only for submerged tests, i.e. when carried out below the water table.

The corrected excess head (*H*) is the sum of the applied pressure (*P*) and the excess gravity head (*h*) minus the head loss (h_e) due to friction losses through the supply pipes and rods. The head loss varies with volume of flow and is usually obtained from standard charts or an on site calibration.

For each test the corrected pressure (*H*) against rate of flow (*q*) curve is drawn and the permeability is calculated at excess heads relevant to the particular problem being investigated. Consideration should be given in interpreting the test results to the form of the pressure/flow diagram (Muir Wood and Caste 1970).

REFERENCES

ASTM Standard D4366-86. Standard Test method for stress wave energy measurement for dynamic penetrometer testing system. *ASTM standards*, Vol. 04.08, 872-875.

British Standard 1377, British Standard Specifications

Briaud, J.L., 1979, The Pressuremeter: Application to Pavement Design, PhD dissertation, Civil Engineering, University of Ottawa, Canada.

Briaud, J.L, Lytton, R.L. and Hung, J.T., 1982, Using a Pressuremeter for Pavement Design and Evaluation. *Int. Symposium on the Bearing Capacity of Roads and Airfields.* Civil Aviation Administration, Oslo. The Norwegian Institute of Technology, Trondheim, Norway.

Briaud, J.L., Cosentino, P.J. and Terry, T.A., 1987, Pressuremeter moduli for airport design and evaluation. *Final Report No. DOT/FAA/PM-87/10.* Program Engineering and Maintenance Service, Washington D.C. 20591.

Briaud, J.L. and Shields, D.H., 1982, A Special Pressuremeter and Pressuremeter Test for Pavement Evaluation and Design. *Geotechnical Testing Journal ASTM.* Vol. 2, No. 3.

CSMRS, 1996, Krishnamurthy, K.S., Bora, D.N. and Venkatachalam, K.A. Case Study of Foundation Design on Silty sound Deposits using Cone Penetration Test Data. *Terzaghi 96 Conference on Problematic Soil.* Kakinada.

Companella, R.G. and Alex Sy., 1994, Recent Development in Energy Calibration of Penetration Tests at UBC. *Proc. 13 Int. Conference on Soil Mechanics and Foundation Engineering,* New Delhi, Vol. 2. 151-156.

Cadling, L. and Odenstad, S., 1950, The Vane Border; An Apparatus for Determining The Shear Strength of Clay Soils Directly in the ground. *Royal Swedish Geotechnical Institute Proc. 2.*

Clough, G.W., Briaud, J.L. and Hughes, J.M.O., 1990, The Development of Pressuremeter Testing Special Lecture 2. *Proc. of 3rd Int. Symposium on Pressuremeters.* British Geotechnical Society, Oxford University.

Earth Manual, 1968, Department of the Interior. Bureau of Reclamation, USA.

Flaate, K., 1965, Field Vane Shear Tests with Delayed Shear *5th Pacific Area National Meeting of ASTM* (Seattle).

Freeme, C.R., 1983, Evaluation of pavement behaviour for major rehabilitation of roads. *Technical Report RP/19/83.* Pavement Engineering Technology, Division of Roads and Transport Technology CSIR, Pretoria.

Gibbs, H.J. and Holtz, W.G., 1957, Research on Determining the Density of Sands by Spoon Penetration Tests. *Proc. 4th Int. Conf. Soil Mechanics and Foundation Engineering*, London.

Jamiolkowski, M. and Presti C.C.F. Lo, 1994, Validity of In-situ Tests related to real behaviour. *Proc. 13th Conference on Soil Mechanics and Foundation Engineering*, Vol. 5, 51-55.

Menard, L., 1957, Mesures Insitus des Proprietes Physiques des So/s. Annales des Ponts et Chausees No. 3.

Meyerhof, G.G., 1956, Penetration Tests and Bearing Capacity of Cohesionless soils. *Journal of Soil Mechanics and Foundation Engineering*, Paper No. 866, Vol. 82.

Marsland, 1971, The shear strength of Stiff Fissured Clay *Proc. Roscoe Memorial Symposium Foulis.*

Meigh, 1964, Discussion on Sutherland. Instt. Of Struct Eng. Vol. 42 No. 2.

Muirwood & Caste, 1970, In-situ testing for the channel Tunnel. *Symposium on In-situ investigations in Soils and Rocks British Geotechnical Society*, Paper 10.

Meigh, A.C. and Hobbs, N.B., 1973, *Soil Mechanics.* Chapter of Civil Engg. Reference Book. Butterworths.

Powell, J.J.M. and Uglow, I.M., 1985, A Comparison of Menard, Self boring pressuremeter and push in pressuremeter tests in a stiff clay fill. *Proc. Conf. On Advances in Underwater Technology and Offshore Engineering*, London.

Palmer, D.J. and Stuart, J.G., 1957, Some Observations on the Standard Penetration test and a correlation of the test with a new Penetrometer. *Proc. 4th Int. Conf. Soil Mechanics and Foundation Engineering*, London. Vol. 1.

Robertson, P.K. and Campanella, R.G., 1986, Guidelines for use, interpretation and application of the CPT and CPTU. *Soil Mechanics Series no. 105*, Deptt. Of Civil Engg., University of British Columbia.

Strydom, J.H. and Sauders, P.J., 1994, The Pencel Pressuremeter: A useful tool for routine site investigation. *13th International Conference on Soil Mechanics and Foundation Engineering.* Vol. I, p. 129-132.

Schultze, E. and Menzenback, E., 1961, Standard Penetration Test and Compressibility of Soils. *Proc. 5th Int. Conf. Soil Mechanics and Foundation Engg.*, Paris. Vol. I.

Serota, S., 1971, Discussion on Paper by Ireland, Moretto and Vargas (1970). *Geotechnique,* Vol. 2.

Schmertmann, J.H. & Palacios, A., 1979, Energy Dynamics of S.P.T. *Journal of Geotechnical Engineering,* ASCE 105 (GT8) 909-926.

Schmertmann, J.H., 1978, Guidelines for Cone Penetration Test-Performance and Design. US Department of Transportation. Report no. FHWA-TS-78-209.

Sanders, P.J., 1991, A proposed Subgrade Design Procedure using the PENCEL shear Pavement Pressuremeter (PSPP). *Contract Report DPVT/ C. 1962. 1* Division of Roads and Transport Technology CSIR, Preloria, July 1990.

Sully, J.P. and Campanella, R.G., 1994, Evaluation of Field CPTU Dissipation data in over-consolidated fine-grained soils *13th International Conference of ISSMFE,* Vol. I. 201-204.

Terzaghi, K. and Peck, R.B., 1948, *Soil Mechanics in Engineering Practice.* John Wiley and Sons, New York.

Thorburn, S., 1963, Tentative Correlation Chart for the Standard Penetration Test for non-Cohesive Soils. *Civil Engg.* and Public Works Review.

Tomlinson, M.J., 1969, *Foundation Design and Construction.* Sir Isaac Pitman, London 2nd Edition.

Vucetic, M., 1994, Cyclic Characterization for seismic regions based on PI. *13th International Conf. of ISSMFE,* New Delhi, 1994. 329-332.

Vucetic, M. and Zorapapel, G.T., 1990, The effect of liquefaction process on the spectral content of strong ground motion. *Proc. Third U.S.-Japan Workshop on Soil Liquefaction,* San Fransisco. 403-418.

Wroth, C.P., 1982, British experience with the self-boring pressuremeter. *Proc. Symp. on the Pressuremeter and its Marine Applications Institute du Petrole,* Laboratories des Ponts et Chausees Paris.

Windle, D. and Wroth, C.P., 1977, In-situ measurement of the properties of stiff clays. *Proc. 9th Int. Conf. On Soil Mechanics and Foundation Engineering,* Tokyo.

Wroth, C.P. et. al., 1979, A Review of the engineering properties of soil with particular reference to the shear modulus. *Cambridge University Engineering Department Internal Report.*

Walker, L.K. and Jewell, R.J., 1979, The selection of design parameters in weathered rock. *Proc. 7th Eur. Conf. On Soil Mechanics and Foundation Engineering.*

Zorapapel, G.T. and Vucetic, M., 1990, The effect of porepressure increase in sandy liquefiable deposit on the spectral content of strong motion. *Proc. 4th US National Conference on Earthquake Engineering.* Vol. 3. Palm Springs. 793-802.

2

Characterization of Strength of In-Situ Soils

W.H. Ting[1]

INTRODUCTION

In contrast to the superstructure where the construction material can be selected and then manufactured, in geotechnical problems the material is as found in-situ and often non-homogeneous.

Due to the wide range of soil and rock materials that are often encountered on site, to implement a geotechnical design it is necessary to determine the soil properties from soil testing and deduce design parameters that are *characteristic* of the deposit encountered as well as providing a *safe bound* for the designed geotechnical structure or treatment.

The procedure of characterisation is considered as one of the crucial steps in the implementation of geotechnical structures. To achieve a rational scheme for characterisation, it will be necessary to treat the subject by placing the procedure of characterization within the overall plan for solving geotechnical problems. The subject is herein introduced against the background of geotechnical construction, followed by stating the general principles of characterisation before focusing on specific examples in the characterisation of soil strength.

GEOTECHNICAL CONSTRUCTION

Geotechnical structures may be conveniently separated into various categories according to their basic loading support methods. For instance, division permits the sharing of knowledge between different types of construction within each category, as well as between different categories. In this way it is possible to benefit from transfer of technology from one to the other. Geotechnical construction involves the ground as well as structural solutions applying concrete and steel and other building materials. Ground improvement or treatment is also placed within the

[1]Geotechnical Engineer, Dr. WH Ting, Consultants 18 Jalan SS20/10 Damansara Kim, 4700 Petaling Jaya Selangor, Malaysia

context of the construction. The categories of support systems are as shown below:

1. *Direct Support System* for External Loading
   (a) Embankment and fill ground directly support water, traffic and building loads. The examples are bund, dam, road and airfield, and reclamation. The loads are in turn *transmitted* to a natural ground foundation.
   (b) It is to be noted that such support systems are readily replaceable by other structural systems that may be considered as alternatives.
2. *Support to Ground* (as slope, vertical face, basement, cavity, etc.) with associated loading
   (a) The support is distinguished by the need to resist important *lateral earth pressures* in various forms as a result of ground weight loading.
   (b) The support is also typified by the contribution of ground strength to the total support in addition to structural support. (An example where ground strength is sufficient on its own is a stable slope.)
   (c) Additional support is provided by traditional structural systems. The examples are retaining walls, pile walls, structural membranes, cavity lining, culvert, etc.). Part of the received load is transmitted to the natural ground foundation.
   (d) Anchors, bolt, dowel, nail, reinforcement installed in the geotechnical structure may provide direct support; or act through the structural support.
3. *Foundation Support System* by Natural Ground
   (a) The foundation support is in the way of a *final stable support* of the loadings transmitted to the natural foundation by the two previous support systems or by direct superstructure loading.
   (b) *Passive earth pressure* has an important role in providing the required resistance.
   (c) Typical examples are embankment on soft ground, and when loading is transmitted by spread foundation and piles to the natural foundation.
   (d) The foundation support can also be in the form of an improved ground.
4. *Ground Improvement / Treatment*
   (a) The ground itself may be directly improved by an appropriate treatment. A wide variety of methods exist and typical methods are briefly mentioned hereinafter as illustration.
   (b) Densification by static and dynamic compaction (such as for embankment and reclamation fill).
   (c) Consolidation by surcharge and drainage by pre-formed vertical drains.
   (d) Columns (sand, stone, cement, lime, etc.).

(e) Dewatering and de-saturation
(f) Cement grouting and chemical injection
(g) Freezing

PRINCIPLES OF CHARACTERISATION

General

The characterisation of soil properties for the design of structures follows basic engineering procedures that take into account the inherent limitation of soil mechanics in the definitive explanation of geotechnical processes.

The following sections briefly describe the limitations and the specific role of characterisation of soil properties within the context of the solution of geotechnical problems as a whole. The soil properties are determined by in-situ tests as well as by laboratory tests on field samples.

Design Methodology

General principles

The successful solution of all geotechnical problems entails the implementation of a suitable theoretical framework to manage the construction during the planning and design stage. Design parameters are input into a *computation model* as the framework, and results obtained either aided by an application program or by calculations with a spreadsheet. The computation of algorithms in a spreadsheet has gradually replaced charts and tables. The spreadsheet is appealing as changes in the variables are displayed, and progress in physical processes may be followed during computation. The results obtained from the model are applied to design, as dimensions of structural components with the required material properties, capable of supporting the specified loading with a distortion that is within the safe bound.

In geotechnical engineering the design as implemented on the site, inevitably necessitates the modification of the theoretical framework to take into account the variability of soil properties and other conditions encountered on the site. The modification is carried out by empirical procedures in the form of site observations and testing and back-analysis of the original framework.

Computation models

The building of an appropriate computation model involves challenging problems. The particular geotechnical process is modeled as the response of a body with a given configuration and material property to a specified loading. The basic link between loading and response is the constitutive equation for the strength and stress-strain relationships, which may also involve other physical processes (e.g. hydro-mechanical).

Various physical and mathematical models have been proposed as computation models. The models are idealisations that approximate to the actual physical processes. Simplified models are often introduced if computation procedures become intractable. Full-scale physical models may be implemented in the form of in-situ load testing and measurements and monitoring of loads and movement of the site structure.

Numerical techniques such as finite difference or finite element are then applied to the given boundary value problem, to solve the set of equations that are generated. The mathematical model may have analytical or numerical solutions. More often than not numerical solutions have to be obtained, as closed form solutions are not easily available for complex boundary conditions. Where available, analytical solutions provide a useful visualization of the mechanisms involved in a problem.

Simplified failure mechanisms can also be proposed to complement the computation model (Ting, 1987). In soil mechanics the ultimate state is usually assumed. The limitation of the available computation models naturally causes great concern to geotechnical engineering practitioners. Voluminous technical literature has therefore been devoted to the field performance of geotechnical structures, where various computation models have been evaluated.

An appreciation of the problem of the construction of geotechnical structure has to begin with an understanding as of the limitations of computation models. The limitation of concern herein is the difficulty in characterising the soil properties as input design parameters into the models.

Empirical and observational methods

Where a suitable computation model is not easily available, empirical procedures are often employed. A rational empirical method is nonetheless based on observed physical processes that realistically describe the problem.

The empirical methods are closely related to load testing, measurement and monitoring of full-scale structures in what is known as *observational methods*. In geotechnical engineering, observational methods serve not only towards the design review process, but they also provide the database to be applied to similar site problems in the future as well as for further development of theoretical methods.

Soil Models

Basis

Background

In geotechnical engineering the selection of appropriate design parameters that are characteristic of a given deposit, poses daunting problems. The technique central to characterisation is the development of soil strength and stress-strain *soil models* that incorporate fundamental physical principles. The soil models are an integral part of the computation models as well as empirical methods. The strength model is applied to computation of stability whilst the stress-strain model is required when in addition to stress; the deformation has to be calculated.

The soil properties after characterisation become input design parameters to the computation models. The soil model serves as a theoretical framework to analyse and to characterise soil properties. Besides the models to be applied here other classical models and those of recent origin are also briefly described.

In the soil strengths considered, it is implicitly assumed that effective stress parameters are applied, unless expressly stated otherwise.

SHEAR STRENGTH PARAMETERS

In selecting shear strength, the appropriate strength parameter to be applied has to be determined. For example, Schofield (2000) expressed the view that in design problems, disturbed soil properties may be used with undrained action on soil and $c = c_u$, or with drained action on soil and $\phi = \phi_d$. Worth (1984) preferred the use of the symbol s_u in place of c_u for *undrained shear strength*, with the same definition of half the deviator strength. There is a distinction between the defined undrained shear strength, and the shear strength (s) and the cohesion component (c') of the shear strength obtained by using the Mohr-Coulomb approach. In practice s is approximately the same and smaller in value than s_u and is not a major practical issue when undrained condition is assumed. In drained analysis with low confining stress, cohesion as a distinct component becomes important.

SCHEME FOR CHARACTERISATION

In view of the many issues affecting characterisation, a broad scheme has been introduced under which this presentation is made. The scheme is however not exclusive but determined by the needs in the practice of characterisation. Other methods of the division of the subject are available and acceptable. The two schemes employed are interdependent.

In Table 1, the division of the subject is according to the type of natural ground and its use as a fill.

The conditioning of sediments would usually include drying out, self-consolidation and nominal compaction. The use of the ground as a natural formation could still result in a certain amount of remoulding during eathwork operation.

In Table 2, the subject is divided according to the soil models that will be introduced to analyse soil behaviour for the purpose of characterisation:

Table 1
Soil Types

Soil Type	*Use*	*Processing*	*State*
In-situ sediments	Formation	Natural	Undisturbed
	Formation	Consolidation	Remoulded
Excavated soils	Fill	Nil	Disturbed
	Fill	Conditioned	Remoulded
	Fill	Compacted	Disturbed
	Mining	Mined out	Disturbed
Weathered formation	Formation	Natural	Intact
	Colluvium	Removed	Disturbed

Table 2
Soil Models

Model	*Interpretation*	*Soil Type*	*Processing*	*State*	*Parameters*[1]
Cam clay	Worth (1984)	In-situ sediments	Natural	NC/OC	s_u
Cam clay/	Ting (1999, 2000)	In-situ sediments	Natural	NC/OC	(c′, φ′)
Mohr-coulomb		Weathered formation	Compacted	OC	(c′, φ′)
Empirical	Nithiaraj et al. (1996)	Weathered natural formation	Natural	OC	(c′, φ′)

NC—Normally consolidated
OC—Over-consolidated
[1](c′, φ′) Mohr-Coulomb parameters.

Elastic and elasto-plastic models

The best known classical relationship is the elastic model. Although the elastic model does not perfectly describe soil behaviour, it is nonetheless widely applied as a theoretical framework. The elasto-plastic model is an addition to the elastic model in the way of the introduction of a limiting strength.

A vast array of solutions has been made available based on elastic methods (e.g. Poulos & Davis, 1974; Randolph & Wroth, 1978). Within the safe bound of the performance of geotechnical structures, the "elastic state" is a valid approximation as small deformations are involved. In lieu of perfect linearity, a secant modulus may be also applied, or piece-wise linearity may be assumed. It remains to suitably characterize the elastic modulus of the in-situ soils (e.g. Tatsuoka & Shibuya, 1991).

Mohr-Coulomb

The Mohr-Coulomb failure criterion is a strength model that has enjoyed a vast success. The Mohr-Coulomb shear strength comprise cohesion and a strength component derived from confined friction. Many geotechnical problems such as slope stability have been solved with the failure criterion. A limitation is that being a strength model, the calculation of deformation is not included.

Cam-clay model and other recent developments

The Cam clay theory (Roscoe et al., 1958), based on Critical State Soil Mechanics (CSSM), has significantly advanced our understanding of soil behaviour. It is the most satisfactory formulation that has been developed so far. The model provides a useful theoretical framework not only as a computation model for the design of geotechnical structures but also as a soil model to analyze soil behaviour. The model has found many uses in day-to-day soil mechanics and is relied herein as the basic analytical tool for soil characterisation. It can also be applied for the interpretation of other existing models and testing procedures.

Other varieties of constitutive relationships are continually being proposed. They are often of great complexity but have found specific applications.

Empirical models

In real life, simplified constitutive relationships are widely used. Often various normalisation procedures are introduced to handle the soil parameters; only fundamentally based empirical procedures are proposed herein.

Bounds in Values of Site Parameters

Statistical Approach

The non-homogeneity of soil deposits is greater than other widely used construction materials that are manufactured such as concrete and steel. Following traditional engineering procedure, the application of the principles of statistics to geotechnical engineering has been attempted to deal with the non-homogeneity.

Simons (1981) delivered a lecture on "Statistics, reliability theory and safety factors". He raised the question of the safety of given soil mechanics characteristic values and affirmed that the scatter of soil properties as determined are the highest when compared with other engineering materials. The uncertainties involved are higher than those of assumed loading and computation methods. The following salient points concerning the relevance and limitations of the application of the principles of statistics to geotechnical engineering have been largely deduced from his lecture:

1. The design soil parameters are to be obtained from test values obtained from "representative random samples" of the soil deposits.
2. Well-known statistical techniques such as the determination of the mean values, standard deviation and the coefficient of variation by the assumption of normal distribution can apparently be applied to the evaluation of the scatter of test results on the soil samples.
3. The difficulty begins with deciding whether in the first place, the soil samples are "representative random samples".
4. The random samples may not be representative, due to well-known deterministic factors such as:
   (a) Depositional history (that are geological processes such as weathering, metamorphism, lithology, erosion, etc.)
   (b) Influence of depth (resulting in variations in saturation, degree of weathering, consolidation, etc.)
   (c) Man-made processes (compaction, mining, excavation, etc.)
5. This means that it has to be ensured that the statistical techniques are only applied to batches of soil samples that are obtained from a stratum known to have been formed with consistent deterministic factors enumerated above, varying only in a random manner within that strata. Alternatively, variance analysis may be carried out to test the homogeneity of different batches of samples.
6. Furthermore, in order to apply the statistical techniques the sample size must be statistically significant. This means sample sizes of not less than 20 numbers. This is clearly not practical in most situations in geotechnical investigation. However the Student's *t*-distribution may be applied to evaluate the significance of using small smaples size. (As an example if it is practicable to sink 3 numbers of boreholes for a given structure, according to the *t*-distribution analysis, 3 numbers boreholes will produce test results that are 18.8% of the accuracy of the ideal sample size of 20 numbers.)
7. Statistical techniques, such as regression analysis and the determination of the coefficient of correlation may be applied to determine causal correlation between different variables where there is reliance on different testing methods.
8. The relationship between the Probability of Failure—a statistical viewpoint and the Factor of Safety—a deterministic viewpoint has been established. The probability approach is naturally to be preferred. Due to the practical difficulties experienced in obtaining "representative random samples" of a statistically significant "sample size", the deterministic approach continues to be the dominant method relied on in practice.

Approach in Practice

In classical soil mechanics (e.g., Terzaghi & Peck, 1967), for strength values, it is sometimes simply recommended that the lowest value of strength parameter obtained should be used. In BS 8002 (1994) for earth retaining structures, it is recommended that "representative" soil strength should be values of soil parameters, which are more adverse than the most likely values. The "representative" soil strength need to be "conservative" values, and the most likely values are presumably the mean values. It is apparent that further rationalisation will be in order. The afore mentioned recommendations illustrate that the principles of statistics have always been applied in a notional way to geotechnical engineering so as to conform to basic scientific methods.

The following illustrates how the principle of statistics may be applied in a practical manner in the characterisation of soil strength.

1. The superstructure, being relatively well defined, is known to be designed to a probability of failure of 1 in 10^6. It follows that the foundations supporting the structure should not have a greater probability of failure.
2. It would be useful to relate the characterisation of soil strength with the characterisation of in-situ concrete strength as presented in BS 6089: 1981 "Assessment of concrete strength in existing structures". This is of relevance as the same probability of failure is targeted. In that standard, the characteristic. In-situ concrete strength (f_{cu}) is taken as the strength that is 1.64 times the standard deviation below the mean strength. This is to ensure that with a normal distribution, 95% of the concrete structure sampled will have a strength that is better than the characteristic concrete strength.
3. It will be reasonable to expect that the characteristic strength of a soil in-situ should be selected so that more than 95% of the soil formation has a strength that is better than the characteristic strength. The 95% value need to be further increased if the tests are not on "representative random samples" and the sample size is not statistically significant. Under normal practice the 95% is inevitably increased, as there is real difficulty in determining whether the samples are representative and random, and it is often not practical to obtain sample sizes that are statistically significant.
4. Another factor that cannot be discounted is the reliability of commercial quality soil tests. For the in-situ concrete, BS 6089 recommends that the design strength should be the characteristic in-situ strength further reduced by a factor of 1.5 to reflect differences between site and laboratory conditions.
5. In order to realise optimum structures, the methodology above is confirmed by an engineering judgmental decision that is further discussed hereinafter.

Steps in characterisation

In-depth knowledge of soil mechanics

The necessary skill for characterisation of soil properties would begin with an in-depth knowledge of the fundamentals of soil properties.

In the first place, a sound textbook *knowledge of techniques* of characterisation will be required. A recommended standard text, that efficiently covers the topic of the characterization of soil properties for many situations is that authored by Professors Lambe and Whitman ("Soil Mechanics"). Textbook knowledge by itself will not be sufficient. In the topic of selection of design parameters, it is necessary to be to date with developments in research and analyses of case histories. An appreciation of earlier developments can be obtained from the Proceedings of the VII ECSMFE 1979,—"Design Parameters Geotechnical Engineering" (British Geotechnical Society, 1981). In that conference a variety of geotechnical materials such as soft and stiff clays, granular soils, weak rocks, fill materials, improved soils and special soils have been separately dealt with. The effects of processes, such as degradation, weathering, total and partial saturation, remoulding, and chemical and physical changes on exposure, have also been discussed in great detail for the various types of soils. Several soil models were presented and discussed.

Next an intimate *knowledge of the ground* to be characterized should be acquired. The nature and geology of the deposit had to be understood; the broad classification for construction purposes being natural and prepared (or filled) ground. The ground, depending on their usage is excavated and formed into slope, underground cavity, basement, channels, trenches vertical face, etc., and by compaction into embankments and reclaimed land.

Further knowledge required would include an understanding of the methods cited hereinbefore (computation models, application of statistical principles, etc.) and the field performance and method of construction cited hereafter.

Field performance and methods of construction

As mentioned hereinbefore geotechnical engineering has inherent problems with:

1. Computation models
2. Selection of an appropriate characteristic soil property

As a solution, the performance of similar structures on similar formations needs to be determined in one of the following ways:

1. From personal experience
2. By examining case histories recorded by others
3. By initiating performance studies and applying observational methods when implementing the construction.

The method of construction has to be carefully planned to avoid temporary ground instability. The safety of one part of the structure as affected by construction of other parts has to be ensured when selecting the appropriate geotechnical solution to a construction problem.

Engineering judgement

Finally an engineering judgement has to be exercised to select the appropriate characteristic after having applied the following procedures:

1. The evaluation of the limitations of the relevant computation models
2. The application of the principles of statistics in a practical manner
3. The input of in-depth knowledge of soil mechanics
4. The acquisition of the knowledge of the performance of similar constructions in similar formations, and an awareness that method of construction has to be taken into consideration.

Peck (2000) affirmed, "that the successful practice of engineering requires a high degree of engineering judgement".

Recommendation on characterization

In the following sections the process of characterization of soil strength is presented in detail. Although concerned with characterization of strength, the principles are equally applicable to characterization of other soil properties. The presentation begins with the introduction of the soil model to be applied. In-situ deposits of normally consolidated and over-consolidated soils are first considered. The discussion will then proceed to soil strength in both in-situ and related laboratory testing for soil sediments as well as for weathered formations.

SOIL MODELS FOR CHARACTERIZATION

Wroth's Interpretation of In-Situ Soils

Wroth (1984) in his Rankine Lecture presented the interpretation of in-situ soils within the general framework of the CSSM and the deduced Cam clay model. He presented for engineering application the fundamental basis of the relationship:

$$s_u = S_{ru} p_o' \qquad (1)$$

where s_u is the undrained shear strength equal to half the deviator strength,
p_o' is the effective vertical stress on the tested sample,
S_{ru} is the undrained strength ratio that has a particular value depending on the over-consolidation ratio and nature of the deposit of the sample.

Wroth's interpretation is applicable to in-situ soil sediments. The method may also be applied to investigate built compacted embankments where in-situ shear tests have been carried out. A detailed study of different types of laboratory and in-situ strength tests was also made by Wroth and reference should be made to his lecture for further understanding of differences in the testing results obtained.

The simple relationship proposed is a powerful analytical tool for characterization. The S_{ru} values usually range between 0.2 and 0.25 for normally consolidated clayey soils. Higher values of S_{ru} will denote that a soil is over-consolidated (Wroth's Fig. 31). When p_o' varies linearly with depth, as Eq. (1) is a linear relationship, deviation from linearity will indicate departure from a normal sedimentation process. As an example, crusting may occur on upper soil layers or there may have been erosion after deposition.

Previous to Wroth's efforts, the undrained strength ratio has been widely used as an empirical tool for soft soil (e.g. Jamiolkowski, 1981). The ratio has also been applied to undisturbed weathered formations, but the interpretation of the results would be fundamentally different from that of sediments.

Model for (c',ϕ') Soils

Cohesion as a soil property

In laboratory tests on field samples, a cohesion intercept is obtained as a component of shear strength in the Mohr-Coulomb plot in drained tests on over-consolidated soils and on samples from weathered formations. In the site total confining stress is often reduced (as in excavations), resulting in changes to effective confining stress over the long-term. In the simple physical sense, cohesion may be viewed as a component of shear strength that is important even after the effective confining stress of the sample is reduced to a small quantity. Examples of the maintenance of long-term shear strength after reduction of confining stress can be observed in the field as the sustained vertical banks (Terzaghi & Peck, 1967) and the ability of desiccated surface clay stratum to support traffic loads. Cohesion has a basic role in the maintenance of long-term shear strength.

Some further observations of interest that needs to be accounted for are listed below:

1. The contribution of cohesion (c') in a dominantly frictional (ϕ') soil is significant in the analyzed behaviour of earth structures. This is particularly so for the analysis of slopes and walls where a small quantum of c' has an important effect on the factor of safety.
2. In natural sediments, soil may occur as lightly to heavily over-consolidated deposit. This would be reflected in the range of (c',ϕ') values obtained in triaxial shear tests and a procedure for characterizing the strength behaviour would be pertinent.
3. Compacted clay has over-consolidation features and retains adequate

strength in low confining stress situations such as in support of traffic load in road pavements.

4. In the drained analysis of slopes of compacted clay embankments, the assumption of a totally friction material (Schofield, 2000) means that the shear strength mobilized in lower slopes with low confining stress would be low. Such an assumption may be unduly conservative. Due to the over-consolidated nature of compacted clays, at low confining stress the available sear strength is mainly accounted for by "cohesion" as a component of strength that would be underestimated by assumption of a fully frictional material.
5. The measured strength parameters of a given soil may be altered due to stress release during sampling and sample preparation. It should be possible to relate the altered soil strength to the original expected property.
6. It is to be noted that under field condition cohesion may be subject to degradation over the long-term or reduction for various reasons. In the assignment of cohesion as a component of strength in design, possible decrease in value of the assumed cohesion has to be investigated.

The notion of cohesion as defined in the Mohr-Coulomb criterion may be "embedded" in the Cam clay model (Scofield, 1998), but the recognition of cohesion as a distinctive component of strength will serve to explain and characterize observed shear strength as earlier explained.

In the following sections an attempt shall be made to separate strength into effective cohesion and friction components within the framework of CSSM under triaxial compression test conditions. The study accepts the validity of the Mohr-Coulomb relationship as a tool to analyze the role of the various components of shear strength; the uniqueness of the components needs however to be further investigated.

Cohesion in Relation to Friction

Soil strength is likely to be derived from friction, and cohesion could be regarded as a component of the strength of a modified friction soil. A specific instance of how the property of cohesion can be rendered to a frictional soil is introduced.

Schlosser (1982) succinctly demonstrated that sand (from Fontainebleau) could be transformed into a material with cohesion behaviour by the introduction of reinforcement in the sand. He made the statement that "...at failure caused by the breaking of the reinforcements the samples ("tested") possessed a cohesion....". His observation is based on a series of laboratory tests of samples with and without reinforcement carried out by Long et al. (1972). The relationship in Fig. 10 of their test report is re-plotted here as Fig. 1.

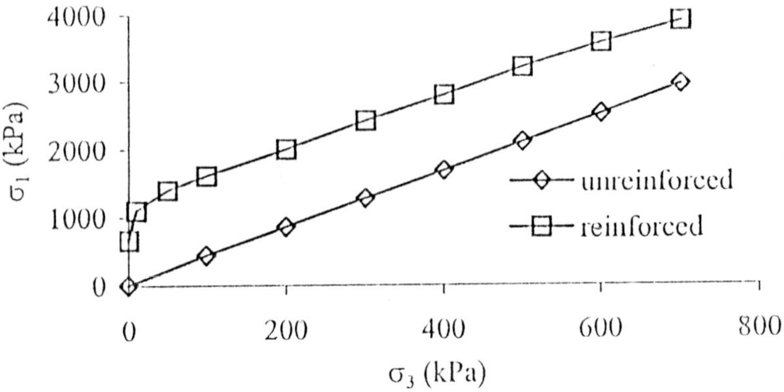

Fig. 1 *Reinforced sand in triaxial test.*

A relationship has been deduced from the findings and presented as:

$$c = K_p / 2R_t / \Delta H \tag{2}$$

where c = cohesion
K_p = Rankine passive earth pressure coefficient
R_t = tensile resistance of reinforcement per linear meter
ΔH = verticle spacing between two adjacent reinforcements

The Eq. (2) is applicable to the reinforced sand at failure ("in the Mohr-Coulomb condition") when the reinforcement is at breaking point. A relevant point to note in Fig.1 is that the friction angle (represented by the slope of the two lines) has the same value both in the un-reinforced and the reinforced case.

Penman (2000), in the course of his presentation, drew attention to the early interest in the ability of a cylindrical soft clay sample to remain intact before an unconfined compression test (presumably as a result of "cohesion" sustained by suction acting as the effective confining pressing, p').

Besides the action of reinforcements, the other "agents" capable of rendering the property of "cohesion" to "frictional" soils in the field are the:

1. Over-consolidation process
2. Suction of partly saturated soils
3. Cementation material:
   (a) as a weathering product of the less resistant components of fresh rock
   (b) derived from subsurface solution processes, such as laterite (from soluble iron oxide), organic compounds and calcareous material.

The agencies responsible for effects of over-consolidation are geological processes such as rise and fall of sea levels, glacial movements, erosion, desiccation, etc., and man-made processes such as compaction.

It is possible that cohesion arising from cementing and mechanical reinforcement effects differs in important aspects from the over-consolidation processes. A clue may be obtained from Eq. (2) where it is shown that the strength and spacing of the reinforcement (in addition to the Rankine passive pressure coefficient) affect the value of the cohesion obtained. The results may be qualitatively interpreted to mean that the measured cohesion value is affected by the strength and distribution of cementing agents and reinforcement elements; but seemingly confining stress is not involved; whereas cohesion from over-consolidation effects will be shown later to be a function of the confining stress. The friction value is an independent but important parameter, and it establishes the "packing" of the material before it is cemented or reinforced and also enters into the Rankine coefficient. In practice the findings will be translated to mean that, in strength characterization it is necessary to differentiate between over-consolidated soils (sediments and compacted clays) and weathered natural formations, especially in relation to the role of the cohesion component.

Ting (1999) derived the cohesion component of strength from the basic assumption of a frictional soil and the formulation is reproduced hereinafter. The derivation is then applied as an analytical tool in the characterization of in-situ soils (Ting, 2000) that will also be presented.

Formulation

DEDUCTIONS INVOLVING CAM CLAY MODEL

Presentation

The Cam clay model representation for normally consolidated (NC) and over-consolidated (OC) soils are shown in Fig. 2. AC and CR are respectively the normal consolidation and swelling lines. Physically the NC and OC soils at the same void ratio (represented by initial states A & R respectively in Fig. 2), have related properties in the Cam clay model as shall be further elaborated in the next section.

(In line with Roscoe et al. (1958), the symbol for the mean effective stress, together with akin quantities, is p and not p'.)

Cam Clay relationships

The Cam clay model assumed that undrained stress paths are 'similar' for NC soils. Wroth (1984) presented the findings of Loudon (1967), in his undrained triaxial shear tests, that the 'similarity' may be extended to stress paths of OC soils. It can be demonstrated by combining the various Cam clay findings that for a given soil the critical state line may be represented by:

$$q_m = q_r = S_r p_e = S_r p_o R^A \qquad (3)$$

where

q_m is the deviator strength of the NC soil

q_r is the deviator strength of the OC soil

$S_r = q_m / p_e$ (= constant for 'similar' paths but p_e may vary)

$p_e = p_o R^A$ (= $p_o \exp((e_o - e)/\lambda)$, Ting; 1968)

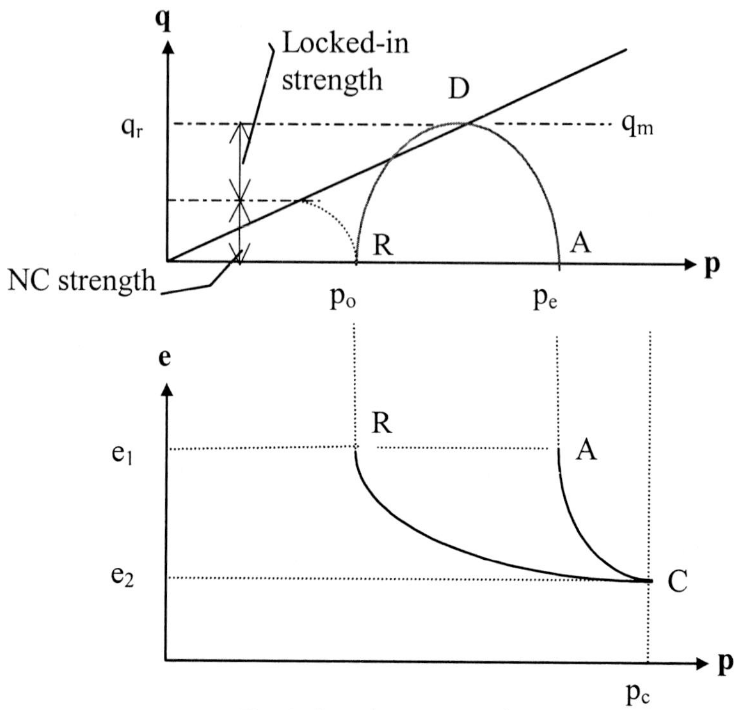

Fig. 2 *Cam clay representation.*

$R = p_c \cdot p_o$ = over-consolidation ratio
$\Lambda = 1\text{-}\kappa/\lambda$ (by definition)
κ = slope of *e*-In *p* swelling relation assumed linear in the model
λ = slope of *e*-In *p* compression relation assumed linear in the model
p_e, p_c and p_o have the meanings shown in Fig. 2

In site problems S_r, can be determined from triaxial shear tests, and p_c determined from consolidation tests. The CSSM undrained shear strength, $q_r/2$, derived from Eq. (3) is plotted against p_o in Fig. 3 for Weald clay.

Locked-in strength (Assumption I)

The deviator strength (q_r) for the OC soil at *R*, (Fig. 2) may be considered to be the sum of two components. It is assumed that the first component, the locked-in strength (q_l), is as a result of a locked-in compression (p_e-p_o) after the soil is compressed from A to C and then swells to *R*, where the soil is at the same void ratio as at A. The second component (q_n) is as a result of the behaviour of the soil as a Cam clay type NC soil under initial confining pressure (p_o). Eq. (3) can then be recast as:

$$q_r = q_l + q_n = S_r(p_e - p_o) + S_r p_o = S_r p_o(R^{\Lambda}) \qquad \text{(3a)}$$

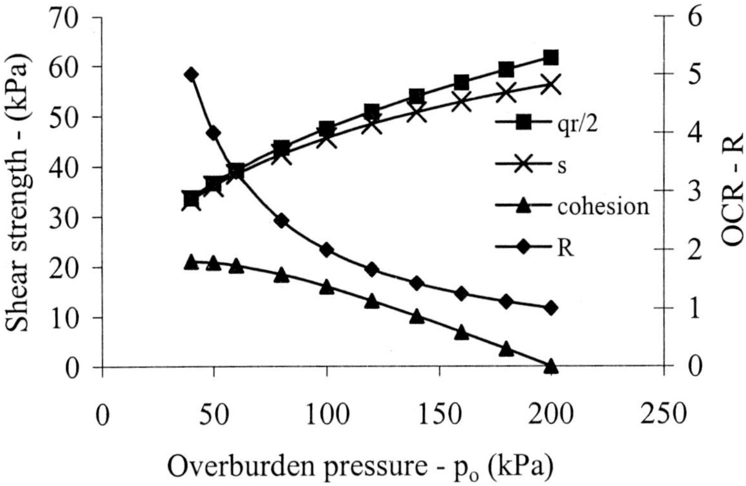

Fig. 3 *Shear strength vs Overburden pressure.*

In the above equation, according to the assumption, the first component (locked-in strength) and the second component (NC strength) are given respectively by:

$$q_1 = S_r(p_e - p_o) \tag{4}$$

$$q_n = S_r(p_o) \tag{4a}$$

Deductions involving Mohr-Coulomb failure relationship

Shear strength

Since the Mohr-Coulomb relationship is assumed to be valid for OC soils, in a triaxial compression test at point S_1, in Fig. 4, the shear strength:

$$s = (q_r/2) \cdot \cos\phi' = (q_1/2+q_n/2) \cdot \cos\phi' = (S_r p_o R^{A}/2) \cdot \cos\phi' \tag{5}$$

The deviator strength q_r, is defined in Eq. (3) and q_1 and q_n defined in Eqs. (4) and (4a).

Cohesion component (Assumption II)

The Mohr-Coulomb equation is $s = c' + \sigma' \tan\phi'$ (Fig. 4).

A follow-up of Assumption I, is to assume that the locked-in strength component (q_1) in Eq. (4) is entirely responsible for the cohesion (c') component of strength. Thus:

$$c' = (q_1/2) \cdot \cos\phi' = [S_r(p_e - p_o)/2] \cdot \cos\phi' = [S_r p_o(R^{\wedge}-1)/2] \cdot \cos\phi' \tag{6}$$

From the Mohr-Coulomb equation and combining Eq. (5) and Eq. (6):

$$\sigma' \tan\phi' = (S_r p_o/2) \quad \cos\phi' \tag{6a}$$

Effective normal pressure (σ')

The pore water pressure (u) at failure has to be determined and is here assumed as Skempton's pore water pressure parameter (A_f) multiplied by the deviator stress (q_r). Worth (1984) provides a theoretical relation for

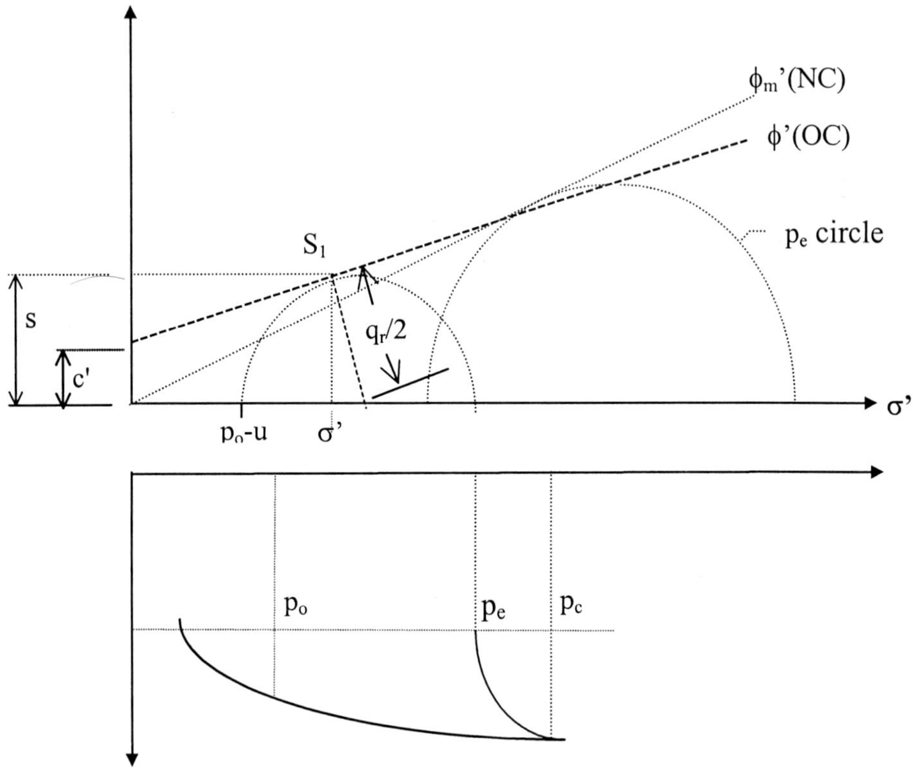

Fig. 4 *Mohr-Coulomb representation.*

A_f (his Eq. (41)) deduced from the Cam clay model that will be applied. Published data on the parameter may also be used. The formulations are further developed as below:

From Fig. 4—Mohr-Coulomb representation:

$$\sigma' = p_o - u + (q_r/2) \,.\, (1\text{-}\sin\phi) \tag{7}$$

where, $u = A_f.q_r$

$$A_f = 1/M.[(R/r)^{-\wedge} + M/3\text{-}1] \text{ (Worth)}$$

$$q_r = S_r p_o R^{\wedge} \text{ Eq. (3)}$$

Solution of equations

Available equations

Together with the Mohr-Coulomb equation there are three other equations:

Eqs. (5), (6) and (7) with four unknowns s, c', $\sigma'$and ϕ'.

Solution

Given:

$$c' = (q_1/2) \,.\, \cos\phi' = [S_r p_o(R^\Lambda-1)/2] \,.\, \cos\phi' \text{ Eq. (6)}$$
$$\sigma' = p_o-u+(q_r/2).(1-\sin\phi') \text{ Eq.(7)}$$

Substituting the above relationships in the Mohr-Coulomb equation: $s = c' + \sigma'\tan\phi'$:

$$s = [S_r p_o(R^\wedge-1)/2] \,.\, \cos\phi' + [p_o-u+(q_r/2).(1-\sin\phi)].\tan\phi'$$

Also,

$$s = (q_r/2) \,.\, \cos\phi' \text{ from Eq. (5)}$$

Equating the two equations for s and inserting and expanding Eq. (7):
$$(S_r p_o R^\wedge/2).\cos\phi' = [S_r p_o(R^\wedge-1)/2].\cos\phi' + [p_o-u+(S_r p_o R^\wedge/2).(1-\sin\phi)].\tan\phi',$$

$$\cos\phi'/2 = [1/S_r - A_f R^\Lambda + (R^\Lambda/2).(1-\sin\phi)].\tan\phi'.$$
$$(1/\sin\phi' - \sin\phi')/2 - R^\Lambda(1-\sin\phi)/2 = 1/S_r - A_f R^\Lambda$$

Thus,

$$(R^\Lambda - 1).\sin^2\phi' - [2/S_r - R^\Lambda(2A_f-1)].\sin\phi' + 1 = 0 \qquad (8)$$

Eq. (8) is a quadratic equation in sin ϕ for a given value of R. Knowing Λ, S_r, and A_f, the equation can be explicitly solved to deduce ϕ'. Once ϕ' is deduced, c' can be calculated explicitly from Eq. (6), and σ' from Eq. (7) to give all the Mohr-Coulomb parameters.

Results and Discussion

Results from formulation

Baisc input parameters

The input parameters required in the presentation of results in the following sections are shown in Table 3. The magnitudes of the parameters need to be consistent with the theoretical relationships of CSSM as presented by Wroth (1984). Random combination of the parameters (some of which are interlinked) would lead to results that would be in conflict with the Cam clay model. The over-consolidation ratio has been restricted to a range of 1.0 to 5.0 to avoid numerical difficulties. The restriction is reasonable as site values usually fall within the range. It is also difficult to obtain reliable test results from heavily over-consolidated soils.

The Cam clay values have been obtained from tests carried out by Ting (1968) and the Weald clay results fromWroth's (1984) paper. The site test data have been obtained from a drainage and irrigation project in Sarawak, Malaysia.

Table 3
Values of basic input parameters

Properties	*Symbol*	*Unit*	*Cam clay*	*Weald clay*	*Site (all)*	*Site (local)*	*s/p*$_o$
Deviator strength ratio	S_r		0.48	0.62	0.72	0.77	0.39-0.96
Swelling index	κ		0.04	0.035	0.03	0.08	0.045-0.03
Compression index	λ		0.16	0.093	0.095	0.22	0.02-0.09
	Λ		0.75	0.624	0.68	0.64	0.78-0.67
Spacing ratio	r		2	2	2	2	2
	M^*		0.8	0.95	1.16	1.2	0.67-1.5
Max. past consolidation pressure#	p_c	kPa	200	200	200	150	200
Equivalent NC pressure#	p_e	kPa	180	50	na	na	na
Friction angle (NC)	ϕ_m'	degrees	21	24.2	29	30	17.5-37.5
Average slope (Hvorslev plot)	ϕ'	degress	12.3	16.6			

*deduced; # when constant; na: not applicable.

Mohr-Coulomb plot

In conventional triaxial shear test on over-consolidated soils three samples are usually tested at initial consolidation pressures (p_o) close to the existing overburden pressure. The samples have the same maximum past consolidation pressure (p_c) with the overburden pressure (p_o) varying along a single swelling line as in Fig. 4. The conventional testing path has been applied to the cases of Cam and Weald clays. The resulting value of ϕ' deduced from the solution of Eq. (8) has been applied to Eqs. (5) and (7) and the resulting relation between s and σ'presented in Fig. 5.

It can be seen that within the range of over-consolidation ratio considered, the (s, σ') relationship is approximately linear. It has to be noted that the stress path has been chosen to suit conventional test procedures. The slope angles of the lines are close to the friction angle value for NC soils and do not represent the friction angle values as calculated from Eq. (8). As will be shown in the next section, the (s, σ') relation is dependent on the path selected for the tests.

The test path dependency is one of the reasons for the existing difficulties encountered in the evaluation of the strength of over-

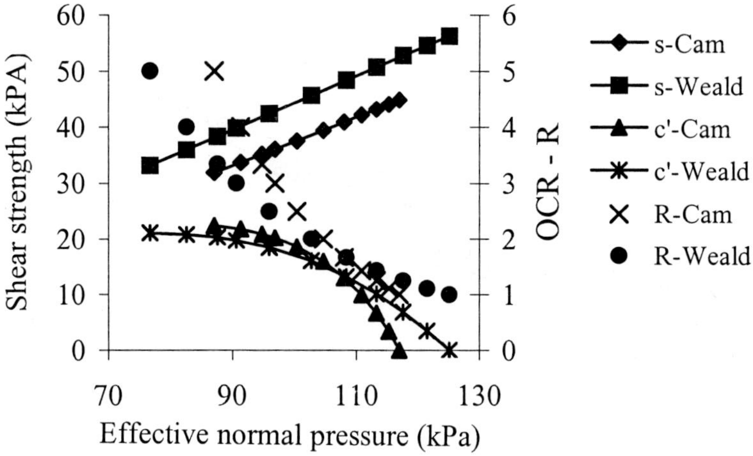

Fig. 5 *Mohr-Coulomb relationship.*

consolidated soils on the Mohr-Coulomb plot. The conventional procedure for the triaxial compression test as described earlier in this section remains the most convenient. To minimize the effect to test path dependency, it is recommended that confining pressures for the laboratory test, of not more than 20 kPa greater and smaller than the existing overburden pressure, is to be applied. Attention however has to be paid to interpretation of the plot obtained from the tests. The friction angle needs to be deduced as the average of the three Mohr circles rather than attempting to pass an envelope as tangent to the circles.

Hvorslev plot

To gain better understanding of the behaviour of the OC soil, the (s, σ') relationship at four constant values of void ratio each (represented by a constant value of p_e in the Hvorslev plot) has been investigated and presented in Figs. 6a and 6b.

The (s, σ') relation on the plot is slightly curved. The average inclinations of the lines are the same (parallel) as expected for the Hvorslev plots. The angle of the average slopes is as shown in Table 3. The plot differs from the Mohr-Coulomb plot presented in Fig. 5 as it follows a constant void ratio path.

Shear strength vs. confining stress

To study the role of the confining stress on soil strength, the shear strength (s) and the cohesion component (c') are plotted against the overburden pressure (p_o) and shown in Fig. 3. Cohesion (c') is in turn plotted against the effective normal pressure (σ') and shown in Fig. 5.

It can be seen from Fig. 3 that c' increases as p_o decreases (as when a soil sample swells on unloading). The point of particular interest is that c' does not vary much with p_o at lower values of p_o which is at higher

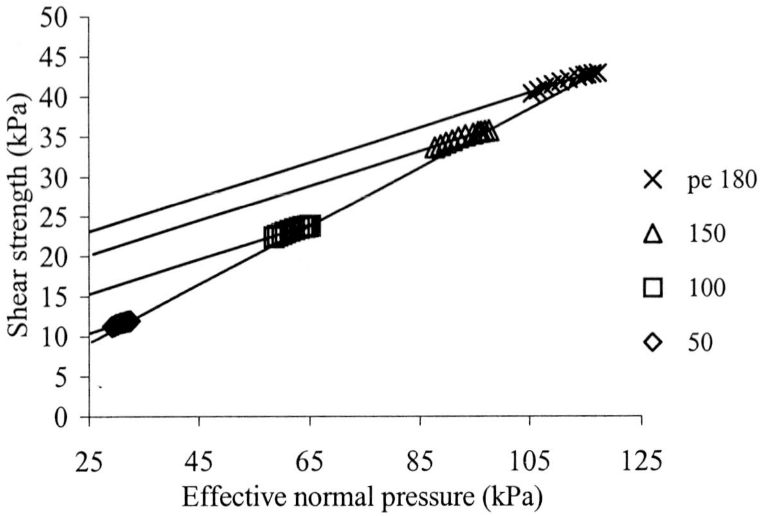

Fig. 6a *Hvorslev plot (Cam clay).*

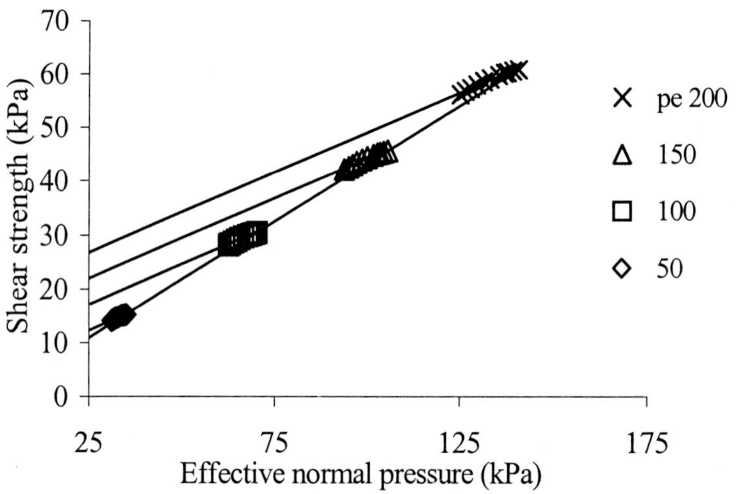

Fig. 6b *Hvorslev plot (Weald clay).*

values of over-consolidation ratios. The same pattern applies to the (c', σ') plot in Fig. 5. In contrast, the variation increases when p_o increases as the over-consolidation ratio approached 1.0 near the NC state.

It is also possible to directly deduce the effects of confining pressure on shear strength from CSSM. This can be done by determining the differentials in the following equation:

$$s_{utc}/p_R' = (M/2) \cdot (R/r)^{\Lambda} \text{ (Eq. 22, Wroth, 1984).}$$

where s_{utc} is half the deviator strength and $p_R' = p_o$.

The resultant relationship will be complex whereas the Mohr-Coulomb approach herein provides greater simplicity and clarity.

It can also be seen in Fig. 3 that the difference in shear strength as defined by CSSM (as half of the deviator strength) and by the Mohr-Coulomb equation are small especially at higher over-consolidation ratios.

Strength components

Model soils

The value of ϕ' deduced from Eq. (8) has been applied to Eq. (6) to obtain c' for the cases of the model soils (Cam and Weald clays) and the results are presented in Figs. 7a and 7b.

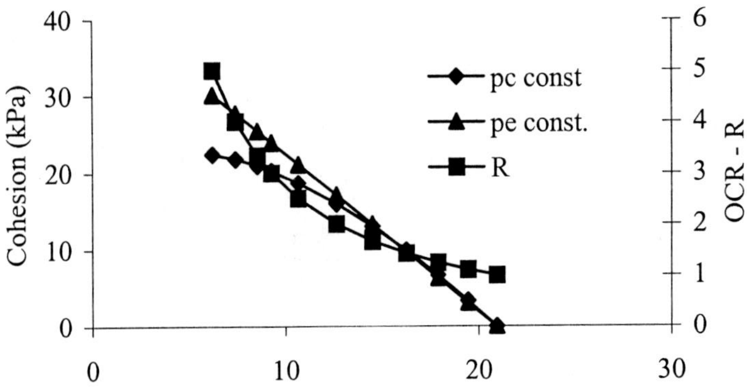

Fig. 7a *Cohesion vs. Friction (Cam clay).*

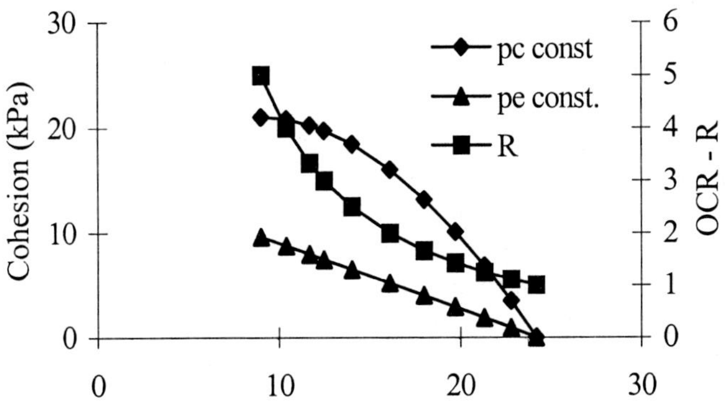

Fig. 7b *Cohesion vs. Friction (Weald clay).*

The test path dependency in conventional triaxial shear test (p_c constant, p_o decreasing by swelling) and in Hvorslev plots (p_e constant, p_o varying at constant void ratio) is revealed in Figs. 7a & 7b in the different (c', ϕ') relations obtained. Linear relationship has been obtained in the figures for (c', ϕ') for the cases of p_e being constant (and constant void ratio) as assumed in the Hvorslev plot.

Site soils

It is usually difficult to obtain good quality site data and limited test data has been obtained from the site in Malaysia (Table 3). The effective cohesion and friction values obtained in laboratory triaxial strength tests on site samples are plotted in Fig. 8. There is some scatter in the results. The variation is possibly due to particle size distribution, mode of formation and quality of sampling and testing.

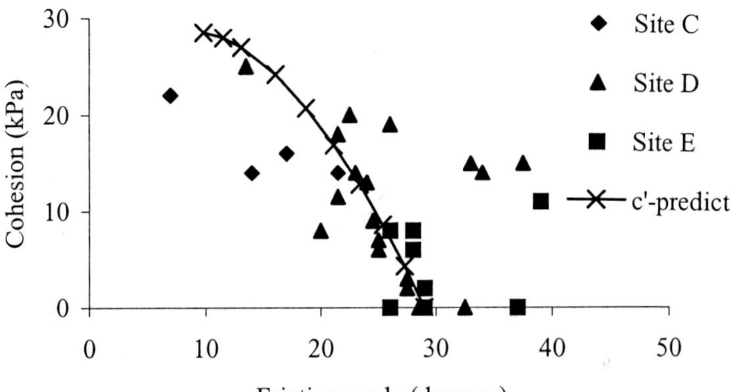

Fig. 8 *Cohesion vs. Friction (site).*

A predicted (c', ϕ') relation representing the best fit of the test data is also presented in the same plot. It can be seen that the predicted relation with an assumed input q_m' value of 29 degrees provides a reasonable representation of the site data.

In strength characterization, the parameters of interest are the strength components (s, c', ϕ') of soil samples from various depths with varying initial confining pressure (p_o). Some limited test data from a localized site within the area presented have been analyzed. The technique of characterizing the strength parameters using the formulations herein is illustrated by comparing site and predicted results in Fig. 9. It can be seen that the predicted values reasonably described the limited data obtained from site.

Shear strength ratios

In natural soft deposits the ratio of shear strength (s) to overburden pressure

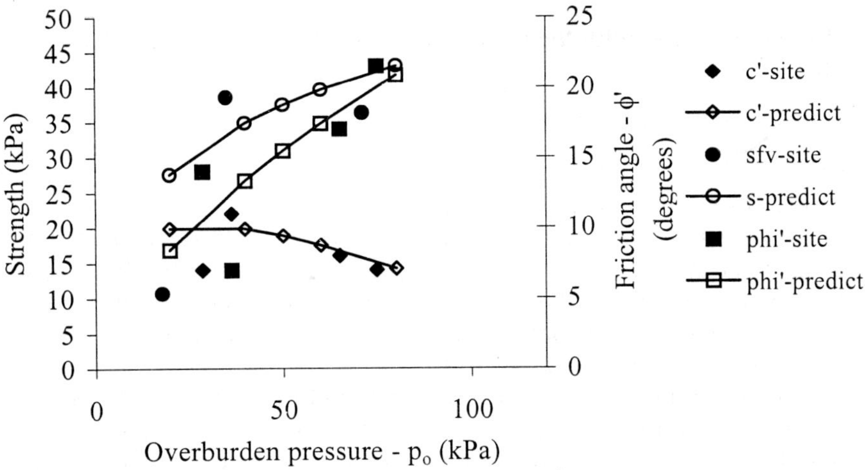

Fig. 9 *Strength characterisation—Sarawak site.*

(p_o) is of interest to characterize the variation of strength with depth of the deposit.

Since $s = (q_r/2) \cdot \cos \phi'$,

$$q_r = S_r p_o R^{\Lambda}$$

$$s/p_o = (S_r R^{\Lambda}/2) \cdot \cos \phi' \tag{9}$$

The ratio from the equation is plotted against friction angle for NC soils (ϕ_m') for values of consolidation ratio from 1 to 5 in Fig. 10. The value of the basic input parameters have been varied to obtain the input ϕ_m'parameter as in Table 3.

The predicted values are close to known values for site samples. It can

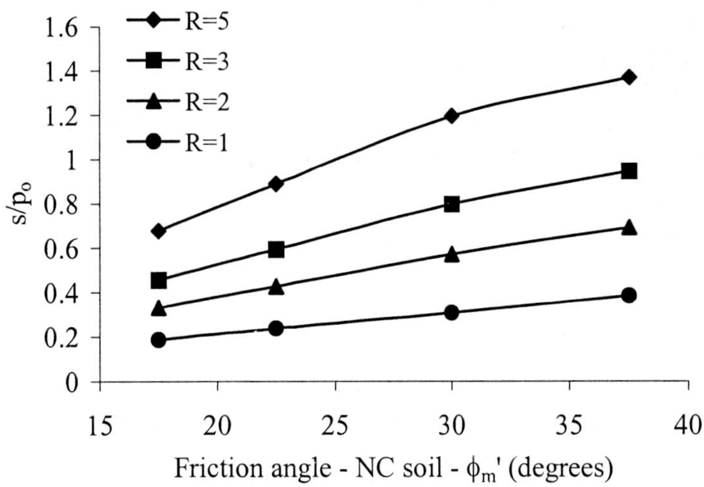

Fig. 10 *Strength ratio variation.*

also be compared with the presentation for s_u (= $q_r/2$) by Wroth (1984) in his Fig. 31.

DISCUSSION ON FORMULATION

Assumptions and deductions

Locked-in deviator strength from Cam clay model

The supposition that within the framework of CSSM the deviator strength of all soils (normal or over-consolidated) are the same on the q/p_e vs. p/pe plot (see Loudon, 1967) is an approximation. The assumption may be modified, but any deviation may be considered as a secondary effect at this stage of the theoretical development.

In Assumptions I and II, the deviator strength is separated into cohesion and friction components [Eqs. (4), (4a), (6), (6a)]. The assumptions are based on an observation on the effects of the change in confining pressure of a soil from an initial state on the normal consolidation line, compressed, and then allowed to swell to an over-consolidated state at the same void ratio as the initial state. The assumption has the merit that it is derived from the observed compression behaviour (p_e–p_o) of a real soil.

Shear strength and cohesion component

Since the Mohr-Coulomb relationship has been applied to the over-consolidated soil, the definition of shear strength will differ from that of Wroth (1984) who assumed that the undrained shear strength is half of the deviator strength ($q_r/2$). The difference has been observed to be small (Fig. 3). Assumption II follows from Assumption I. It has to be noted that the expression for cohesion [Eq. (6)] involves the friction angle.

Effective normal pressure and pore water pressure parameter A_f

To arrive at an explicit solution that accounts for the effective normal pressure (σ'), it is necessary to introduce Skempton's pore water pressure parameter (A). Wroth's theoretical relationship for the pore water pressure parameter derived from CSSM can be conveniently applied. The parameter can also be obtained from measured values.

Cam clay and Mohr-Coulomb relationship

The deduced components of strength (cohesion and friction angle) are path dependent quantities. Their deduction was implemented by applying the Mohr-Coulomb relationship within the framework of the Cam clay model.

Application

The formulation deduced herein is a convenient tool to analyse the cohesion characteristics of natural deposits of over-consolidated and compacted clays. It is often required to examine the strength characteristics of natural sediments deposited over large areas. With the available formulation, combined with sufficient in-situ and laboratory shear tests, it is only necessary to carry out consolidation tests to determine the over-consolidation ratio and other consolidation parameters, to determine the basic CSSM input parameters. Examples of initial site application to assess the theoretical formulation are presented in Figs. 7a, 7b, 8, and Fig. 9. The detail applications in site practice are presented in the next main section.

The interpretation of triaxial compression test results needs to take into account the test path dependency of the (s, σ′) relation and the deduced strength components (c', ϕ'). With better knowledge of a deposit, it may be advantageous to deduce cohesion and friction from the theoretical formulation, which is then applied to analyze the results obtained from laboratory test on prepared or in-situ samples.

Conclusion on model formulation

1. An example of how cohesion behaviour can be rendered to a frictional soil has been presented.
2. It has been possible to separate soil strength into cohesion and friction components for NC and OC soils within the framework of the CSSM. The separation has been carried out by assuming the validity of the Mohr-Coulomb relationship as a tool in deducing the strength components. The success in the separation will afford a better understanding of the individual roles of the components when subjected to changes inconfining pressure.
3. Though (c', ϕ') may be separated, they are not independent of each other and of the confining pressure (p_o). The relationship is path dependent, and examples of paths with constant p_c and p_e (with void ratio constant) are shown Fig. 7a and 7b.
4. The deduced shear strength and effective normal pressure relation (s, σ′) appears reasonable on the Mohr-Coulomb plot and when presented as the Hvorslev plot. The relation (s, σ′) is not unique and is test path dependent as explained for the plots. The conventional laboratory triaxial compression test procedure has to take into account the test path dependency. Confining pressures for the laboratory test, of not more than 20 kPa greater and smaller than the existing overburden pressure is recommended.
5. The effect of confining pressure on shear strength can be directly deduced by methods of CSSM. The advantage of applying the Mohr-Coulomb relationship as proposed herein, is that the relationship is well known and conveys a simple and clear concept of the role of cohesion and friction. The relationship furthermore provides a convenient and direct method of determining physical values of the individual components of strength in laboratory tests.
6. Assumptions I and II have been made besides those derived from the

Cam clay model and Mohr-Coulomb relationship. They are applied to formulate relationships that lead to an explicit solution of the friction angle (ϕ') in Eq. (8). The angle obtained can in turn be applied to Eq. (6) and (6a) to obtain the separate cohesion and friction components of strength.

7. The Assumption I made [in Eq. (4)] that the observed compression (p_e-p_o) is responsible for the cohesion component of strength implies that cohesion is frictional in origin being dependent on confining stress.
8. The effects of the changes in confining pressure (p_o) and effective normal pressure have been considered and it was shown that cohesion is not significantly affected by the changes of pressures at higher over-consolidation ratios encountered at lower pressures (Fig. 3).
9. The deduced (c', ϕ') relation reasonably described some overall site data on cohesion and internal friction (Fig. 8). A strength characterization exercise on limited localized site strength data within the same area has also been carried out (Fig. 9) to test the applicability of the formulation.

PRACTICE OF CHARACTERIZATION

In-Situ Soils (Wroth's Interpretation)

The undrained strength ratio has been applied as an analytical tool to the field vane shear test results at two sites to characterize the strength of the deposits (Fig. 11 and Fig. 12). At the first site the soil has been identified as marine clay with a smaller scatter in strength values than in the latter, which is a floodplain deposit.

In Fig. 11, the plot of the theoretical lines 0.2 p_o and 0.4 p_o indicates the following characteristics:

1. The 0.4 p_o line passes approximately through the mean of the undisturbed strength values. The values of S_{ru} = 0.4 shows that the soil is over-consolidated as the value of S_{ru} exceeded the theoretically expected maximum value of 0.25 for normally consolidated soils.
2. The remoulded strength values fall below the 0.2 p_o line as an upper bound, and S_{ru} = 0.2 is a typical lower value for normally consolidated soils (Wroth's Fig. 31).
3. In this particular instance the earthworks operation in reclaiming the ground did cause remoulding and the characterization of the deposit with $s = 0.2\, p_o$ was found appropriate. Although the line is the upper bound of the observed remoulded values, it is acceptable because the degree of remoulding of the earthworks operation is not complete.

In Fig. 12, as a floodplain deposit the soil is expected to be normally consolidated. The following points are noted:

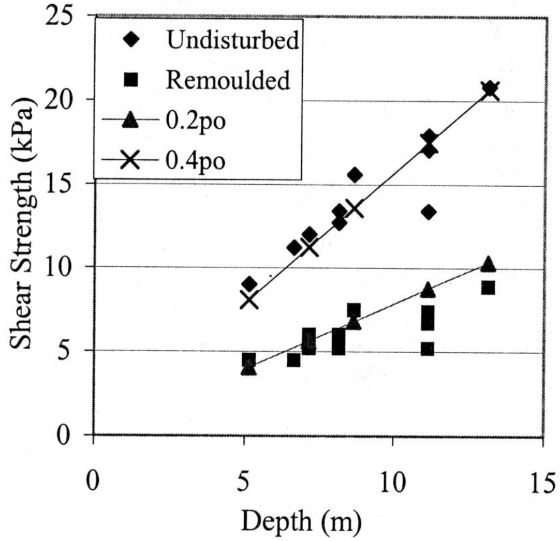

Fig. 11 *Field vane shear tests (Klang).*

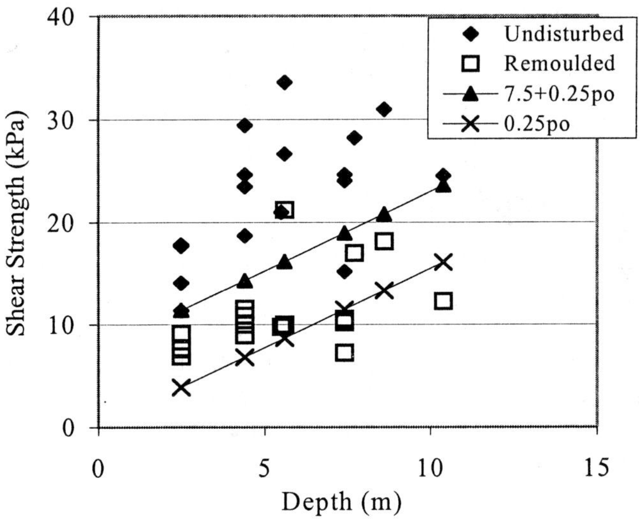

Fig. 12 *Field vane shear tests (Sarawak).*

1. The normally consolidated line: $s = 0.25p_o$ was applied with a higher S_{ru} value of 0.25, as the soil being a river bank deposit contains larger size particles such as silt and some sand besides clay.
2. The base line is $s = 7.5 + 0.25\ p_o$, with the offset of 7.5 kPa being introduced, as an upper layer may have been recently removed by erosion after consolidation.
3. The base line forms the lower bound of the undisturbed and the upper

bound of the remoulded strength values. This has often been the observation for normally consolidated site deposits that were characterized using the proposed method.

4. The base line in this instance has therefore been selected as the characteristic undrained shear strength for estimating the stability of embankment on the deposit and found satisfactory in practice.

Bjerrum (1972) advocated the introduction of a correction factor to the field vane shear strength before it is applied in the analysis of the stability of embankment on soft ground. With the correction factor, a factor of safety of 1.3 is also recommended. He cited rate of loading, anisotropy and progressive failure as factors justifying the correction. In a given site it is not simple to investigate whether the factors attributed by Bjerrum is important. There is often a scatter of strength values obtained and he has not discussed how the scatter affects the choice of correction factor. The factor is to be introduced only when investigation justifies its applicability. In the method proposed here where the undrained strength ratio is used as a theoretical framework for evaluation and where a lower bound has already been applied, the Bjerrum correction factor should not be included. Instead consistent with Bjerrum's approach, where the selection of strength is linked with the choice of factor of safety; the latter should be from 1.25 to 1.5. Based on observation and knowledge of performance of similar constructions the choice of factor of safety can then be refined.

Interpretation for (c′, ϕ′) Soils

Procedure of characterization

GOVERNING EQUATION AND RELEVANT PARAMETERS

The following presentation is directed towards the application of the (c', ϕ') soil model to site problems.

In the section on soil models, Eq. (8) has been derived for which an explicit value of ϕ' can be obtained for a given value of R (over-consolidation ratio). With each value of ϕ', the cohesion component of strength (c') can be obtained from Eq. (6), the shear strength (s) Eq. (5) and the deviator strength (qr) from Eq. (3).

When normalized with p_o,

$$c'/p_o = S_r[(R^{\Lambda}-1)/2] \,.\, \cos\phi'$$

$$s/p_o = (S_r R^{\Lambda}/2) \,.\, \cos\phi'$$

$$q_r/2p_o = S_r R^{\Lambda}/2$$

and the normalized Mohr-Coulomb strength components are uniquely related to ϕ' and R, whilst the normalized deviator strength is uniquely related to R and independent of ϕ'.

EXAMPLES OF STRENGTH CHARACTERIZATION

Basic parameters

Typical values

The typical values of the parameters involved in the normalized relationships for a soil being considered are presented in Table 4:

Table 4

Basic prameters

Properties	*Symbol*	*Typical Values*
Swelling index	κ#	0.045
Compression index	λ#	0.2
1-κ/λ	Λ#	0.775
Spacing ratio	r#	2
Friction angle (NC)	ϕ_m'#	17.5-37.5 degrees
6* sin ϕ'_m (3-sin ϕ'_m)	M#	0.668-1.527
Deviator strength ratio: $M(I/r)$	S_r#	0.391-0.893
Overburden pressure	p_o	varies
Max. past consolidation pressure	pc#	200kPa
Friction angle	ϕ'	varies
Over-consolidation ratio (pc/po)	R	varies

constants.

Method of selection

In Table 4 the parameters to be applied, taken in order, are to be determined as follows:

1. The consolidation parameters κ and λ may be obtianed from laboratory consolidation tests. In the chart, κ = 0.045 and λ= 0.2 are constant parameters and the assumed values are typical of those applied in CSSM. With further experience other relevant values may be selected. $\Lambda = 1\text{-}\kappa/\lambda$ = constant. Wroth (1984) suggested that the spacing ratio (r) might be assumed to have a value of 2.
2. ϕ_m' is a fundamental parametric constant of the charts that needs to be carefully selected by inspecting the range of (c', ϕ') values obtained in the laboratory tests. A suitable value of ϕ_m'may be selected close to the value of ϕ' that is coupled to a very low value of c'. With increasing experience in characterization it may be necessary to consider whether the selected value of $\phi_m'$is consistent with the classification property of the soil such as particle size distribution and density. The parameter may also be linked to other parameter such as κ and λ with their implication as a function of Λ. On a longer term it will be advantageous to link the parameter $\phi_m'$to typical particle size distribution of a given deposit where higher and denser sand contents would be expected to lead to higher values of the parameter. The next parameter, $M(\phi_m')$=6* sin $\phi_m'/(3 - \sin \phi_m')$ = constant.

3. The constant $S_r = M(1/r)^{\Lambda}$ is from Wroth's Eq. (22) for $R = 1$ at the NC state.
4. The effective overburden pressure (p_o) is a site characteristic of the given soil sample with an implicit assumption that $K_0 = 1$. A constant value of p_c is assumed in the plots thus fixing R for the given sample. The parameters s, q_r and c' correspond to the given value of R and when normalized with p_o will be independent of the assumed value of p_c. In the charts for convenience the parameter $p_c = 200$ kPa is assumed. A manual check was made by assuming that $p_c = 100$ kPa. It was confirmed that the normalization of the strength components (c, s, q_r/2) as stress ratios with the quantity p_o produces relations that are independent of the effect of varying p_c.
5. In some site applications, the undrained shear strength may be obtained as unconfined compressive strength (UCS), as vane shear strength; as strength in triaxial undrained shear tests (UU tests) or indirectly as N values in standard penetration tests (SPT). In such applications it may then be desired to obtain the appropriate characteristic range of (c', ϕ') values corresponding to the undrained shear strength for design purposes. The related parameters are s and $qr/2$.

Plotted charts

The charts as plots of c/p_o, s/p_o and $q_r/2p_o$ against ϕ' for various values of the parameter ϕ_m'are presented as Fig., 13a, 13b, 14a, 14b, 15a and 15b respectively. The strength components s/p_o, c/p_o, when normalized by p_o are unique functions of R and ϕ'. The strength $q_r/2p_o$ is a unique function of R. Since the parameters R and $\phi'$are also uniquely related to each other as in Eq. (8); the normalized strength components are unique functions of ϕ' in the plotted figures. Discussion and application of the charts follow.

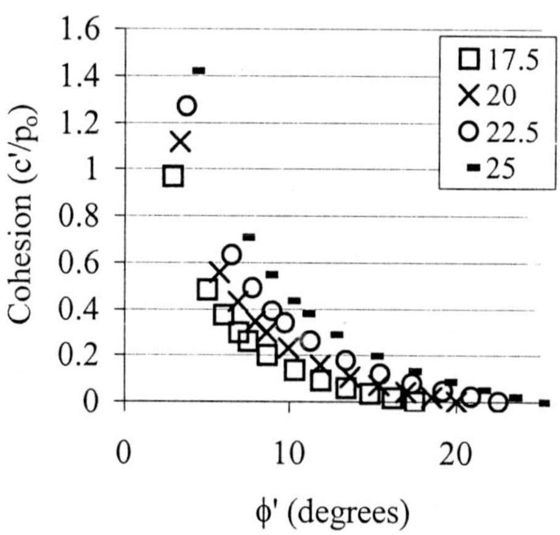

Fig. 13a *Cohesion (ϕm': 17.5-25).*

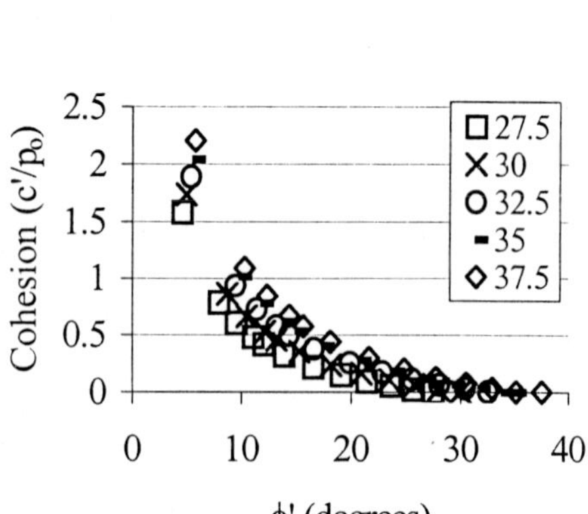

Fig. 13b *Cohesion (ϕm': 27.5-37.5).*

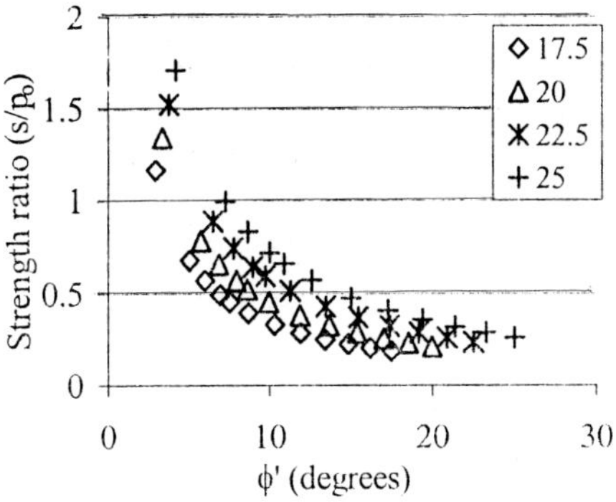

Fig. 14a *Shear strength (ϕ_m': 17.5-25).*

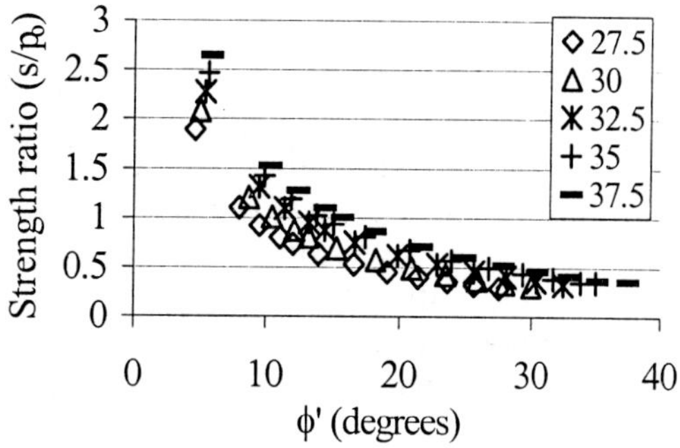

Fig. 14b *Shear strength (ϕ_m': 27.5-37.5).*

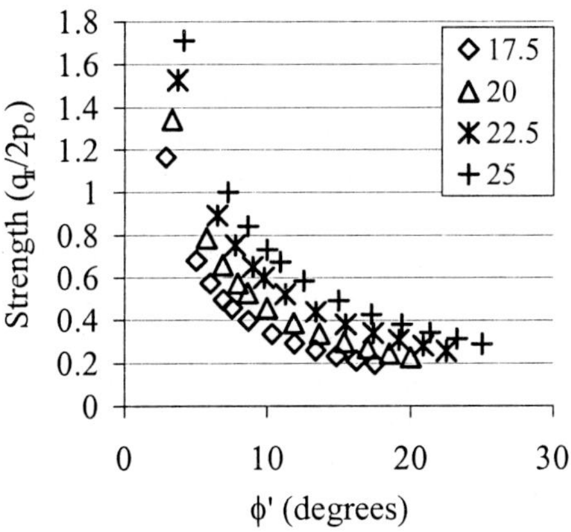

Fig. 15a *Deviator shear strength (ϕ_m': 17.5-25).*

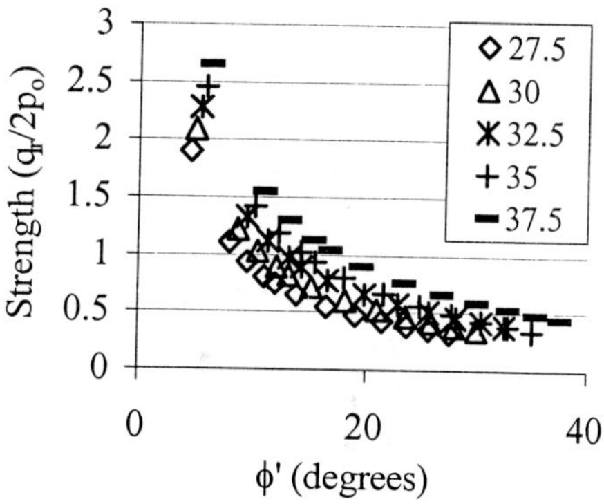

Fig. 15b *Deviator shear strength (ϕ_m': 27.5-37.5).*

Input parameters for the relations

1. For a given relation p_c is constant (= 200 kPa) and p_o is decreased in unloading.
2. The parameters (c', s, $q_r/2$) are calculated as functions of ϕ' as p_o and thus R ($=p_e/p_o$) is varied.
3. The quantity ϕ' is uniquely related to the parameter R [as the input parameter ($S_r\Lambda$) are constants and the quantity A_f is uniquely related to R for each solution of the governing equation]. Therefore R also varies as ϕ' is varied.
4. The quantity $q_r/2p_o$ is close to the value of s/p_o because effect of $\cos\phi'$ therein is not important.

Inherent implications

1. It is pertinent to point out that relationships in the charts are path dependent and represents the given swelling path on unloading.
2. One of the reasons for finding a range of (c, ϕ') values from the site is that the given soil is derived from the same deposit, but is subjected to varying degree of unloading due to variation of the confining pressure p_o as well as sample disturbance.
3. The plotted quantities tend towards a constant value as ϕ'approaches the parameter ϕ_m' with R approaching 1.0 as full frictional behaviour is attained.

Interpret (c′, ϕ′) from laboratory testing of in-situ soils

1. The in-situ soil would have been interpreted by Wroth's method after field vane shear tests or other in-situ shear tests have been carried out and characteristic undrained shear strength values obtained as required.
2. The problem here is to interpret the (c', ϕ') values obtained in additional laboratory effective strength tests. The effects of sample disturbance, and swelling into an over-consolidated state have to be accounted for.
3. An example has been presented in Fig. 8 where a best-fit theoretical (c', ϕ') relationship is constructed for the site data.
4. The over-consolidation ratio may be incorporated in the (c', ϕ') plot (Fig. 7). This helps to assess the extent to which the sample has swelled in the course of implementation of the laboratory test.
5. Finally, characteristic shear strength profiles can be selected as in Fig. 9, if a shear strength parameter is desired. If separate (c', ϕ') strength components are required for the computation model, then pair of values has to be assigned (also as in Fig. 9) for a range of overburden pressure (p_o).

Determine ϕ$_m$′: given (c′, ϕ′) of in-situ soils

1. ϕ_m' is the basic characteristic friction parameter of a given soil that is always of interest. ($\phi_m'=\phi_d$ is the friction parameter recommended by Schofield to be applied directly for drained action.) Whether the parameter is to be applied directly is a matter of judgement.
2. Calculate the values of c'/p_o. The overburden pressure (p_o) is from a site measurement of the soil sample.
3. The values of $(c'/p_o, \phi')$ are located on Figs. 13a and 13b to determine the line for which the pair of values is the closest. The parameter ϕ_m' lies on the intersection of the selected line and the $c'/p_o = 0$ axis.
4. It may be desirable to look at past records to determine whether the selected value of ϕ_m' is reasonable. The selection of ϕ_m' would be aided by examining the particle size distribution in for sand and silt contents

of the soil samples, as a higher coarse content is known to lead to a high value of the parameter.

Deduce basic parameters from (s, c′, ϕ′) of embankment soils

1. Ting (2000a) proposed a simplified model (Fig. 16) to interpret the behaviour of compacted partly saturated soils when fully saturated. In that model partly saturated soil samples compacted to satisfactory densities would, when fully saturated (path "*fd*"), have an identified state "*d*" on a normal consolidation line of the same soil.
2. In most embankments in practice, it is assumed that the compacted soil would eventually reach a fully saturated state. Laboratory shear tests are therefore conducted on back-saturated samples of the soil that has been initially prepared by compaction at lower moisture contents.
3. A range of (s, c′, ϕ′) values are obtained in laboratory strength tests that needs to be characterized for the embankment slope stability. From the test values the over-consolidation ratio (*R*) can be calculated from the relationship: $R^{\Delta} = s/(s\text{-}c)$ which can be derived by combining Eqs. (5) and (6).
4. The ratio s/p_o in Eq. (5) can be calculated once *R* is known, as ϕ′and S_r can be obtained from the test results. An appeal can then once more be made to Fig. 14 to determine the relevant value of ϕ_m' that is

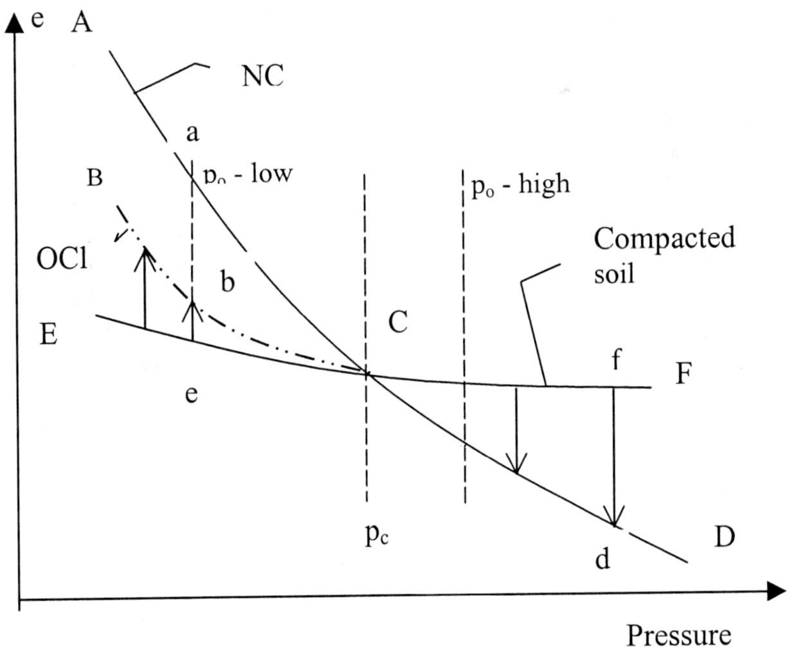

Fig. 16 *Simplified model for compacted soil.*

characteristic of the state of the given soil sample on the normal consolidation line.

5. The characteristic strengths can then be determined from the plotted charts corresponding to the relevant values of p_o for the embankment.

Determine (c', ϕ') from ($q_r/2p_o$) of in-situ soils

1. The values of $q_r/2p_o$ obtained from in-situ shear tests are located on Figs. 15a and 15b. The appropriate value for the parameter ϕ_m' is estimated from particle size in previous records.
2. With a knowledge of ϕ_m' and the given value of $q_r/2p_o$, the relevant value of ϕ'can be established.
3. The selected values of ϕ' and of the parameter ϕ_m' are then applied to the c'/p_o vs. ϕ' charts (Figs. 13a and 13b) to obtain the characteristic range of (c', ϕ') values for design.

Characterization of (c', ϕ') soils in general

In the drained stability analysis of site problems involving slopes and excavations, an assumption can initially be made of a purely frictional (ϕ_m') soil with $c' = 0$ using the method hereinbefore. A quantum of cohesion (c') can then be progressively added; this would mean that ϕ' that is uniquely related to c' would be gradually reduced from the initial value of ϕ_m'. Cohesive and cohesionless behaviour can be analyzed as distinct but related components of shear strength, and a rational method of assigning a cohesion component can then be made; taking into account the sensitivity of factor of safety to the quantum of cohesion.

Weathered Natural Formation

Background

In earlier sections it was explained that the cohesion component of a given weathered natural formation could be similar to that found in reinforced sand, with the cementation material acting as a reinforcement element. The friction component would represent the material "packing" that is related to changes in confining stress during the formation of the weathered soil. The characterization of weathered natural formations thus differs from that of the over-consolidated soils (that may include compacted weathered material), where a range of (c', ϕ') values are uniquely related to each other, and to the existing overburden pressure. However for the weathered formation, the effect of the confining stress due to the friction component in the computation model is still relevant.

In practice, although the characterization of weathered natural formations will be based on empirical methods, to advance our knowledge, theoretical methods have to be kept in mind. For example the undrained shear strength ratio may be applied as a normalization procedure for the strength parameters obtained.

The treatment of the subject refers to conditions in Malaysia although the principles involved may be applied to other environments that have experienced similar geological processes. The formations in Malaysia as a tropical country, with the high temperatures and rainfall (of 300 cm per annum), are subject to deep weathering. The depth of weathering in sedimentary rock formations may reach about 20 m (and more in granite formations in view of its greater permeability).

Colluvium sometimes overlies the weathered formation. The colluviums are usually loose deposits ranging from clayey silt to boulder sizes and their only treatment is often complete removal or avoidance.

Site investigation

The weathered natural formation to be considered consists of material of weathering grades of IV (Highly weathered) to VI (Residual soil), although exceptionally grade III (Moderately weathered) material may be included. Detailed topographical and geotechnical investigations have to be carried out. In view of the complexities of the formations, besides the detailed investigation of the construction site, macro-surveys of the site and surrounding areas are also essential. Frontal, lateral and back slopes and adjoining grounds should be given equal attention. The survey could be carried by geophysical methods calibrated with detail boring. The construction site may then be properly placed within its overall setting with the natural slopes and drainage pattern providing useful conceptual information.

It is difficult to obtain good quality undisturbed sample of the hard weathered materials and the Mazier sampler (that reduces the possibility of flowing drilling fluid eroding the core) has been used. Laboratory test on extracted field samples is carried out in addition to in-situ strength testing. Pressure-meter test is highly recommended as it provides a fundamental insight into strength characteristics. A field shear box test would be relevant where feasible. In Malaysia, the most convenient field-testing method is the Standard Penetration Test.

Strength and stability

TYPICAL STRENGTH VALUES

Due to the non-homogeneity of the weathered deposits, there will be a wide variation in the strength characteristics within a given site. The non-homogeneity is due to geological differences in lithology, metamorphism and degree of weathering of the parent rock.

Typical effective strength results for some weathered sedimentary rock deposits are shown in Table 5. The results include those presented by Nithiaraj et al. (1996), studies made along the Seremban-Ayer Keroh

Highway (Ting et al., 1990) and from the Slim River Area (Ting, 1998).

Table 5
Weathered sedimentary rock deposit strength results

Location	*Weathering Grade*	*Clay*	*Silt*	*Sand*	*Gravel*	*Effective Strength*		*Data Type*
		%	%	%	%	c' kPa	ϕ_m' (degree)	
Kampar, Perak	VI	5-58	9-40	7-49	0-74	8-30	18-45	Range
	V	0-35	0-69	0-94	0-55	9-71	13-42	
	IV	0-26	0-55	0-83	2-60	7-9	16-41	
	VI	NA	NA	NA	NA	15	20	Typical
	V	NA	NA	NA	NA	30	22.5	
	IV	NA	NA	NA	NA	0	45	
Seremban	Varies	NA	NA	NA	NA	10	37	Typical
Ayer	VI	10	69	21	0	38	24	
Keroh	V	NA	NA	NA	NA	10	36.5	
	V	8	50	29	13	50	21.5	
Slim River Perak	Varies	4-43	18-88	1-38	0-5	5-60	11-46	Range
	VI	43	18	38	1	19	42.5	Typical
	V	4	72	23	1	5	31	
	IV	NA	NA	NA	NA	60	29	

NA=Not Available.

It can be seen from the table that the friction angle can be up to 45 degrees and cohesion up to 71 kPa.

Strength Variation

The variation of soil strength, for sediments due to changes in effective confining stress, and for weathered natural formation due to cementation, has been clearly identified. Other causes of variation of soil strength also need to be kept in mind. The time frame of the problems considered is within the life span of the structures. The variation of soil strengths of the weathered natural formation as a result of various factors in geological time scale will require separate consideration.

Some shear tests on weathered soils have revealed peak and residual strengths indicating that the deposits can have characteristics similar to over-consolidated soils. The non-uniform mobilization of shear strengths, (of the residual strength due to development of higher strains as opposed to peak strength at small strains) along a potential failure surface leading to progressive slope failure is well established.

An important factor that require careful consideration is the quantum of the effective cohesive strength selected for design, and its potential decrease over the long-term. The effects of the lowering of cohesive strength (c') on factors of safety is well established. Chandler and Skempton (1974) in their study of slopes in brown London clay and Upper Lais clay have argued that the assumption of $c'=0$ in the design of the slopes is unduly conservative. Even though the measured values of c' in-situ is wide ranging, in practice as a matter of prudence of c' value not exceeding 10 kPa is usually applied in slope stability analysis.

The loss of strength with time, for the formation encountered in the Slim River area of Malaysia (Ting, 1998) leading to delayed failure of road cuttings, is a well-known occurrence. The mechanism of the failure is little understood although the loss of strength due to the weakening of bond by weathering processes as proposed by Bjerrum is one possibility. Chandler (1972) in his study of the effects of weathering on shear strength, whilst recognizing that weathering is associated with a progressive reduction of the effective cohesion and frictional strengths, reasoned that the actual weathering process has to be identified. For example, a process such as oxidation is too slow to account for failures within the normal life span of the structures. In the case of the Slim River slopes, the relevant processes with the appropriate time scale could be changes in temperature and moisture content. Progressive failure due to equilibration of pore water pressures has been ruled out because of the time scale in the failure of the permeable slopes.

Ting et al. (1990) deduced that the slope failures along a highway that they studied could be attributed to failure occurring along planes of discontinuities that existed in the parent rock. The shear strength of the discontinuities could be less than that of intact samples, as failure could not be accounted for by the strength of the intact samples.

METHOD OF CHARACTERIZATION OF STRENGTH

Because of the variation in strength results due to differences in geological processes within a given formation, there will be difficulty in the selection of a characteristic strength for the design of the slope in weathered formations.

In their paper Nithiraj et al. (1996) have proposed a method for the selection of strengths of weathered soils in the stability analyses. The relevant findings of the paper are summarized in Fig. 17 and Fig. 18. (Fig. 8 and 7 in the paper).

It can be seen from Fig. 17 that there is a broad correlation between c' and ϕ' values for the weathering grade V material that are encountered at the given site. The nature of the correlation needs further investigation, as it may not be due to over-consolidation effects that was proposed for

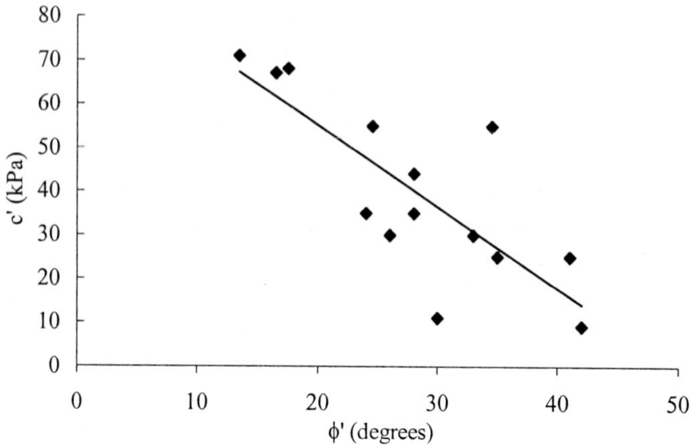

Fig. 17 *Variation of c′ with ϕ′ for weathered schist.*

the sediment in Fig. 8. For the formation, given *N* values from Standard Penetration tests; ϕ′ may be deduced from Fig. 18. The value of *c*′ may then be deduced from Fig. 17 for the deduced value of ϕ′. In the context of the discussion hereinbefore, ϕ′ would represent the "packing" of the weathered material whilst *c*′ would represent the strength and distribution of the "cementing" agent. A given formation may first be divided into various strata according to the characteristic *N* values measured. A single set of (*c*′, ϕ′) values may then be assigned to each of the characterized stratum, and the required computation then carried out for design of the structure.

The proposed method may be applied in the determination of characteristic strength parameter. Together with experience built up with

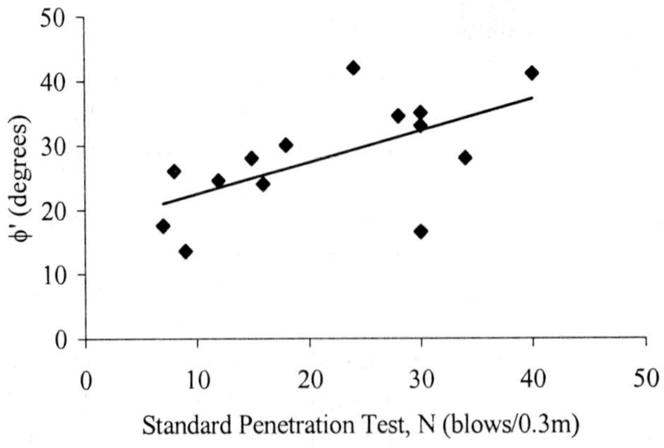

Fig. 18 *Variation of ϕ′ with SPT 'N′ for weathered schist.*

performance of the constructed geotechnical structure, the method is not only relevant as an empirical practice in characterization, but it can also form the basis for further research to provide a fundamental solution to the problem.

CONCLUSION ON CHARACTERIZATION

In Table 1 the soils types to be considered have been presented. The soil models to be applied have been proposed in Table 2.

In conclusion to summarize the findings of the characterization procedure, the soil parameters recommended to be applied, corresponding to the geotechnical structure load and support are shown in Table 6.

Table 6
Summary of recommended procedure for parameters

Deposit	*Soil Model*	*Load/Support*	*Parameters*
In-situ sediments	[1]Cam clay	Natural	s_u
In-situ sediments	[2]Cam clay/ Mohr-Coulomb	Interpretation	[3]Mohr-Coulomb
Compacted sand	Mohr-Coulomb	Embankment/fill	ϕ_m'
Compacted soil	Cam clay/ Mohr-Coulomb	Interpretation	ϕ_m' Mohr-Coulomb
Compacted soil	Cam clay/ Mohr-Coulomb	Embankment/fill	Mohr-Coulomb
Weathered formation	[4]Empirical	Natural	Mohr-Coulomb

[1]Cam clay interpretation (Wroth, 1984).
[2]Cam clay/Mohr-Coulomb interpretation (Ting, 1999, 2000).
[3]Mohr-Coulomb strength parameters are (c', ϕ').
[4]Empirical method (Nithiaraj et al., 1996).

The Table shows that the proposed Cam clay/Mohr-Coulomb soil model may be applied directly to the interpretation of soil parameters, or for correlation with parameters from other soil models if they are to be preferred. Some examples of the application of the soil models introduced have been given, but other useful application of the models can still be found.

Despite some limitations of the soil and computation model, their role in providing a theoretical framework is still essential in the analysis and characterization of soil properties for in-situ applications, and in the design and construction of geotechnical structures. The necessary adjustments to the construction under site conditions may be made by implementation of the well-established observational methods supported by back-analysis and adjustments to the original model.

REFERENCES

BS 8002, 1994, *Code of practice for Earth retaining structures*. British Standards, Institution, London. pp. 7, 22.

Bjerrum, L., 1972, Embankment on soft ground. *Proc. ASCE, Spec. Conf. Performance of earth and earth-supported structures*, Lafayette, Vol. 2:1-54.

British Geotechnical Society, 1981, Design Parameters in Geotechnical Engineering. *Proceeding of the VII ECSMFE*, Brighton, 1979, Vol. 1 to 5.

Jamiolkowski, M., Lancellotta, R., Marchetti, R., Nova, R., and Pasqualini, E., 1981, Design Parameters for soft clays. *Proc. of the VII ECSMFE*, Brighton, 1979, Vol. 5:27-58.

Lambe, T.W. and Whitman, R.V., 1969, *Soil Mechanics, SI Version*. John Wiley & Sons, New York, pp. 301, 306, 427, 452.

Long, N.T.; Guegan, G; and Legeay, G., 1972, Etude de la terre armee a l'appareil triaxial. *Rapport de recherché No. 17*, Laboratories des Ponts et Chausees.

Loudon, P.A., 1967, Some deformation characteristics of kaolin. *Ph. D. thesis*, University of Cambridge.

Nithiaraj, R., Ting, W.H., and Balasubramaniam, A.S., 1996, Strength Parameters of Residual Soils and Application to Stability Analysis of Anchored Slopes. *Geotechnical Engineering*, Vol. 27, No. 2, Dec. 1996, SEAGS, Thailand.

Penman, A.D.M., 2000, Geotechnical engineering and building research. The early days of soil mechanics at BRS, (An essay on the foremost British government soil mechanics laboratory from earliest time to 1957), *Proc. of the Geotech - Year 2000*; Developments in Geotechnical Engineering, AIT, Bangkok, pp. 7-28.

Poulos, H.G. and Davis, E.H., 1974, *Elastic solutions for soil and rock mechanics*. John Wiley & Sons, New York.

Peck, R.B., 2000, Words of Wisdom. *Geotechnical News*, September, Vol. 18, No. 3:37.

Randolph, M.F. and Wroth, C.P., 1978, A simple approach to pile design and the evaluation of pile tests. *Behaviour of Deep Foundations*. ASTM STP 670, 484-499.

Roscoe, K.H., Schofield, A.N. and Wroth, C.P., 1958, On the yielding of soils. *Géotechnique*, pp. 9, 71-83.

Schlosser, F., 1982, Reinforced earth mechanism, behaviour and design methods. *Symposium on Soil & Rock Improvement, etc.* AIT, Bangkok: F-1, 1-26.

Schofield, A.N., 1998, Mohr-Coulomb error correction. *Ground Engineering*, August, 1998.

Schofield, A.N., 2000, Rankine's earth pressure fallacy. *Proc. of the Geotech-Year 2000*, Developments in Geotechnical Engineering, AIT, Bangkok: pp 1-2.

Simons, H., 1981, Statistics, reliability theory and safety factors. *Proc. of the VII ECSMFE,* Brighton, 1979, Vol. 5:81-90.

Tatsuoka, F. and Shibuya, S., 1991, Deformation characteristics of soils and rocks from field and laboratory tests. *Proc. of the IXth ARC on SM & FE,* Vol. 2, December, Bangkok, pp. 63-66.

Terzaghi, K. & Peck, R.B., 1967, *Soil Mechanics in Engineering Practice.* John Wiley and Sons., New York, pp 145-146, 197, 490, 499.

Ting, W.H., 1968, Some effects of history on the stress-strain behaviour of kaolin. *Ph. D. thesis,* University of Cambridge.

Ting, W.H., 1987, Conceptualization by mechanism in stability problems. Panellist Contribution, Discussion Session 6, *Proc. of 8th Asian Reg. Conf. on SM & Fn. Engr.,* Vol. 2, Kyoto, Japan.

Ting, W.H., 1998, Slim River Soil Properties. *Private computer file.*

Ting, W.H., 1999, The failure behaviour of over-consolidated soils. *Geotechnical Engineering,* Vol. 30, No. 3, December: 153-165.

Ting, W.H., 2000, Characterization of components of shear strength. *Geotechnical Engineering,* Vol. 31, No. 3, December: 165-170.

Ting, W.H., 2000a, A survey of embankment construction practice and future developments. *Proceedings of the Geotech-Year 2000,* Developments in Geotechnical Engineering, AIT, Bangkok: 232-233.

Ting, W.H., Garib, Z., and Ramanathan, N., 1990, Study on the stability of slopes along Seremban-Ayer Keroh expressway. Seminar on *Geotechnical Aspects of North-South Expressway,* Kuala Lumpur, 275-281.

Wroth, C.P., 1984, The interpretation of in-situ soil tests. Rankine Lecture, *Géotechnique* 34, No. 4, pp. 449-489.

3

The Use of In-Situ Tests for Foundation Design

Barry M. Lehane[1]

INTRODUCTION

The first evidence of the use of in-situ tests to characterize the ground dates to over 2000 years ago when the Chinese measured the force required to push various types of penetrometers into the ground (Broms & Flodin 1988). Developments since then have taken the form of the design of devices fulfilling one or a combination of the following functions:

(a) provisions of a repeatable and sensitive measure of the ground's consistency
(b) facilitation of small-scale load tests of the ground
(c) direct measurement of a specific soil parameter.

The most popular in-situ tests (performed to assist foundation design) are separated into these three categories in Table 1. Part 3 of the European Standard, *EuroCode 7*, provides a comprehensive account of both the operation and interpretation of all these in-situ tests.

The tests listed in Table 1 have their own particular advantages and the initial part of this paper discusses a number of issues relating to the quality of parameters determined using popular devices of each category. This review provides the background for the second half of the paper which examines the application of some in-situ tests based foundation design procedures. Four examples, taken directly from the author's experience, are selected for this purpose and are used to highlight problems associated with design procedures that are based solely on in-situ test data.

CHARACTERIZATION OF IN-SITU TESTS

Category (A) Devices

These devices provide a measure of the consistency of the ground and are the most popular in general use around the world. The qualitative nature

[1]Dept. of Civil & Resource Engineering, University of Western Australia (Formerly Trinity College, Dublin, Ireland)

Table 1
Categorisation of in-situ tests

	Quantity measured
Category A (Qualitative tests)	
Standard Penetration Test (SPT)	N blows/300 mm
Dynamic Probe Testing (DPT)	M blows/mm
Hand-held MacIntosh probe	M blows/75 mm
In-situ Californian Bearing Ratio (CBR)	CBR value
Weight Sounding Test (WST)	N turns/0.2 m
Category B (Miniature load tests)	
Cone Penetration Test (CPT)	q_c, f_s, u_c
Plate Load Test (PLT)	E_{plt}, q_{ult}
Screw plate test	E_{plt}, q_{ult}
Pressuremeter	G_{ph}, p_{lim}
Dilatometer (DMT)	E_{dl}, $p_{1.1}$
Category C (Soil parameter measurement)	
Pressurement	σ_{h0}, $(c_u)_{ps}$, ϕ'_{ps}
Vane test	$(c_u)_{vane}$, S_t
Seismic cone/Crosshole tests	v_s and hence G_{max}
CPT	c_h

of the data obtained is such that the tests represent a relatively small increase in sophistication to what has been used in site investigations over 2000 years ago.

Standard Penetration Test (SPT)

The SPT is a robust and versatile test and remains very popular in many parts of the world. Recent recognition of the need to correct for energy losses and other effects (e.g. see Skempton, 1986) has improved the reliability of design correlations involving SPT blowcounts (*N*). These correlations must, however, be treated with caution given the relatively poor repeatability of the SPT *N* value, particularly in cohesionless soils, e.g. see Robertson & Fear (1995).

Providing a clean base at the bottom of a borehole for the SPT tool, avoiding over-driving of the casing and ensuring sufficient water head within the casing are some of the practices essential to the measurement of a meaningful *N* value. Measured *N* values are, however, affected in many other less well-known ways. For example, as discussed at a variety of papers presented at the First Internationl Symposium on Penetration testing held in Orlando, USA in 1988:

- Larger borehole diameters (above 125 mm) lead to reduced *N* values
- Use of a cone fitted to the base of the SPT tool (a practice commonly employed to avoid damage to the sampler head in coarse sands

and gravels) when testing in sands can lead to N values which are about 50% greater than if the standard sampler is used.

- Non-standard rods affect the absorption of energy, e.g. heavy rods lead to reduced N values.

Dynamic Probing (other than SPT)

Dynamic cone testing using small-scale driving rigs or using portable devices such as the 'MacIntosh probe' are popular, particularly for small-scale projects and where access problems do not permit other in-situ tests. The blowcount profiles provided by these tests are a good qualitative description of the consistency of the ground and, unlike the SPT, are not sensitive to drilling related disturbances. There are, however, relatively few established and reliable correlations for the blowcounts recorded and designers often find themselves attempting to translate their blowcounts to equivalent SPT N values.

Category (B) Devices

These devices may be considered as miniature load tests and with proper interpretation can be used effectively to extrapolate the data obtained to fullscale conditions.

Cone Penetration Test (CPT)

The standard CPT provides a direct measurement of the ultimate end bearing stress (q_c) of a 60° cone with a projected end area of 10 cm^2, as it is pushed into the ground at a steady rate. It also measures the ultimate side friction (f_s) that can be mobilized in the area immediately above the cone tip. These parameters relate to failure conditions and therefore the most satisfactory correlations for the CPT match q_c and f_s with soil parameters at ultimate conditions.

Almost all cones are now equipped with a pore pressure sensor (and are referred to as piezocones) and consequently are capable of detecting variations in soil permeability over short depth intervals. The pore pressures recorded while the cone is stationary (i.e. in what is referred to as a dissipation test) may also be used to measure the in-situ horizontal coefficient of consolidation (c_h).

While the benefits derived from performing CPT's are well advertised, designers should be aware of limitations. For example,

- q_c values in soft soils can be grossly in error without (a) careful instrument calibration and selection of load cell zeros and (b) correction to the measured end resistance due to the presence of the filter stone (known as the 'unequal area effect').
- Pore pressure readings obtained during and after passage of the cone through heavily overconsolidated (dilatant) stiff clays are usually unreliable because of the tendency for filter stone to de-saturate.

- The presence of relatively small cobbles or coarse gravel can severely disrupt the verticality of the penetrometer installation and, in the limit, lead to cone/rod damage.
- The resolution of the sleeve friction measurement, f_s, is low compared to the end resisance and f_s values are particularly sensitive to the roughness of the sleeve which is progressively modified with continued usage.

Plate load test (PLT)

The popularity of PLTs declined in recent years because of perceived scale effects associated with their use and because of the general recognition that a soil's stress-strain relationship is highly non-linear. However, many workers have since shown that PLTs can, with careful interpretation, provide excellent data with which to assess the shear stiffness and strength characteristics of the ground. PLTs must, in this instance, be supplemented by appropriate non-linear Finite Element analyses of their test results. Soil stiffness characteristics may, alternatively, be estimated using an approximate technique proposed by Atkinson (2000). Very accurate measurements of applied loads and plate movements are essential to reliable interpretation of PLTs. Errors associated with bedding-in settlements have to be minimized (for example, by using cast in-situ concrete 'plates') and care should be taken in assessing appropriate loading rates.

Category (C) Devices

These devices were designed to measure specific geotechnical parameters and, as such, may be expected to be of great value in design.

Pressuremeter

Pressuremeters impose simple two-dimensional (axisymmetric) loading to soils and are therefore very amenable to analysis. The self-boring pressuremeter (SBPT) provides a good indication of the in-situ horizontal stress (σ_{h0}) and can be used to measure the soil's plane strain friction angle (ϕ'_{ps}) and undrained strength, $(c_u)_{ps}$. The SBPT also provides a direct measure of what is referred to as a pressuremeter shear modulus (G_{ph}); this may be used to estimate the soil's shear stiffness characteristics with the aid of non-linear finite element analyses or by employing an approximate technique proposed by Jardine (1992). However, despite these advantages, SBPTs are rarely performed because of their expense in addition to the significant expertise required during operation and data interpretation.

It should be noted that the Menard pressuremeter stresses the ground radially from a pre-bored hole (i.e. in ground where the in-situ stress conditions have been altered). As such, it is essentially a Category (A) device and only provides a measure of the soil consistency.

Vane test

As with SBPT, the vane test is another well-conditioned test. It allows direct measurement of the undrained strength of a clay in the vane mode of shearing, $(c_u)_{vane}$ and also provides a good indication of the clay's sensitivity (S_t).

Shear Wave velocity (v_s) measurement

The assessment of the shear wave velocity (v_s) of the ground (and hence its very small strain elastic shear modulus, G_{max}) is an important part of any realistic numerical prediction of ground movements. v_s can be readily obtained using the crosshole technique or the seismic cone. The crosshole technique requires three boreholes, one to house the source and two for the receivers, which are in line with, and at different distances from the source. The seismic cone has all the features of the standard CPT but incorporates geophones above the friction sleeve to receive a shear wave sent from ground level. The assessment of the arrival times of the shear waves (so that velocities may be calculated) requires a certain degree of expertise. It should also be noted that, because of natural and stress anisotropy in the ground, the v_s value depends on both the direction of shear wave propagation and polarization.

EXAMPLES TAKEN FROM CASE HISTORIES

The foregoing characterization and discussion of in-situ tests provides the background for this section of the paper, which presents a number of examples taken from case histories to illustrate some typical issues that arise when using in-situ test based prediction techniques in foundation design.

Shallow Foundation Settlement

The application of in-situ tests to the settlement prediction of shallow foundations is examined for a site located in the outskirts of Dublin, Ireland. The ground conditions were typical of most of the Dublin area and comprised a deep deposit of boulder clay overlying limestone bedrock. The soil, known locally as Dublin black boulder clay (DBC), is a hard, heavily overconsolidated, well graded, low plasticity and low permeability, clay (lodgement) till. Typical index parameters measured at the test site were as follows: water content≈10%, liquid and plastic limits of ≈25% and ≈12% respectively, clay fraction≈15% and silt fraction≈30%. The stoney nature of DBC poses difficulties for cone penetration testing and therefore only SPT *N* data were recorded. The *N* values obtained within the uppermost 4 m of the deposit at the site are plotted on Fig. 1 and indicate considerable variability. Such variability may be partly attributed to the

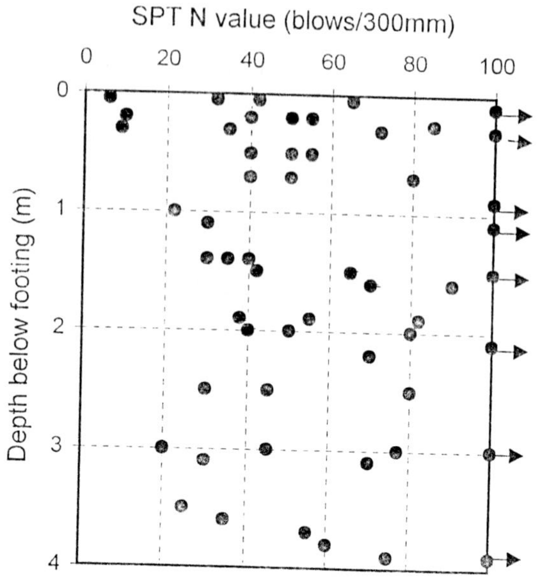

Fig. 1 *SPT N values measured at site of footing test on DBC.*

influence of coarse gravel and cobbles on the measured blowcounts and to possible drilling disturbance due to the addition of water to boreholes to facilitate operation of the shell of the cable percussive rig.

Despite the variability in SPT *N* values, practitioners in Dublin continue to employ empirical correlations with the *N* value, and for shallow foundations, assume an equivalent elastic shear modulus. G_{eq}(KPa) of 600 N (Farrell 1989). A plausible lower bound *N* value for the data on Fig. 1 is about 25, while an average *N* value is about 50. These *N* values equate to design G_{eq} values varying by a factor of two from 15 MPa and 30 MPa.

Rather than arbitrarily select a G_{eq} value of between 15 Mpa and 30 MPa, a number of plate load tests in addition to a fullscale foundation test were performed to improve confidence in settlement predictions for the proposed footings at the site. The average response recorded in two plate load tests, PLT's using 305 mm diameter plates firmly bedded down in the DBC at a depth of 0.15 m is shown in Fig. 2. Each PLT was completed within a period of less than 30 minutes and loading can therefore be considered undrained. The equivalent shear modulus of the ground (G_{eq}) is conventionally interpreted from the linear elastic equation as:

$$G_{eq}=[I_f(D/2)\, q\, (1-\nu)]/\rho \qquad (1)$$

where ν is Poissons ratio (taken=0.5 for undrained conditions), I_f is an influence factor (=π/4 for a rigid circular plate), *D* is the diameter of the plate and ρ is the plate settlement at a bearing pressure of *q*. Application of Equation (1) to the test data on Fig. 2 gives G_{eq}=40 MPa at $q \approx 1000$ kPa.

The footing test was performed within a few metres of the PLT's and entailed loading of a 1.5 m square reinforced concrete pad founded on a 50 mm thick sand blinding layer at 0.1 m depth. A gap was maintained around the perimeter of the footing and, to accelerate consolidation during the test, four 1.5 m deep, 50 mm dia. holes were drilled at 0.7 m centres prior to casting of the footing. Loading to a maximum bearing pressure (q) of 1020 kPa was achieved over a period of 3 months. Creep settlement rates reduced to less than 0.02 mm/day prior to application of the next load increment, illustrating the essentially drained nature of the test. Settlement readings were not obtained until after a dead load equivalent to a bearing measured of 90 kPa had been applied to the footing. The subsequent bearing pressure (q) *vs.* settlement (ρ) response obtained for the footing is shown in Fig. 3 in addition to the interpreted long-term (drained) response. Applying Equation (1) to this long-term response and taking a typical operational Poissons ratio of 0.25 (e.g. see Lehane & Cosgrove, 2000) and $I_f = \pi^{0.5}/2$ for a square rigid footing (with B equated to D) gives an equivalent shear modulus, G_{eq}, at maximum load of 29 MPa.

The G_{eq} values backfigured for the footing test and PLT's of 29 MPa and 40 MPa respectively are somewhat higher than what a designer would select on the basis of the G_{eq} *vs.* N correlation i.e. between 15 and 30 MPa. Moreover, the designer may well attribute the differences in the backfigured values to natural variability of the soil and to scale effects.

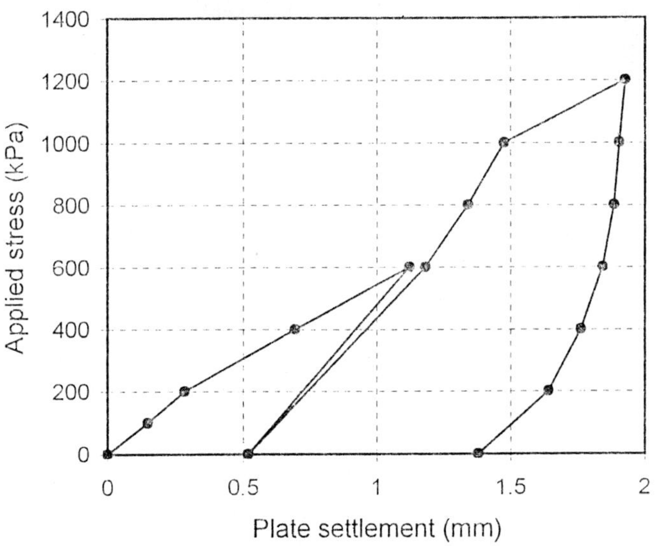

Fig. 2 *Plate load test result at site of footing test on DBC.*

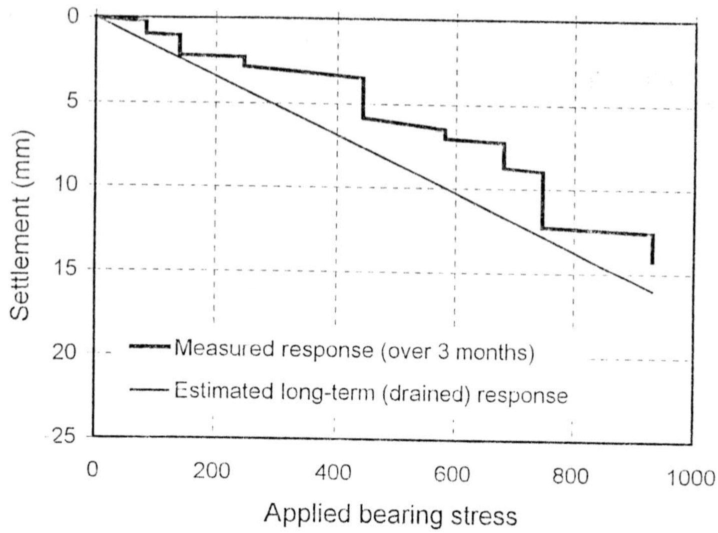

Fig. 3 *Applied bearing stress-settlement relationship measured in footing test on DBC.*

Triaxial tests, which included local strain instrumentation, on rotary cored samples of the DBC were subsequently performed to investigate its stiffness characteristics. These showed that the secant shear stiffness (G_{sec}) varied directly with the mean effective stress (p') at strain levels in excess of ≈0.1% and that, under the conditions pertaining beneath shallow foundations, G_{sec}/p' reduced from ≈230 at a vertical strain of 0.1% to ≈50 at a vertical strain of 1%. The ratio of settlement to footing width (ρ/B) provides an indication of the strain level beneath a shallow foundation (e.g. see Berardi & Lancellotta 1991); this ratio was ≈0.5% beneath the PLT's at q=1000 kPa and ≈1.0% at around the same applied stress in the footing test. On this basis, the discrepancy between the backfigured G_{eq} values from the plate and footing test is fully explained when the differences in the operational strain levels is taken into consideration.

It is clear that the laboratory stiffness measurements facilitated a rational interpretation of the ground stiffnesses measured in-situ. The final solution involved a combination of the findings from in-situ and triaxial tests to deduce a relationship between G_{eq} and ρ/B for a given bearing stress. Moreover, this case history illustrates the very approximate nature of correlations between stiffness and SPT N value as well as the exaggerated variability often implied by N data.

Bearing Capacity of a Shallow Foundation

This case history examines the results from a footing test performed at the UK soft clay test bed site in Bothkennar, Scotland. The site was investigated extensively using the full range of available in-situ tests, sampling techniques and laboratory testing procedures, and is summarized in a full edition of *Geotechnique* (1992). The in-situ tests indicated that the consistency of the clay increased slowly with depth and could be considered relatively constant between depths of 2 m and 6 m.

Table 2 lists the average undrained strengths indicated by in-situ tests between these depths. It should be noted that the N_K value of 15 selected to correlate the cone end resistance with s_u was the mean value suggested by Lunne & Kleven (1981) when related to the peak vane strength of lightly overconsolidated clays. It is evident from Table 2 that N_k defined in this way is ≈9 for Bothkennar clay. Clearly, therefore, site specific N_k values must be derived if CPT data are to be employed to assess undrained strength values.

Table 2
Undrained strengths (s_u) between 2m and 6m at Bothkennar

Field vane (peak)	25.0 kPa
Field vane (remoulded)	10.0 kPa
Self-boring pressuremeter (SBPT)	19.5 kPa
Cone Penetration Test (CPT) with N_k=15	14.5 kPa

The footing was a reinforced concrete, 2.2 m square pad, which was cast at a depth of 0.8 m. Loading to a maximum applied stress of 138 kPa was achieved using 1 m^3 concrete blocks over a period of about three days, by which time the footing had settled by 200 mm; see Fig. 4. Based on a review of pore pressure measurements made beneath the footing, Jardine

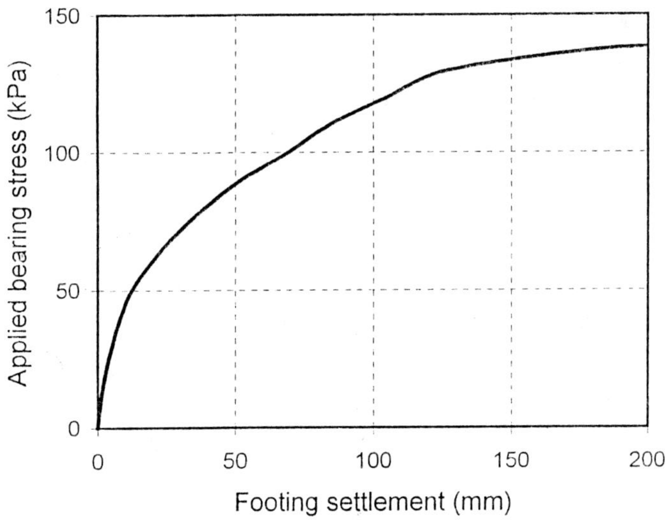

Fig. 4 *Applied bearing stress-settlement relationship measured in footing test on bothkennar clay.*

et al. (1995) argue that conditions from 1 m below the base of the footing were essentially undrained. The ultimate bearing capacity of the footing (q_f) may then be related to the clay's undrained strength by the relationship:

$$q_f = N_{sc}\, s_c\, d_c\, s_u + \sigma_{v0} = N_c\, s_u + \sigma_{v0} \tag{2}$$

where σ_{v0} is the vertical stress at the level of the footing base and N_{sc} is the bearing capacity factor for a surface strip foundation (=2+π). Applying Brinch Hansen (1970) relationships, the shape correction factor, s_c, is 1.2 for a square foundation and the depth correction factor, d_c, is 1.15, giving an overall N_c value of 7. Substituting into equation (2) and accounting for the dead weight of the concrete pad indicates an operational s_u value of 20.3 kPa.

This backfigured s_u value of 20.3 kPa may be compared with s_u values inferred from the in-situ tests in Table 2. It is clear that these tests can only provide a very approximate means for assessing the ultimate undrained bearing capacity of a footing, e.g. employing the peak vane strength in equation (2) would lead to a over-prediction of capacity of 25% while employing the strength inferred from the CPT would lead to an equivalent under-prediction. Even more significant variations in s_u are measured in laboratory tests, e.g. s_u is typically about 7 kPa in triaxial extension tests on conventional piston samples but is 27.5 kPa in high quality Sherbrooke samples tested in anisotropically consolidated triaxial compression tests. The mechanisms associated with the bearing capacity failure of a spread foundation are evidently far more complex than can be inferred from a single in-situ or laboratory test.

Shaft Capacity of Driven Piles

Lehane & Jardine (1994) and Lehane *et al.* (2000) described static load tests performed on jacked 102 mm dia. steel piles at the Bothkennar site referred to previously and on 250 mm square pre-cast piles at Kinnegar, which is located on the outskirts of Belfast, Northern Ireland. The stratigraphy at both sites comprises soft, very lightly overconsolidated, estuarine clay with the CPT profiles shown on Fig. 5; note that these q_t profiles are the piezocone end resistances corrected for the 'unequal area effect' e.g. see Lunne *et al.* 1997. The piles at both sites were installed to a depth of 6 m and the peak local shaft shear stresses (τ_f) inferred from load distribution curves measured in tension tests at both sites are shown in Fig. 6.

The comparable modes of penetration of driven piles and cone penetrometers has prompted the development of many empirical formulae relating q_t and shaft friction. One of the most popular in use is the simple relationship proposed by Bustamante & Gianeselli (1982) which gives:

$$\tau_f = q_c / \beta \approx q_t / \beta \tag{3}$$

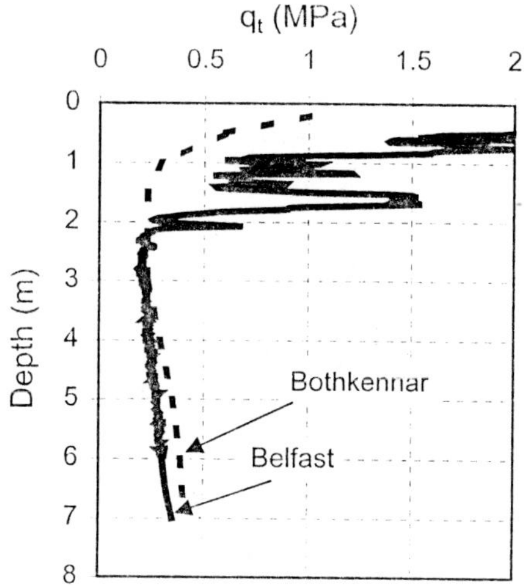

Fig. 5 *CPT q_t profiles at bothkennar and belfast.*

where the value of β recommended for driven piles in soft clay with q_t<1 MPa is 30. Application of this correlation to the q_t profiles in Fig. 5 implies average shaft shear stresses between 3 m and 6 m at Kinnegar and Bothkennar of 8.5 kPa and 10.7 kPa respectively. Measured average values of 12 kPa at Kinnegar and 20 kPa at Bothkennar are higher than these predictions. More significantly, however, the relative difference in capacities is evidently much larger than that of the q_t values, implying that the magnitude of β varies with the nature of the soft clay deposit.

Lehane *et al.* (2000) recently proposed a CPT-based approach for driven piles in clay. This approach builds upon the effective stress methods proposed by Lehane *et al.* (1994) and Jardine & Chow (1996) by employing the CPT q_t value rather than the clay's overconsolidation ratio (OCR) as a measure of clay consistency. The method also allows for the formation of a residual shear surface at the interface between the pile shaft and clay by relating τ_f via Coulomb's friction relationship to the interface friction angle (δ_f) measured in appropriate ring shear interface tests.

The τ_f profiles predicted using the method in Lehane *et al.* (2000) are plotted on Fig. 6 and are seen to provide a good match to the magnitudes and distributions of the measurements. Notably, the approach also predicts the relative difference between the τ_f values recorded at the sites. The calculations indicated that the primary reason for this difference is the lower δ_f value of 23° in the Belfast clay compared to the higher interface residual friction angle of 31° for Bothkennar clay. This example therefore serves to highlight the inadequacy of expressions such as Equation (3) and

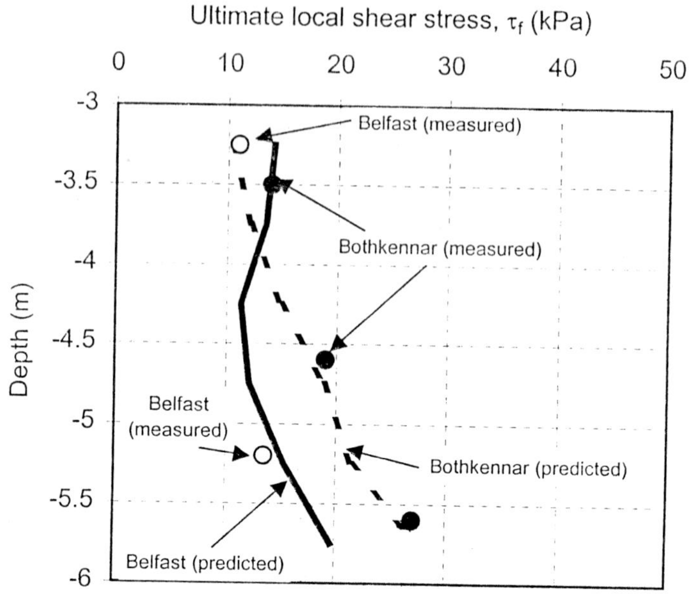

Fig. 6 *Predicted and measured ultimate shear stresses for tension pile tests at bothkennar and belfast.*

the importance of measuring residual interface friction angles for displacement pile design. Ramsey *et al.* (1998) provide a detailed specification on how appropriate angles can be measured, at relatively low cost, using the ring shear apparatus.

In-Situ Rates of Consolidation

The growing popularity of the CPT has recently promoted interest in how it may be used to derive parameters for embankment construction on the lightly overconsolidated Belfast Clay referred to previously. In particular, dissipation tests performed using the piezocone to obtain a direct in-situ measurement of the horizontal coefficient of consolidation (c_h) have been cited as a means of providing a rapid effective means of determining the rates of consolidation beneath embankments.

The times required to achieve varying degrees of dissipation of the excess pore pressures generated by the cone installation are related to c_h using time factors generated by numerical analyses. The most commonly employed numerical analyses are those reported by Houlsby & Teh (1988) who modelled the cone installation process using a modified version of the Strain Path Method and assumed an uncoupled treatment of the dissipation process. Applying the Houlsby & Teh (1988) procedure to piezocone dissipation tests conducted between 3 and 5 m depth in the

Belfast soft clay yielded c_h values of 20 ± 10 m²/year. Although this c_h value is significantly higher than the vertical coefficient of consolidation assessed from 1-D compression tests on laboratory samples, it has been argued by local practitioners that the high c_h value is due to the influence of the macro-fabric within the in-situ material. Laboratory samples are believed to be too small to provide a representative measure of the effect of this in-situ structure.

Fortunately, Davies & Humpheson (1980) conducted a thorough backanalysis of piezometer data recorded following construction of a 6 m high trial embankment on alluvium overlying the Belfast clay. The back-analysed coefficients of consolidation are plotted against effective stress on Figure 7 where they are compared with laboratory measured coefficients and with those derived from the piezocone tests. It is clear that there is general agreement between the laboratory and back-analysed coefficients at the same stress level. The marked reduction in the coefficients as the effective stress increases from pre-yield to post yield values is typical of many clays. In addition, it is clear that the c_h values derived from the piezocone tests are representative of pre-yield coefficients. Baligh &

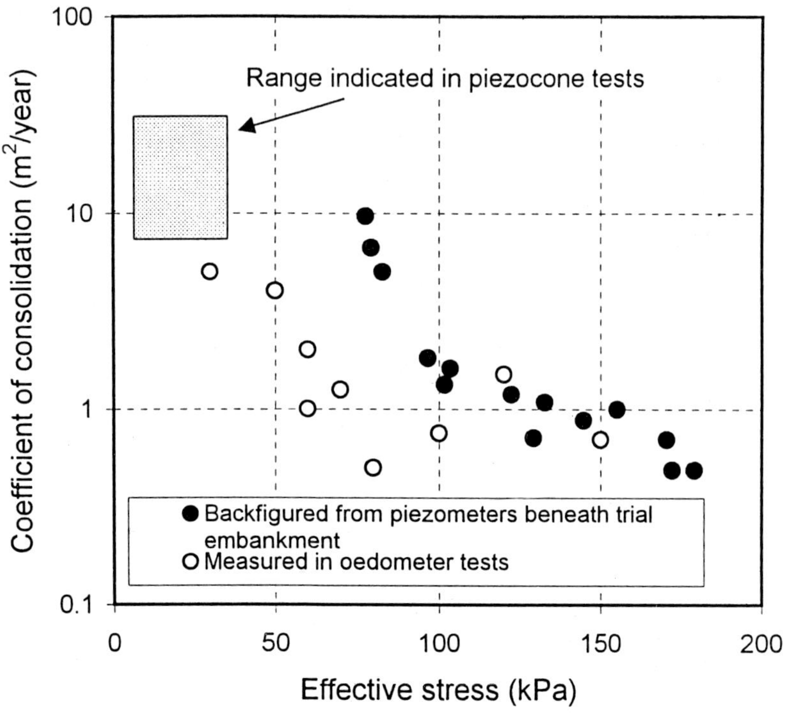

Fig. 7 *Coefficients of consolidation for belfast clay.*

Levadoux (1986), in fact, show that at least 50% of consolidation surrounding a piezocone takes place in the re-compression mode. The piezocone c_h values are therefore not generally relevant to the prediction of consolidation rates beneath embankments on lightly overconsolidated clay. Lehane (1992) shows that they are, however, ideally suited to predicting the time for equalisation around driven piles in both lightly and heavily overconsolidated clays.

CONCLUSIONS

A wide variety of in-situ tests are used to characterise soil and to allow the assessment of parameters for foundation design. Characteristics of the most common in-situ test devices are evaluated within a framework that distinguishes tests on the basis of their capacity to assist geotechnical design. This evaluation reveals that many popular in-situ tests, very often, do not provide a direct measure of the parameter required in design calculations.

The examples selected in the second half of the paper demonstrate that in-situ tests do not, on their own, provide sufficient information about the ground to conduct an efficient design. Interpretation of both in-situ and laboratory tests using a rational soil mechanics framework is essential for the extrapolation to full scale. In particular, it has been seen:

(a) how insights gained from laboratory tests allow improved interpretation of the settlement characteristics observed in plate load and shallow foundation tests;

(b) that the ultimate strength controlling the failure load of a shallow foundation on soft clay cannot be established reliably from in-situ tests alone;

(c) how correlations between CPT data and the shaft capacities of driven piles in clay need to include for the effect of the interface residual friction angle on skin friction;

(d) that the coefficients of consolidation interpreted from piezocone dissipation tests in clay are only applicable to the material's response in the overconsolidated state.

REFERENCES

Atkinson J.H., 2000, Rankine lecture: Non-linear soil stiffness in routine design. *Geotechnique* 50(5), 487-508.

Beradi R. and Lancellotta R., 1991, Stiffness of granular soil from field performance. *Geotechnique*, 41(1), 149-157.

Bustamante M. and Gianeselli L., 1982, Pile bearing capacity prediction by means of static penetrometer CPT. Proc. *2nd European Symp. On Penetration Testing*, Amsterdam, 2, 493-500.

Brinch Hansen J., 1970, A revised and extended formula for bearing capacity. *Danish Geotechnical Institute Bulletin*, 28, 5-11.

Broms B.B. and Flodin N., 1988, History of soil penetration testing. Proc. *1st International Conf. On Penetration Testing*, Orlando, USA, 1, 157-220.

Bustamante M. and Gianeselli L., 1982, Pile bearing capacity prediction by means of static penetrometer CPT. *Proc. 2nd European Symp. On Penetration Testing*, Amsterdam, 2, 493-500.

Davies J.A. and Humpheson C., 1980, A comparison between the performance of the two types of vertical beneath a trial embankment in Belfast. Vertical Drains, 19-31, Thomas Telford, London, UK.

Farrell E.R., 1989, Settlement parameters of Dublin Black Boulder Clay. *Ground Engineering*, 22(5), 33-35.

Geotechnique, 1992, Eight Geotechnique Symposium in print: Bothkennar soft clay test site, 42(2).

Houlsby G.T. and Teh C.I., 1988, Analysis of the piezocone in clay. *Proc. Ist Int. Symp. on Penetration Testing*, 2, 777-783.

Jardine R.J., 1992, Non-linear stiffness parameters from undrained pressuremeter tests. *Canadian Geotech. J.*, 29(3), 436-447.

Jardine R.J. and Chow F.C., 1996, New Design Methods for Offshore Piles. MTD publication 96/103, HSE Books, London, UK.

Jardine R.J., Lehane B.M., Smith P.R. and Gildea P.A., 1995, Vertical loading experiments on rigid pad foundations at Bothkennar. *Geotechnique* 45(4), 573-587.

Lehane B.M., 1992, Experimental investigations of displacement pile behaviour using instrumented field piles. PhD thesis, Univ. of London (Imperial College).

Lehane B.M. and Cosgrove E., 2000, Applying triaxial compression stiffness data to settlement prediction of shallow foundations on cohesionless soil. *Geotech. Engineering*, 143, 191-200.

Lehane B.M. and Jardine R.J., 1994, Displacement pile behaviour in a soft marine clay. *Canadian Geotech. J.*, 31(2), 181-191.

Lehane, B.M., Jardine R.J., Bond A.J. and Chow F.C., 1994, The development of shaft resistance on displacement piles in clay. *Proc. 13th Int. Conf. Soil Mech. and Foundation Engng.*, New Delhi, 2, 473-476.

Lehane, B.M., Chow F.C., McCabe B.A. and Jardine R.J., 2000, Investigating potential relationships between the shaft capacity of closed-ended piles driven in clay and the CPT end resistance. *Geotech. Engineering,* 143, 93-101.

Lehane B.M., McCabe B.A. and Phillips D.T., 2000, Instrumented single and group piles in Belfast soft clay. *Australian Geomechanics*, 37(5), 1078-1088.

Levadoux J.N. and Baligh M.M., 1986, Consolidation after undrained piezocone penetration I: Prediction. J. *Geotech. Engineering*, ASCE, 112(7), 707-726.

Lunne T. and Kleven A., 1981, Role of CPT in North Sea foundation engineering. Proc. ASCE National Convention: Cone Penetration Testing and Materials, 1, 76-107, American Society of Civil Engineers.

Lunne T., Robertson P.K. and Powell J.J.M., 1997, Cone Penetration Testing in Geotechnical Practice, Blackie Academic & Professional.

Ramsey N., Jardine R.J., Lehane B.M. and Ridley A., 1998, A review of soil-steel interface testing with the ring shear apparatus. *Proc. Conf. On Offshore Site Investigation and Foundation Behaviour,* Society for Underwater Technology, London, Kluwer, 237-258.

Robertson P.K. and Fear C.E., 1995, Application of CPT to liquefaction potential. *Proc. Int. Symp. On Cone Penetration Testing,* Linkoping, Sweden, 3, 57-79.

Skempton A.W., 1986, Standard Penetration test procedures and the effects in sands of overburden pressure, relative density, particle size ageing and overconsolidation. *Geotechnique* 36(3), 425-447.

4

In-Situ Characterization of Coarse-Grained Weathered Soils

A.S.R. Swamy[1] *and K.R. Saxena*[2]

INTRODUCTION

The geological features of Northern India differ considerably with those of Southern India, normally known as Peninsula. Geologically India has been divided into two areas the *extra-peninsular area*, that is the country lying to the west, north and east of the Indo-Gangetic plain, and the *Peninsular area*, namely the triangular portion of the country situated on the south of Vindhya range of mountains. The geological history of these two divisions is altogether different. The extra-peninsular area evidently has undergone changes repeatedly due to immersions beneath the sea and has been greatly compressed and disturbed since the Tertiary period. This has resulted in *fossiliferous* beds and contorted and complicated *mountain ranges* as the Himalayas and the hills of the Punjab. On the contrary the Peninsula has not undergone any changes and no great compression since the close of the Paleozoic period.

WEATHERING

The close of the volcanic outburst left the peninsular area of India into a vast basaltic plain. Subsequently due to sub-aerial denudation, the plain got divided into hills and valleys. With the exception of the alluvial deposits, the trap is covered with *laterite*, a porous argillaceous rock, impregnated with iron peroxide, not distributed uniformly through the mass and varying in thickness from 12 m to 60 m. The exposed surface of laterite is irregular due to washing away of softer material, and is traversed by tubes and cavities. The soil produced by the disintegration of the basalt is 'black cotton soil'.

An interesting feature of the peninsular India is the group of boulders with rounded corners standing one on top of the other, presenting a

1. H.No. 1-5-74/7/2, New Maruthi Nagar, Chaitenya Puri, Hyderabad 500 060, India
2. H.No. 7-123, Plot A-16, HUDA Residential Complex, Saroornagar, Hyderabad 500035, India

wonderful example of sub-aerial denudation. Action of heat over the granite gneiss splits it vertically, which reaches the planes of foliation and splits the rock horizontally. The carbonic acid gas in rain dissolves the feldspar and corner of the rock gets rounded off, causing sliding of a block and resting it on the side boulder, fantastically balanced, and giving a picturesque view, as shown in Fig. 1. The direct result of this disintegration is the formation of overlying beds, commonly called *moorum*.

Of the three disintegrated or decomposed, weathered materials, 'Black cotton soil', a fine grained soil, has been studied intensively as its clay content varying in classification from CL to CH, imparts it shrinkage and swelling qualities. Various methods have been developed for in-situ characterization of black cotton soils, the most important being the settlement behaviour under sustained loads. In most cases this property controls the performance of structures.

Laterites and moorum are coarse-grained disintegrated formations. Laterite covers smaller area and is the only formation resting on the trap. *Moorum,* extending over relatively larger area, are the residual soils, formed due to weathering of granitic rocks. Both need altogether different approach for study of in-situ characteristics and evaluation of their in-situ properties, as foundation material and this remains a matter of engineering judgment. The present common in-situ tests do not yield data to fit into existing design practices of foundation engineering. These limitations are illustrated by case histories.

LATERITE

The term 'laterite' is applied to ferruginous and alumino-ferruginous materials, according to Nixon and Skipp (1957), which possess concretionary appearance and lie over soil profile developed in-situ over a parent rock, originally associated with lateritic profile. Thus 'a laterite or lateritic soil is one which is derived from a leached rock, charged with segregated iron nodules', CRRI (1975).

The laterite has been divided into two types, the *high level* and the *low level.* High level laterite is found as a cap formed on the high trap beds of the plateau, altitude above sea level varying from 600 m to 1000 m. Low level laterite; cover a large tract, extending up to both coasts of Peninsula. They contain grains of sand and pebbles, bedded in the ferruginous matrix, and are normally classified as sandy clay or gravel. The laterite resists disintegration. High level laterite, when they get disintegrated, the low level laterite are formed. The age of the laterite, taking into account its occurrence, at different heights and resulting from alteration of the underlying beds in-situ, may be varying from very old to recent times.

In various tropical and sub-tropical regions of the world, laterites and lateritic soils are formed over different types of rock. Laterite has been

Fig. 1 *Formation of boulders—disintegration of rock (with original).*

found very useful as road making material. The Road Research Laboratory, England, has described its use as in road making material in its *Bulletin* (1960) and *Technical Paper* (1962). The Central Road Research Laboratory (CRRI), India in its *Road Research Paper* (1975) has studied its potential use in road and airfield construction as low grade aggregates. In Ghana (1968-69) and elsewhere, the performances of laterites and lateritic soils have been studied and certain co-relations have been established between consistency limits and CBR values. But as foundation material of buildings, perhaps an interesting case history in India, was presented by Saxena and Prabhu (1972).

Profile of Laterites

As mentioned above, the low level laterite, which may be sandy or gravelly clays in ferruginous matrix are partly formed by disintegration of the high level laterite. A typical profile of *high-level laterite* can be conceived as given in Fig. 2. It is the cap-rock, which resists disintegration due to its high compressive strength. This hard mass, though full of cavities, indicating the washing away of softer material, bears a strong resemblance to scoriaceo igneous product, similar to a vesicular soft rock. The different stages of development of profile from parent rock commonly identified are as given below:

1. In the *initial stage*, it is the unweathered parent material.

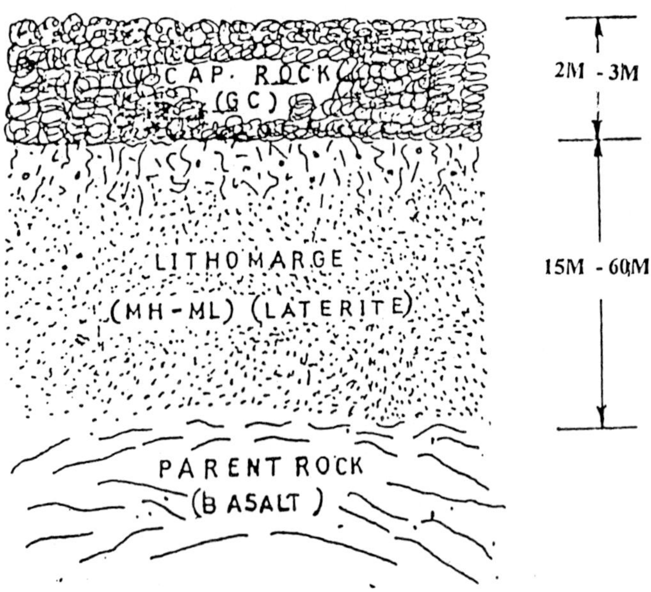

Fig. 2 *A typical lateritic profile.*

2. In the second stage, weathering commences, much of the original material is still found.
3. In the third stage, there is still resistance to weathering by a part of the material. A greater part of weatherable mineral compound gets decomposed and clay content increases, resulting in softness of the material.
4. This is the final stage of decomposition of minerals, leaving aside the weathering resistant mineral parts.
5. In the *final* stage, weathering ceases and the profile is fully developed.

Plasticaly may also indicate the Various Stages of profile. Decrease in plasticity indicates incomplete weathering and increase in plasticity confirms the profile trend. The hard incompressible crust of high strength is followed by layers with less bearing capacities. That is decrease in shear strength.

Characteristics of laterites

The compacted state of laterite mainly depends upon genetical and compositional factors and the pretest conditions. The parent rock and position of soil in profile that is in-situ weathering stage affect the density, besides the gradation characteristics and plasticity of fines and the moisture content. Like the tropical residual soils, laterites also do not conform, in general, to the various concepts of classical soil mechanics, which are based on studies of sedimentary and unweathered soils. The shear strength of residual soils, for example, is influenced by fabric and bond strength and the location of sample within the in-situ profile of the soil formation.The liquid limit, which does not include the micro-fabric or bond strength, is not a useful parameter for correlation with compressibility index. Similarly grain-size of the material has no correlation with permeability and hence the compressibility as determined by routine consolidation tests, can be taken as an indicative quality rather than as an absolute value. Hence characteristics of laterites can be best evaluated, by identifying the location of the sample in the specific profile of the soil formation and tests on the in-situ conditions. It is also not possible to remould the laterite to its field density, which is in order of 2.0 g per cc. or more, depending upon the presence of coarser material.

Building foundations on laterites

Sub-soil exploration

A typical laterite stretch was experienced on the border of Andhra Pradesh and Karnataka States, as shown in plan, Fig. 3, in the southern peninsular part of India at an elevation of 600 to 610 m, above sea level. The sites I and II earmarked for the construction of sugar factories required sub-soil exploration for design of foundations. It was a *high-level laterite formation* and much against the wishes of the owner, due to the expenditure involved,

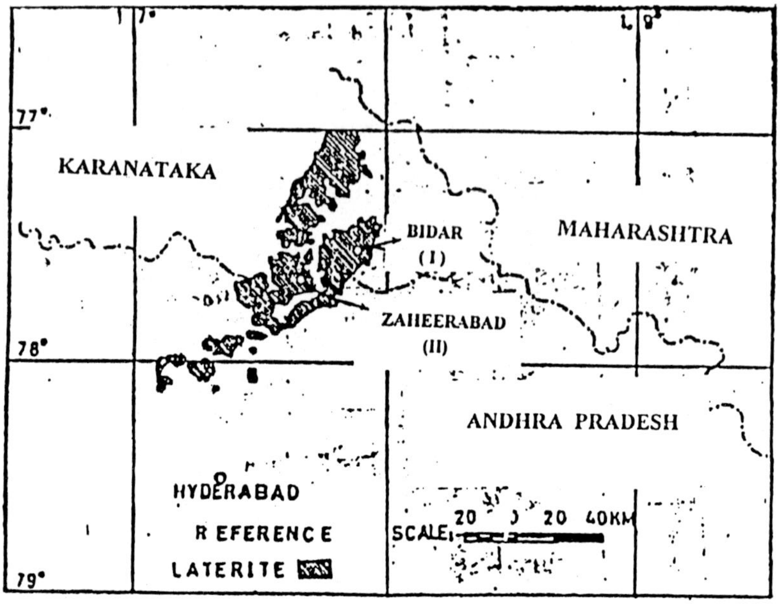

Fig. 3 *Map showing the laterite belt, in southern India.*

sub-soil exploration was conducted up to the parent rock. Table 1 gives the results of the routine 'identification' and 'engineering' tests. Engineering tests were performed on undisturbed or remoulded samples taken from various depths. Table 2 gives the chemical analysis of samples taken from Site II.

Lateritic profile

As shown in Fig. 2, the lateritic profile has been found almost similar at the two sites with variation in thickness of the hard crust and the compressibility characteristics of silts underlying it. The site II is at higher elevation than site I. The sub-soil exploration was limited to 10 m depth, as the bores could not be retained due to caving in of the soils of softer consistency, as shown in Fig. 4 at deeper levels and ground water, besides the difficulty in driving of the casing pipes through the top hard crust. This profile matches with the profile of lateritic deposits found in other tropical and sub-tropical countries. The parent rock has been ascertained to be basalt, a basic igneous rock. The formation of laterite, as mentioned above is due to chemical weathering in regions of alternating wet and dry seasons and where evaporation is also rapid and nearly equal to rainfall. Chemical decomposition of the rocks results in formation of hard crust due to deposition of mineral matters below the surface. The rainwater further

Table 1
Results of identification & engineering tests

Sl. No.	Field Reference of Soil Sample		Liquid Limit %	Plastic Limit %	Plasticity Index %	Gradation Analysis				ISI Classification	Density In-situ	F.M.C. %	Specific Gravity	Void Ratio	Shear Strength Parameters		Remarks
	Location	Depth of Sample (m)				Gravel %	Sand %	Silt %	Clay %						C in kg/cm²	φ in deg.	
1	Bidar-Site I	0-2.0	43	26	18	70	26	1	3	gm/cc							Hard Crust Laterite
		3.25	75	36	39	5	13	37	45	MH	1.19	49.00	3.07	1.585	1.054	12	
		4.00	76	35	41	6	10	40	44	MH	1.14	51.50	3.07	1.690	0.500	21	Softer
		6.00	75	35	40	1	15	42	42	MH	1.25	44.00	3.03	1.430	0.850	16	Strata
		7.50	73	44	29	0	33	47	20	MH	1.22	47.00	2.98	1.440	0.850	16	(lateritic
		9.00	64	39	25	0	42	42	16	MH							Silts)
		10.00	64	40	24	1	40	45	14	MH	Stiff Strata						
2	Zaheerabad-Site II	0-3	44	21	20	89	4	4	3	GC							Hard Crust Laterite
		3.60	47	28	19	26	23	25	26	ML/CI	1.92	18.00	2.84	0.479	0.353	30	
		5.10	46	28	18	31	23	28	18	ML	1.72	25.00	2.86	0.663	0.130	8	Softer
		7.90	50	34	16	14	28	33	25	ML	1.55	31.50	2.76	0.780	0.130	8	Strata
		9.40	45	32	13	21	21	34	24	ML	1.58	29.00	2.81	0.778	-		(lateritic Silts)

Table 2
Chemical analysis of laterites

Constituents	*Site-I Bidar*	*Site II Zaheerabad*			
		3.3-3.8 m	*5-5.34 m*	*7-7.5 m*	*8.85-9.75 m*
Loss on ignition (%)	14.77	12.93	13.17	13.01	12.17
Silica (%)	42.50	30.24	30.62	32.94	32.50
Iron oxide (%)	25.00	31.61	32.94	32.09	28.74
Alumina oxide (%)	11.20	20.79	17.42	16.33	22.66
Silica sesquioxide ratio	1.17	0.58	0.62	0.68	0.63
Silica/al. ratio	3.80	1.45	1.71	2.02	1.44

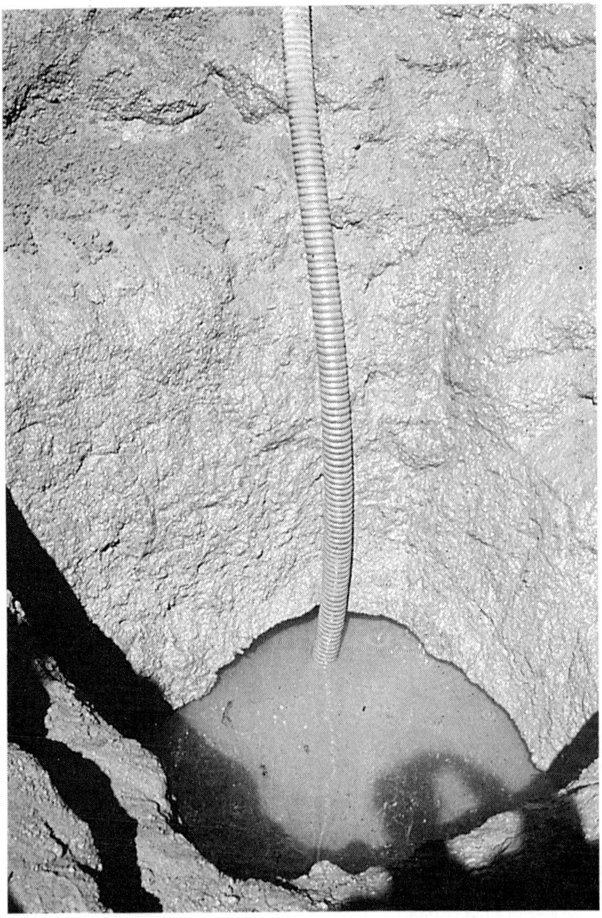

Fig. 4 *Laterite—open pit profile (with transparency).*

carries the dissolved salts only a short distance where they are retained by capillarity and as evaporation takes place a deposit rich in mineral is formed. The only requirement for laterite formation is that the parent rock should contain some iron. The typical red colouration is due to the presence of iron, but the vesicular formation is difficult to explain. There is general agreement about formation of laterites with those reported by Krishnan (1949), Dapples (1958), Novis-Ferreira and Correia (1965) and many others.

The softer layers below the hard crust contain fine grained material, and are normally called *lateritic clays*, devoid of concretions, and are mostly derived from the first cycle of weathering of the parent material as described by Nixon and Skipp (1957).

Pedologists suggest a particular ratio of SiO_2: R2O3 as a guide of laterization. However, if total of sesquioxides exceeds 50 % there is a strong presumption that the material is laterite. Sherman (1952) has reported the presence of montmorillonites and Kaolinites in laterite clays whereas Simpson (1952) has shown the presence of halloysite and kaolin. At the two sites, from the relationship between the liquid limit and clay fraction, it is deduced that based on the *activity* the finer fraction belonged to montmorillonite and kaolin groups.

Laboratory tests

The hard crust material, when pulverized, was found to contain 70 to 90 % of gravel, grain size more than 4.75 mm. The soil beneath was found to contain more of silt and clay size particles, clay fraction decreasing with depth at site I, but more or less it was uniform at site II. According to Vargas (1953), the clay content also reveals the source of laterite as basalt. The weathered over burden, which is of great significance to civil engineers, does not find a place in regional geological maps, unless specially mapped.

The test results given in Table 1, reveal that in general the in-situ densities decreased with depth, due to increase in finer fraction. The soils were also found to be saturated. The ground water table at site I in the month of June was met at 4 m depth and at site II, in the month of November, ground water was not found within the depth of exploration, that is 10 m. The *e*-log *p* curve obtained from the laboratory consolidation test on undisturbed samples of soft stratum indicated, that at site II, the pre-consolidation pressure was less than over burden pressure, giving an apprehension that settlement may occur under normal load.

Field tests

At the two sites, mill house of the sugar factories were to be constructed. A raft of size 17 m × 14.5 m was proposed as foundation for a dead load

Table 3

Test results for bearing capacity of foundations soils of aerodynamics laboratory

Project: B.H.R.L. R&D Unit at Ferozguda

Sl. No.	Field Reference of Soil Sample		% Pass 4.75 mm I.S. Sieve	% Pass 75 micron I.S. Sieve	Gradation Analysis (%)				Consistency Limits (%)				ISI Classi-fication	Specific Gravity	Field Dry Density gm/cc	Field Moisture Content %	Void Ratio	Field Satura-tion Moisture %	Box Shear Parameters of FDD/SMC		Safe Bearing Capacity kg/cm² with f/S=3
	B.H. No.	Depth in M			Gravel	Sand	Silt	Clay	Liquid Limit	Plastic Limit	Plasticity Index	Shrink-age Limit							C in kg/cm²	ϕ in deg	
1	2	3	4	5	6	7	8	9	10	11	12	13	14	15	16	17	18	19	20	21	22
1.	1	1.00	60	15	20	65	8	7	32	24	8	–	SC	2.65	2.04	4.6	0.29	11.3	0.36	46	–
2.	1	2.00	100	11	–	89	7	4	32	22	10	–	SC	2.65	2.04	6.8	0.29	11.3	–	–	–
3.	2	1.00	100	25	–	75	18	7	32	23	9	–	SC	2.65	1.92	8.8	0.3	11.3	0.25	37	–

of 500 tons of the machinery and for 33,000 kW turbo-generator, weighing 34 tons, a framed structure of 6 RCC columns on a raft of 8.5 m × 4.5 m was designed, based on the hard crust 2 to 3 m thick, spread over the area. But the profile of underlying strata and the size of the raft designed required field tests to as certain the safe bearing pressure, assuming the failure of the hard crust either in flexure or by punching. The only pertinent test that was possible, keeping in view the remoteness of the sites, was the plate load test. Though not of direct use, standard penetration tests, were also conducted from 2 m to 8 m depths. The *N* values recorded, varied from 20 to 100, which reflected the variation in strength of the strata.

The load tests were conducted using a 30 cm square plate. Keeping in view the load tests were planned on the hard crust, its thickness over the softer soils was varied. The load-settlement curves are given in Fig. 5. The one metre thick hard crust for a loaded plate of 30 cm width did not fail in shear or settlement for a load of 15-kg/sq.cm. For reduced thickness of hard crust, about 30 cm, failure was noted for a load of 9 kg/cm sq. The softer material showed high settlement under a load of 2 to 2.5 kg/cm.sq. The structural designers were cautioned about consideration of high compressive strength of the hard crust alone for designing the foundations. For laying the raft foundations over the hard crust and to use it as a natural raft, it was necessary that the extent of the crust and its minimum thickness is established over the area involved. The settlement of the softer strata underlying the hard crust and within the bulb of pressure was considered as the governing factor for design of raft.

MOORUM

Moorum or *murum* is more or less a colloquial term, mostly found in Indian literature, used for weathered residual coarse-grained soils, originating from granite and similar rocks. It is often confused with laterites, which have got a typical vesicular concretionary appearance and are porous argillaceous rocks and as mentioned above, are impregnated with peroxide of iron. It has also got a typical profile and is formed at and above a certain altitude; laterization is associated with the chemically active water, which decomposes the rock during percolation. Whereas *moorum* is the weathering product of boulders and is the direct result of disintegration, initially caused by vertical split of granite gneiss through action of heat. These vertical cracks reaching the planes of foliation result in horizontal split in the rock. Rain and wind by dissolving the feldspar, result in blunting of the corners of the rock as they get rounded off, giving rise to a chain of imbalance of the shattered rock, resulting in formation of boulders and the deposition of disintegrated coarse grained material called *moorum.* The coarse and fine-grained content of *moorum* also gives it the nomenclature of hard or soft disintegrated rock in its residual compacted form.

SITE INVESTIGATION FOR AN AERODYNAMIC FACTORY

Site Geology

The area under investigation measured about 35 hectares and topographically was sloping towards west. Geologically granites with

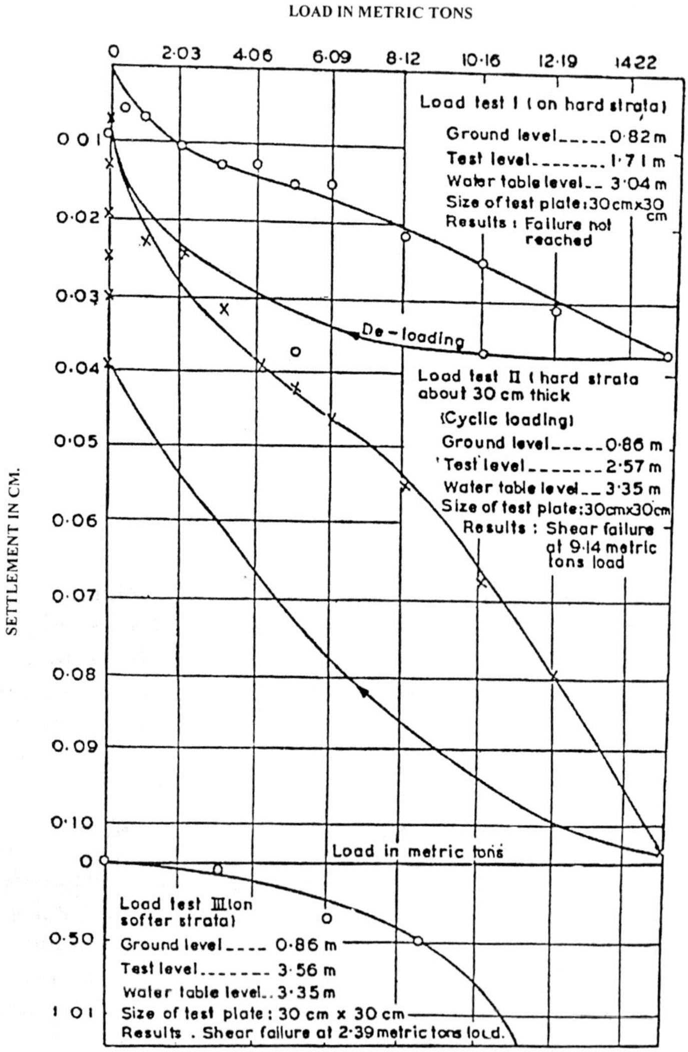

Fig. 5 *Results of load tests on laterite.*

pink and grey outcrops underlie the area. Two sets of joints were observed, one striking N-S and dipping at an angle of 45 degrees and the other striking ESE-WNW and dipping almost vertically. The weathered zone area varied from 0 to 10 metres, minimum on eastern and northern sides and maximum in western and southwesten sides. The stream flowing on the western border served as the source of water.

Subsoil Investigation

Bore logs

Two bore holes, of 150 mm diameter, as shown in Figs. 6, 7, one at the proposed Turbo Machinery and the second at the foundations of the compressors. The depth of the bores varied from 13 m to 15 m, and as required by designers, standard penetration tests and collection of undisturbed soil samples were attempted, alternatively at one-metre intervals. The location of bore No. 2 had to be changed thrice to avoid boulders met with at two to three metres depth to extend the bore up to a depth of 13 m. It was not possible to get true '*N*' values for 45 cm penetration, as required by the ISI standard. Undistrubed soil sampling, even by varying diameter, from 100 mm to 38 mm and with higher area ratio than allowed, was not possible. It was also not possible to remould the sample at the field density of the order of 2 gm/cc for conducting any laboratory test.

Bearing pressure

The estimation of ultimate bearing pressure of the soils, which are relatively in denser state below 2 m depth, was not possible either by SPT or Terzaghi's general shear failure criteria. The in-situ properties clearly indicate that any undesirable settlement was also not expected, so as to derive any inference for arriving at any safe bearing pressure of the soil. It is also experienced that it may not be possible to provide static reaction loading to conduct plate load test to any conclusive stage, either to cause shear failure or excessive settlement. The bearing pressure normally recommended by authors in such geological formations has been 4 to 5 kg/sq. cm or 45 to 50 tons per sq. m. Once the uniformity of sub soil formation is established, the plate load tests with static loadings three to four times the designed load has been performed to check the settlement.

It is clear as mentioned earlier also, that due to mode of formation of weathered soils as described above, the classification by mechanical means or on the basis of index properties itself pose several difficulties. Much more difficult is to predict in-situ strength, deformation and compressibility.

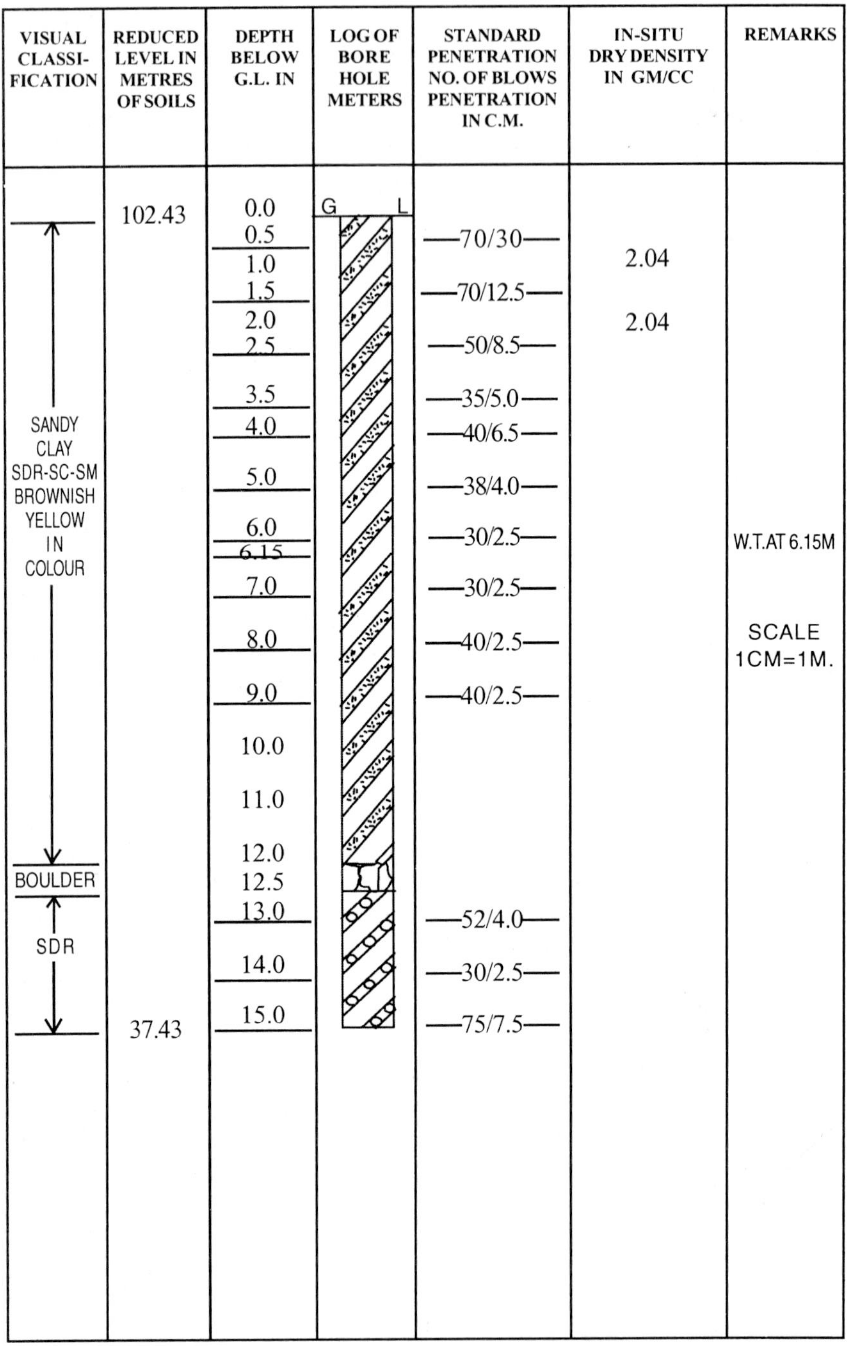

Fig. 6 *Log of borehole no. 1.*

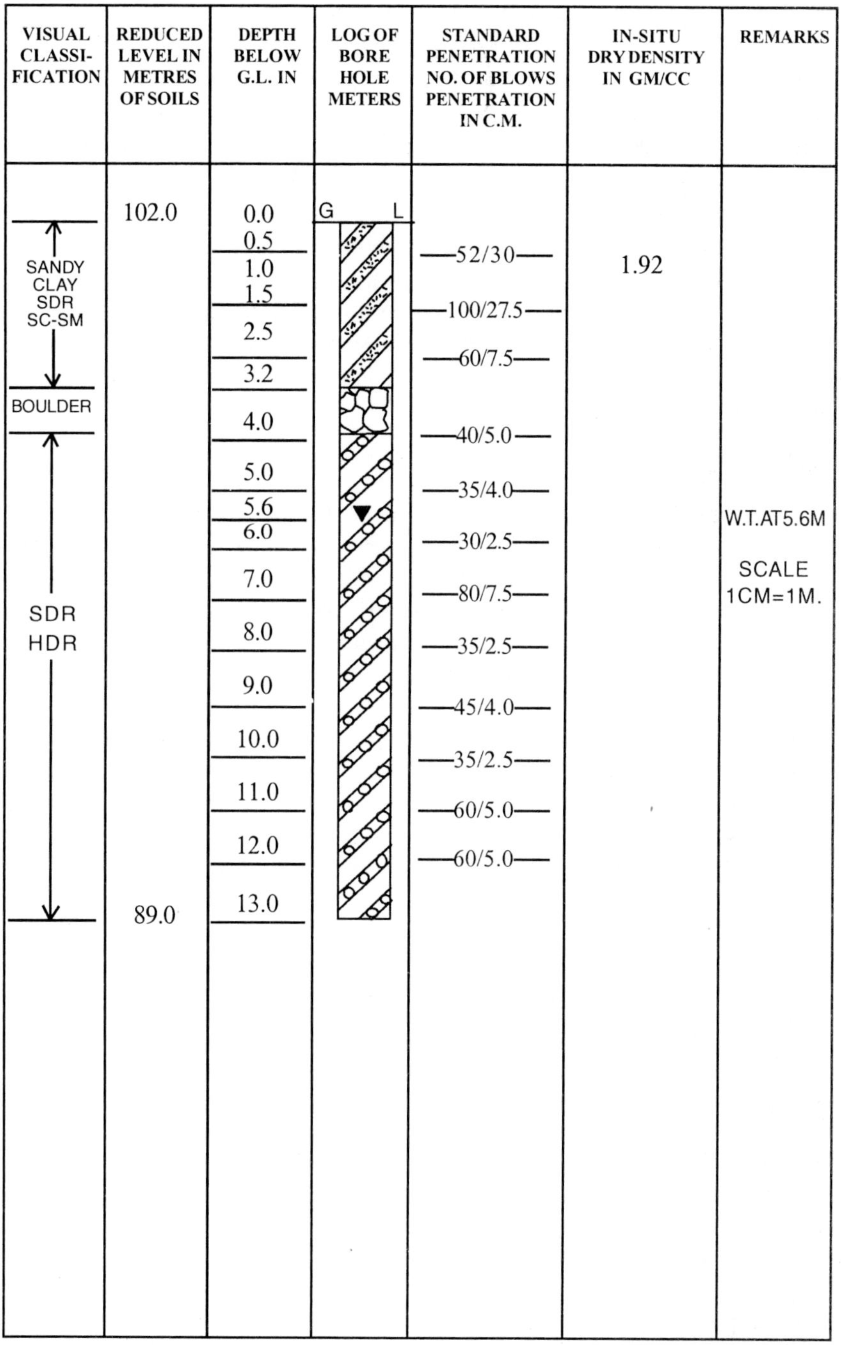

Fig. 7 *Log of borehole no. 2.*

In-situ Dynamic Properties: The feel of the material in disturbed state, when compared to its in-situ highly compressed state and more over in view of the structure planned on it, made it necessary that dynamic tests are also conducted. 'Forced Vibration Tests' and 'Wave Propagation Tests' were performed as per standards, using cement concrete block of size 1.5 m × 0.75 m. × 0.7 m.The following coefficients were obtained:

1. Coefficient of Elastic Uniform Compression (*Cu*) as 5.41 kg/cm^3
2. Coefficient of Elastic Uniform Shear (*Cr*) as 4.98 kg/cm^3

These values with respect to the foundation soils, for the prototype foundations, appeared to be low when compared to those given by Barkan (1962) and others. Obviously the dynamic conditions created destroyed the bonds between the particles, due to very low content of fines in the soils, of the order of 5 to 7%.

CONCLUSIONS

1. Residual soils, formed due to chemical weathering or weathering by atmospheric agents, like rain, frost, wind and day and night temperatures, particularly in arid and semi-arid regions, result in various types of soil profiles. These differ in engineering behaviour due to parent rocks and their environmental conditions. In tropical regions these factors get optimized.
2. Weathered rock names do not indicate mechanised properties. Weathering grades, and index properties may better indicate physical performance. There still exists confusion about the terminology of coarse grained weathered soils, like laterite and non-cohesive residual soils, locally named as moorum, in India.
3. Drainage is the major factor which influences the weathering and the resulting profile of the weathered material, either disintegrated or decomposed.
4. The well established laboratory or field tests cannot be directly used to determine the properties of weathered coarse grained soils for engineering designs. On the contrary, as illustrated it is observed that tests are conducted to verify the soil parameters used for design.

REFERENCES

Barkan, D.D., 1962, Dynamics of Bases and Foundations, McGraw-Hill, New York.

Central Road Research Institute, 1975, Late rite As A Highway Constructional Material, By R. L. Nanda, Road Research Papers, No. 141, Okhla, Delhi.

Dapples, E.C., 1958, Basic Geology for Scientists and Engineers, John Wiley Publication.

Krishnan, M.S., 1949, Geology of India and Burma, Published by Madras Law Journal Office, Madras.

Novais-Ferreira, H. and Correia, J.A., 1965, 'The Hardness of Lateritic Concretions and its Influence in the Performance on Soil Mechanics Test' VIth Int. Conf. Of Soil Mechanics and Foundation Engineering, Toronto, Vol. I.

Nixon, I.K. and Skipp, B.O., 1957, 'Part 5, Laterite' Proc. Of the Institute of Civil Engineers, London.

Road Research Lab. England, 1960, Roadmaking Gravels and Soils in Central Africa, Overseas Bulletin No. 12; Dept of Scientific and Industrial Research, England.

Saxena, K.R. and Prabhu, J.S., 1972, 'Building Foundations on Laterites, Proceedings of Technical Sessions, Vol. I, 13th Annual General Meeting, Indian Geotechnical Society, New Delhi.

Sherman, G.D., 1952, 'Genesis and Morphology of Alumin. rich Lateritic Clays in Problems of Clay and Laterite Genesis', Proceedings of American Inst. Min. Engineers, New York.

Simpson, J.B., 1912, 'Notes on Laterite in Western Australia', Geological Magazine, Decade V, Vol. 9.

Vargas, M., 1953, 'Some Engineering Properties of Residual Clay Soils, occurring in Southern Brazil', Proc. Of 3rd Int. Conf. On Soil Mechanics and Foundation Engineering, Zurich, Vol. I.

5

Ground Investigations and Parameter Assessment for Different Geological Deposits in Sweden

Ulf Bergdahl[1], *Rolf Larsson*[2] *and Leif Viberg*[3]

INTRODUCTION

General

This paper describes how geotechnical engineers in Sweden normally perform their site investigations, both in the field and laboratory, how the results are compiled to obtain a ground model and how the geotechnical parameters for work such as foundation design are assessed. The geological features vary considerably throughout the country and the composition and content of a site investigation program will vary according to the type of ground on the site. The paper has therefore been divided into sections with respect to the main soil types. For each type of ground, a short description of the geological features and the composition of the ground is provided as an introduction. Thereafter, the field investigation methods and laboratory tests with typical results are presented, together with normal practice for parameter assessment.

The content of the paper has mainly focused on the assessment of strength and deformation properties. Thus, classification of soils, permeability, capillarity and chemical parameters are not treated in detail. For soil classification in Sweden, see Karlsson and Hansbo (1984).

In this paper, use is made of a number of empirical relations. The reader should be aware that these relations represent "best fits" of compiled measurements showing clear relations but a certain amount of scatter.

General Geological Features in Sweden

Most soils is Sweden have been formed during the last glaciation and the post-glacial period. The continental ice sheet began to recede from southern

[1]Senior Chief Engineer, Swedish Geotechnical Institute, Linköping, Sweden.
[2]Research Engineer, Swedish Geotechnical Institute, Linköping, Sweden.
[3]Senior Research Engineer, Swedish Geotechnical Institute, Linköping, Sweden.

Sweden about 14,000 years ago and had practically vanished about 8,500 years ago.

The soil forming processes have been several both during and after the glaciation and some of them continue in our time. The most important geological processes behind soil formations in Sweden are glaciation, ice sheet retreat with meltwater discharge, sedimentation of clay and silt particles in seas and lakes, land upheaval and deposition of organic material. Examples of typical geological profiles for parts of Sweden are shown in Figs. 1 and 2.

During the advance of the land ice, formerly deposited soil material was reworked and the bedrock was eroded. Till material was deposited under the ice as it advanced and during the melting phases, sorted

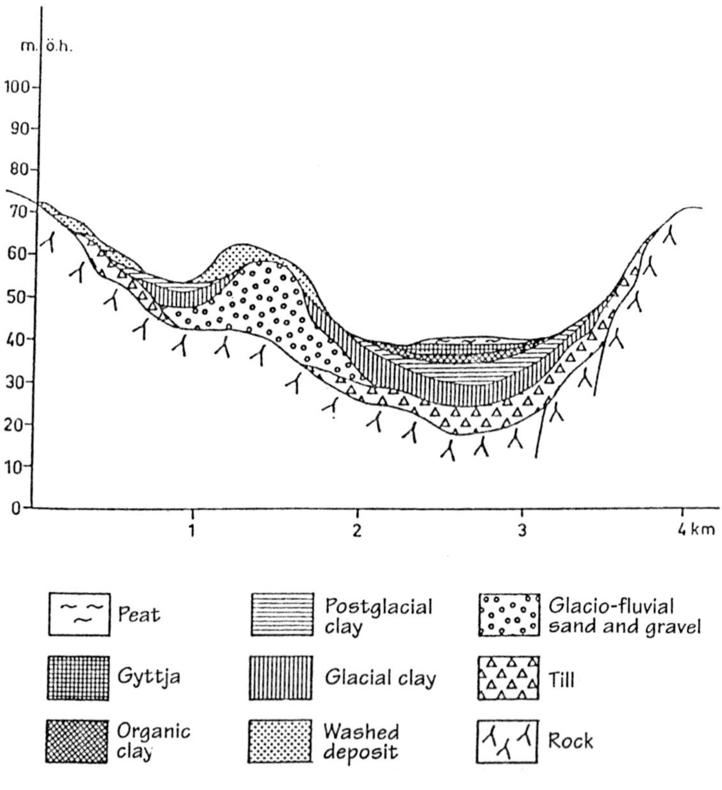

Fig. 1 *Principle geological profile in south-east Sweden. Typically, an esker (glacio-fluvial sand, gravel and pebbles) is partly covered by soft clay, organic soil (peat) is deposited on clay and sand and gravel have been wave washed out on till and clay. The clay is soft and compressible in many places, and piling down into the till layer is a common procedure in construction (from Ahlberg, P. 1979).*

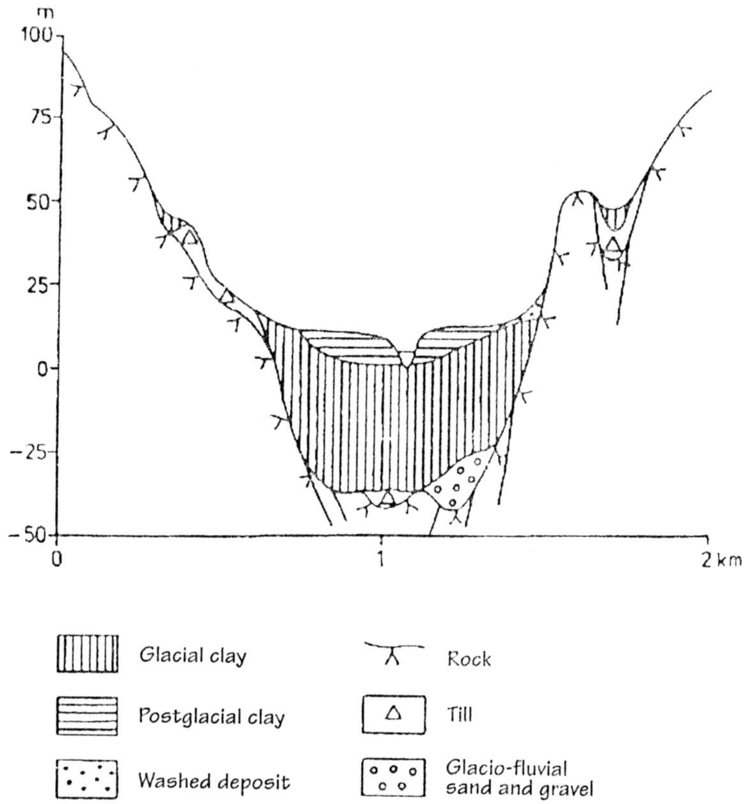

Fig. 2 *Principal geological profile in south-west Sweden. Bedrock topography is steep and deep clay layers fill out the valleys. The clayey areas suffer from numerous landslides due to river erosion in combination with artesian pressure and quick clays. Friction piles in clay is common in the deep clay layers when constructing buildings. (From Ahlberg, P. 1979).*

sediments of cobbles, gravel and sand were deposited by meltwater rivers as eskers and deltas. Finer material sedimented further away from the outlets of the meltwater rivers. The vanishing overburden pressure from the ice sheet resulted in land upheaval and former sea bottoms were raised above the shoreline. During this process, deposited material on the bottom was reworked and transported by sea waves, resulting in new deposits. New rivers were also created by erosion of formerly deposited material.

General Layout of Site Investigations in Sweden

The general purpose of a geotechnical site investigation is to obtain knowledge of the ground regarding soil, rock and groundwater conditions, and to determine the properties of the soil and rock. Site investigations may also include evaluation of existing structures (Bergdahl 1984).

The site investigation normally starts with a desk study of existing maps, aerial photos and previous investigations. The field investigation then usually starts with some type of penetration test depending on existing knowledge and the type of ground at the site. These tests provide an indication of the relative density of the soil and the thickness of different layers down to very dense bottom layers or bedrock. Based on these preliminary sounding results, appropriate methods and equipment for the continued investigations, samplings and groundwater measurements are chosen. The purpose and phase of the investigation and the design and construction requirements should also be considered in this choice of complementary investigations.

Because of the variety of geological features and composition of the ground, a number of tools and boring equipment have been developed in Sweden for sounding and penetration testing, in-situ testing and soil sampling. A number of foreign methods and equipment have also been tested and adopted in Sweden.

The commonly used field investigation methods described in the Swedish Geotechnical Society handbook on geotechnical field investigations (1996) are:

- Sounding and penetration testing methods:
  —Cone and piezocone tests, CPT(U), similar to ENV 1997-3 (1999).
  —Mechanical static sounding test, in which the total penetration resistance is measured continuously and the skin friction at intervals, Dahlberg (1974).
  —Weight sounding test. See ENV 1997-3 (1999).
  —Dynamic probing test HfA. This equipment falls between DPH and DPSH according to ENV 1997-3 (1999). The hammer mass is 63.5 kg, the height of fall 0.5 m and the cone diameter 45 mm.
  —SPT test. C.f. ENV 1997-3 (1999).
  —Soil-rock drilling. Heavy rock drilling machines provided with rock bits and MWD system (Measuring While Drilling) are used in very dense or stony soils and rock.
- Sampling methods:
  —Standard piston sampler ϕ50 mm. See Swedish Committee on Piston Sampling (1961).
  —Peat sampler type SGI ϕ99 mm.
  —Test pits.
  —Auger samplers ϕ36-100 mm.
  —Side intake sampler ϕ35-50 mm.
  —Drive-in piston sampler ϕ25-50 mm.
  —Open tube drive sampler for casing ϕ90 mm.
  —Rotary core drilling.
- In-situ testing methods:

—Field vane test, C.f. ENV 1997-3 (1999).
—Pressuremeter tests of the Menard type (MPM). ENV 1997-3 (1999).
—Flat dilatometer test. ENV 1997-3 (1999).

- Groundwater measurements:
  —Open systems with or without a filter tip.
  —Closed system with filter tip and some form of pressure transducer and recording system.

After the site investigations, it is important to restore the ground in and around the boreholes in such a way that feet or other objects are prevented from penetrating the borehole. In multi-aquifer formations, it is necessary to grout the borehole with a low permeable material in order to avoid leakage between the layers or into a forthcoming underground structure or out of the ground in the case of artesian groundwater.

The results of the site investigations are normally compiled in a site investigation report summarising the results of all investigations in a text with tables, diagrams and drawings, Fig. 3, and the design assumptions for a planned structure and its construction. The report is often divided into two parts. Part 1 presents the investigation results and Part 2 interpretations and evaluations of the investigations, Orre (1979).

Establishing Ground Models

The results from a single test or a single borehole should not be studied on its own, but should as far as possible be integrated in a model for the soils and their properties in the area. The first piece of information in this model is the geological history of the area. This indicates what type of soils and what layering sequence can be expected. It also indicates the age of the different layers and their natural loading history. Later changes in the history in terms of erosion, large landslides and changes in the chemical composition of the pore water, all of which affect the present conditions, may also be found in the geological model.

Based on this information, a rough picture may be drawn indicating what preconsolidation pressures, densities and strengths may be expected and how they may vary along and across the area. Informations on the location of possible anomalies in the general pattern due to earlier landslides or leaching, for example, can also be obtained.

The model of the soil in the area and its properties will be modified and increasingly refined and detailed as more information in terms of test results becomes available. In the model, all the information should be consistent and compatible. When contradictions occur, the reasons must be investigated and the picture clarified. Construction of the model continues until sufficient information is obtained for solving the problems at hand.

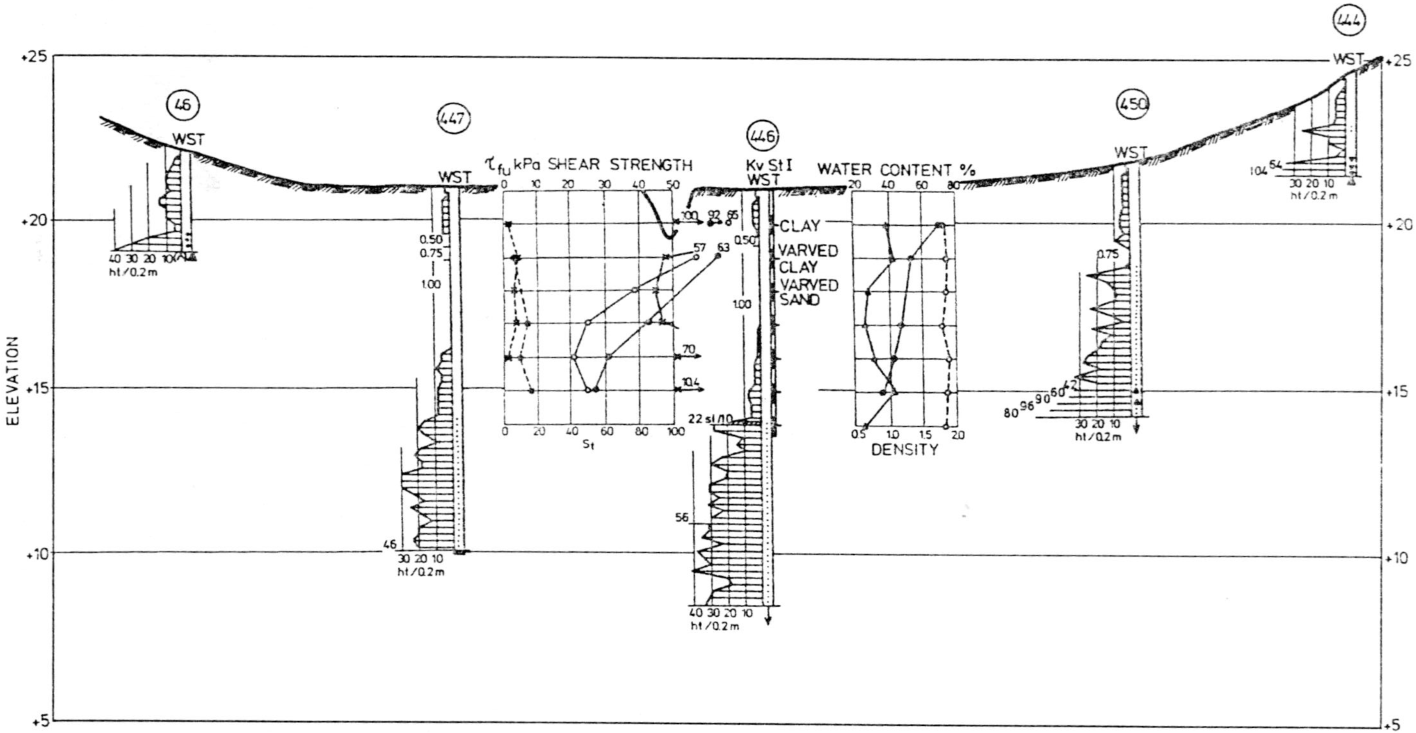

Fig. 3 *Results of a number of penetration tests are normally presented in sections together with results from other field and laboratory tests to illustrate the ground conditions.*

The model may thus vary greatly in complexity, from checking that the shear strengths evaluated in a borehole are compatible with what is normal for a soil of that type and the current preconsolidation to very elaborate models when, for example, the stability conditions for a long bank along a river valley are being considered. However, modelling always increases the reliability of the results. Large savings can thus be achieved both in terms of the required volume of investigations and tests and a more accurate assessment of characteristic values for the design. The latter is related to the level of confidence that can be achieved in the assessment of design parameters.

CLAY AND ORGANIC SOILS OF GYTTJA TYPE

Geological features and composition

As the land ice retreated, large volumes of clay were deposited as sediments in former lakes and seas outside the ice. This clay is called glacial clay because it was formed in close connection with the melting land ice. The meltwater carried material from the ice and the finer particles were deposited in calm water over large areas. In freshwater lakes, the seasonal variation in meltwater production resulted in a thicker, light coloured summer layer and a thinner, dark winter layer with a relatively high clay content. One pair consisting of a light and a dark layer forms an annual varve. The varved clays were formed mainly in fresh water environments. An example of a cutting in varved clay is shown in Fig. 4.

Fig. 4 *Varved clay on till. The thickness of the varves increases downwards because of increasing thickness of silt (white tones). There is a distinct borderline between the soft clay-silt layer and underlying firm till. Photo: L. Viberg.*

In salt water, the glacial clay did not form pronounced varves because of aggregation of particles—flocculation—and a more rapid sedimentation. The organic content originating from plants and animals is very limited in glacial clay.

When the soil-forming process was no longer influenced by the land ice, the post-glacial period began. Post-glacial clay is formed by wave-washed sediments and river deposits in lakes and the sea. The organic content is higher in the post-glacial clays than in glacial clays. On lake bottoms and in bays, the deposition of organic matter may create gyttja, a soil consisting of more than 6% of organic material by dry weight, mixed with mainly silt and clay particles. Transition soils such as organic clay and clayey gyttja are common. Gyttja is common as a top layer on former and present bottoms of lakes and bays. Peat bogs are normally underlain by gyttja and clay.

The thickness of gyttja deposits seldom exceeds a few metres. However, layers of organic clay and clayey gyttja may in many cases be more than 10 m thick.

Clay till is a special, fine-grained type of till. It was formed in the same way as other tills, see Section 6.1.

Investigation Methods and Typical Results

Field investigations

(a) CPT test

CPT tests in clay (and gyttja) have to be performed with simultaneous measurement of the generated pore pressure during penetration, i.e. as piezocone tests. Probes with measuring ranges and accuracy adapted for the particular clay soils should be used in order to enable utilisation of all the possibilities for interpreting the test results. The tests should be performed in accordance with the procedure recommended by the Swedish Geotechnical Society (1993), which is similar to the Reference Test Procedure recommended by the Technical Committee TC 16 of the International Geotechnical Society (1989), but with higher demands on accuracy.

Following the extended possibilities for interpretation, the CPT test has become increasingly used as a tool for determining both stratigraphy and properties in profiles with fine-grained and medium-grained soils. In profiles with normally consolidated or slightly overconsolidated clays, the tests can be performed to depths of about 50 m with normal equipment. For deeper penetrations, some method of reducing the rod friction has to be introduced and when using acoustic signals through the rods, a signal amplifier may be required. In clay till, the use of CPT tests is restricted to deposits with a limited content of cobbles and stones.

The tests are performed at a constant rate of penetration of 20 mm/s. Normally, only short stops are made for adding new rods and taking new grips with the drill rig. Exceptions occur when embedded coarser layers are found within the clay mass and when penetration is stopped in coarser layers below the clay. These occasions provide a suitable opportunity to stop the penetration and follow the dissipation of excess pore pressure with time so that the stabilised in-situ pore pressure can be read off. Such stops normally last about 5 minutes. These values are later used in the interpretation of the test results, where the in-situ pore pressure profile is an important factor. Perfectly hydrostatic pore pressure conditions in profiles with fine-grained soils are an exception rather than the rule in Sweden.

Stops for measuring pore pressure dissipation in the clay and evaluating a coefficient of consolidation are not employed in Sweden. The time for dissipation in these normally very soft and very low permeable soils is too long to be a rational way of obtaining this rather crude estimate. On the other hand, the seismic cone test with short stops at regular depth intervals for determining the travel time for a shear wave from the ground surface down to the probe, i.e. a seismic cone test, has proved to be a very rational way of measuring the variation in shear wave velocity and initial shear modulus in clay profiles.

The results from the CPT tests are interpreted using empirical charts and correlations. The test results provide continuous profiles of depth, cone resistance, sleeve friction and generated pore pressure. A minimum requirement for interpretation is that the in-situ pore pressure profile, u_0 versus depth, is supplemented. This can be measured by piezometers and is often done in the CPT tests as described above. An assumption of hydrostatic pore pressure conditions from an observed free ground water level in an open borehole is normally insufficient.

The charts and correlations used vary with the local geology and soil composition. A number of such charts and relations have thus been proposed for different geographical areas, and in very specific projects a local calibration may have to be made. Swedish experience is employed in a computer program named CONRAD (Larsson et al., 1995) for presentation and evaluation of CPT tests. Some of the correlations used in the program can also be modified to permit local calibration such as that used for clay till in southern Sweden (Larsson, 2001).

The test results are presented in diagrams with the parameters total cone resistance, q_T, sleeve friction, f_T, measured and generated pore pressure, u and Δu ($=u-u_0$), friction ratio, R_f ($=f_T/q_T$), and differential pore pressure, DPPR ($=\Delta u/q_T$), plotted versus depth, see Fig. 5. This presentation does not involve an interpretation, but only calibrated corrections for the data and correlations between them. The presentations are first made in

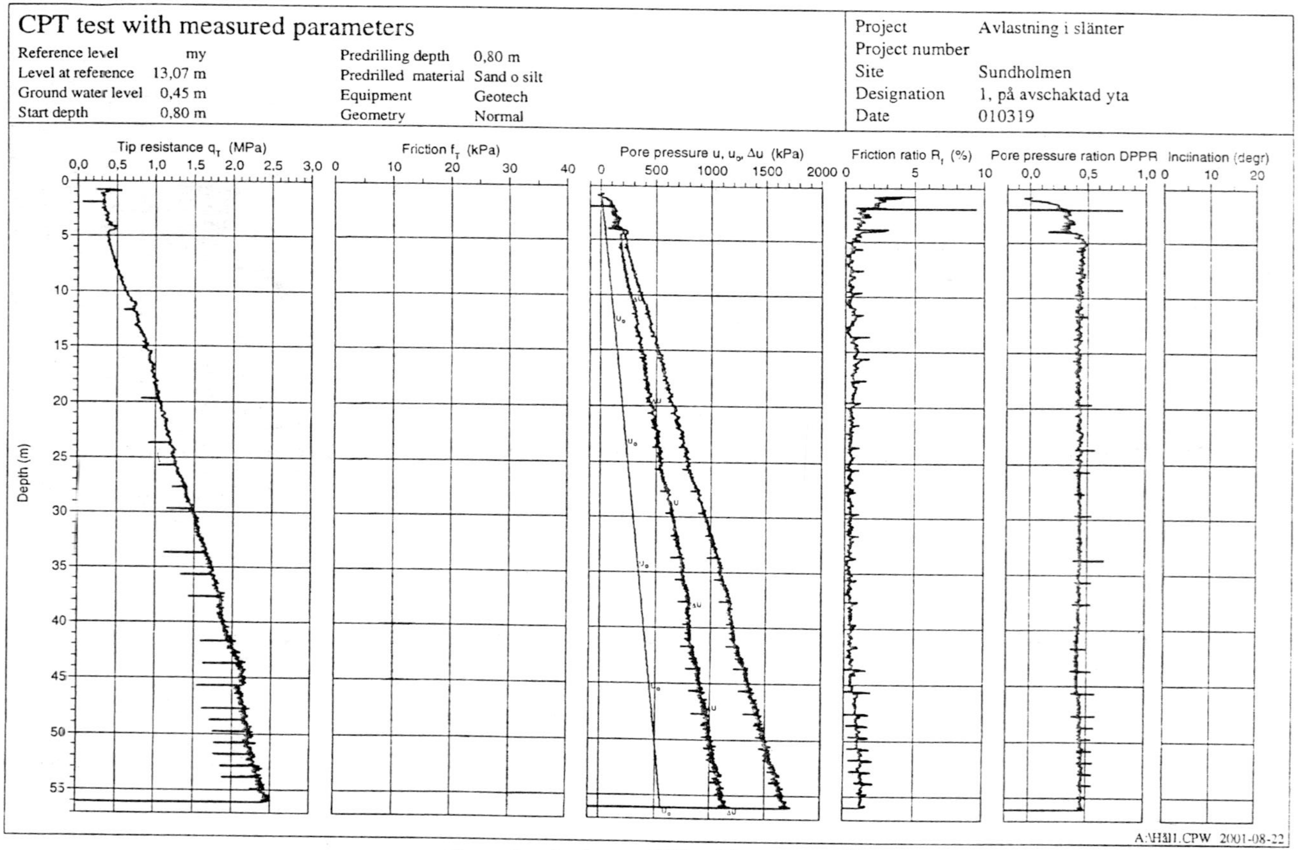

Fig. 5 *Results from a CPT test in a clay profile.*

internationally agreed scales to provide a quick overview. However, these scales do not allow a closer inspection of the results from tests in soft soils. The results are therefore normally also presented in scales adapted to give a better resolution for the particular test.

The interpretation of the test starts with a classification of the soils in the profile based on the test results. In a preliminary chart, see Fig. 6, the profile is divided into sands (and coarser soils), silts and clays and organic soils. This is built up mainly from the relation between the normalised net cone resistance $(q_T\text{-}\sigma_{v0})/\sigma'_{v0}$ and sleeve friction $f_T(q_T\text{-}\sigma_{v0})$. Certain limits for cone resistance and generated pore pressure are also employed in this classification.

When the soil is classified as sand or silt, it is also classified with regard to relative density in a scale from very loose to very dense based

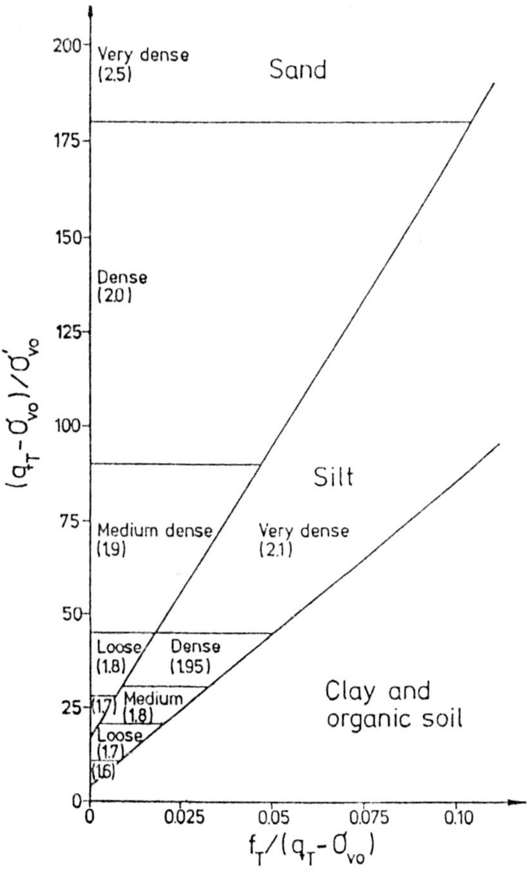

Fig. 6 *Diagram for evaluation of silt and sand and separation of cohesive soils. Numbers in brackets represent estimated water saturated density of the soil in t/m³ (Larsson, 1993).*

on the normalised net cone resistance. It is also assigned a typical density for this type of soil, from which the overburden pressure for the next depth interval is calculated. When the soil is classified as clay or organic soil, special charts for this type of soil are used, see Fig. 7. These charts are based on the net cone resistance, (q_T -σ_{v0}), and the pore pressure parameter B_q (=$\Delta u/(q_T$ - σ_{v0})). From these charts, the soils are classified with regard mainly to overconsolidation and stiffness, and corresponding typical densities are assigned.

This step by step procedure yields a preliminary classification of what is often called "soil type behaviour". This means that the registered data are typical for a certain type of uniform soil, but instead of consisting of a silt, for example, it may well be a clayey sand yielding similar relations. Unless the general soil conditions are very well known, the preliminary soil classification should only be used as information on layering. This will serve as a guide for the selection of depths at which soil samples should be taken and for the choice of complementary investigations.

A preliminary evaluation of the undrained shear strength in the fine-grained soils can also be made at this stage. This is done using the formula

$$c_u = \frac{q_T - \sigma_{v0}}{N_{KT}}$$

where the cone factor N_{KT} is set to 16.3 as an average for Swedish clays and 24 for gyttja. For clay till, a cone factor of 11 is used.

The evaluation can be improved when additional data from laboratory investigations and other field tests become available. Thus, the preliminary classification and densities may be modified as the results are obtained from the routine tests in the laboratory. The evaluation of the undrained shear strength can also be improved with supplementary data on the liquid limit in the soil. The value of the cone factor for clay and gyttja is then calculated from the empirical relation

$$N_{KT} = 13.4 + 6.65\ w_L$$

where w_L is the liquid limit in decimal numbers.

When the liquid limit is determined, an estimate can also be made of the preconsolidation pressure in the soil using the formula

$$\sigma'_c = \frac{q_T - \sigma_{v0}}{1.21 + 4.4 w_L} \quad \text{see Fig. 8.}$$

The evaluation of the preconsolidation is very sensitive to the liquid limit and should only be used as support for the trends evaluated from oedometer tests. When sufficient results are available from oedometer tests and advanced laboratory tests, such as direct simple shear tests or triaxial tests, a local calibration can be performed for the cone factors for both

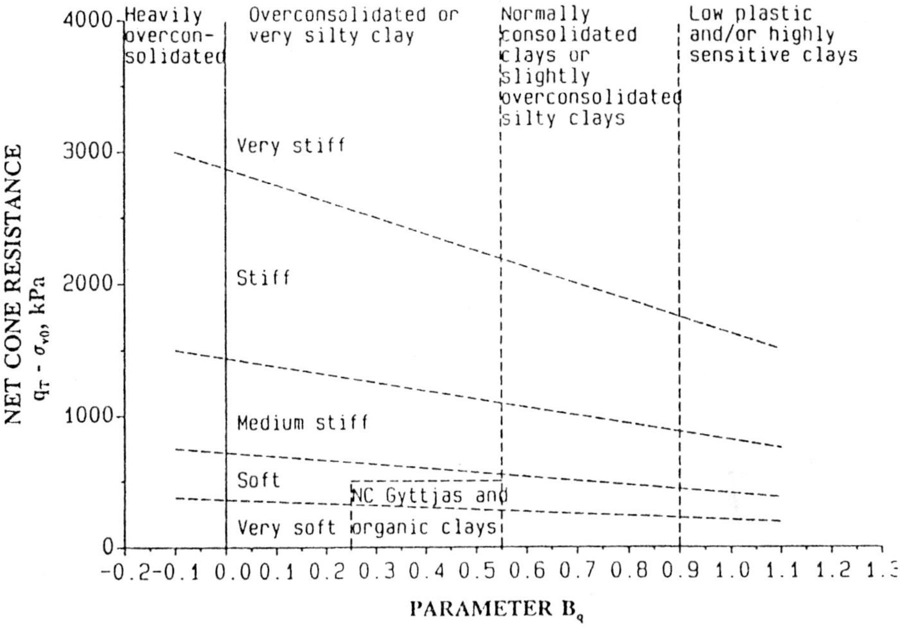

(a)

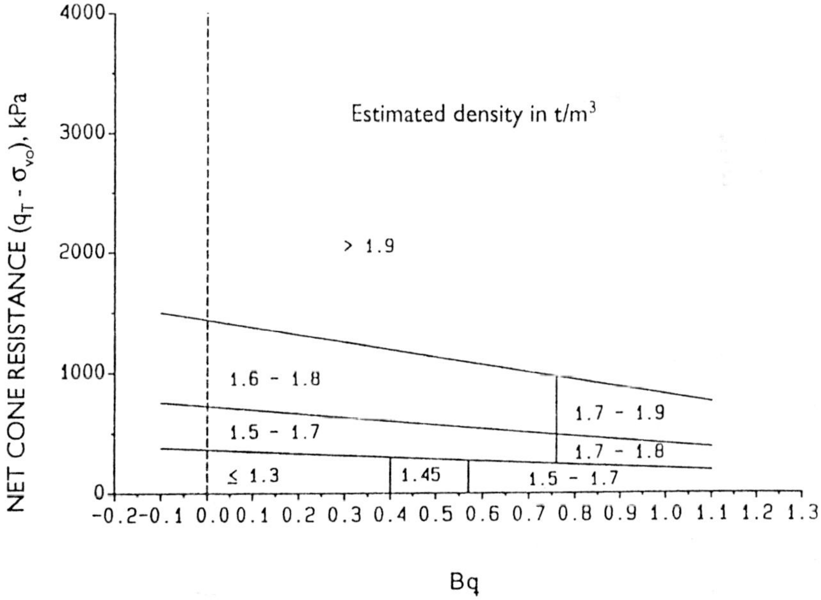

(b)

Fig. 7 *Charts for classification of clays and organic soils (Larsson and Mulabdic, 1991). a) Designation, b) Density.*

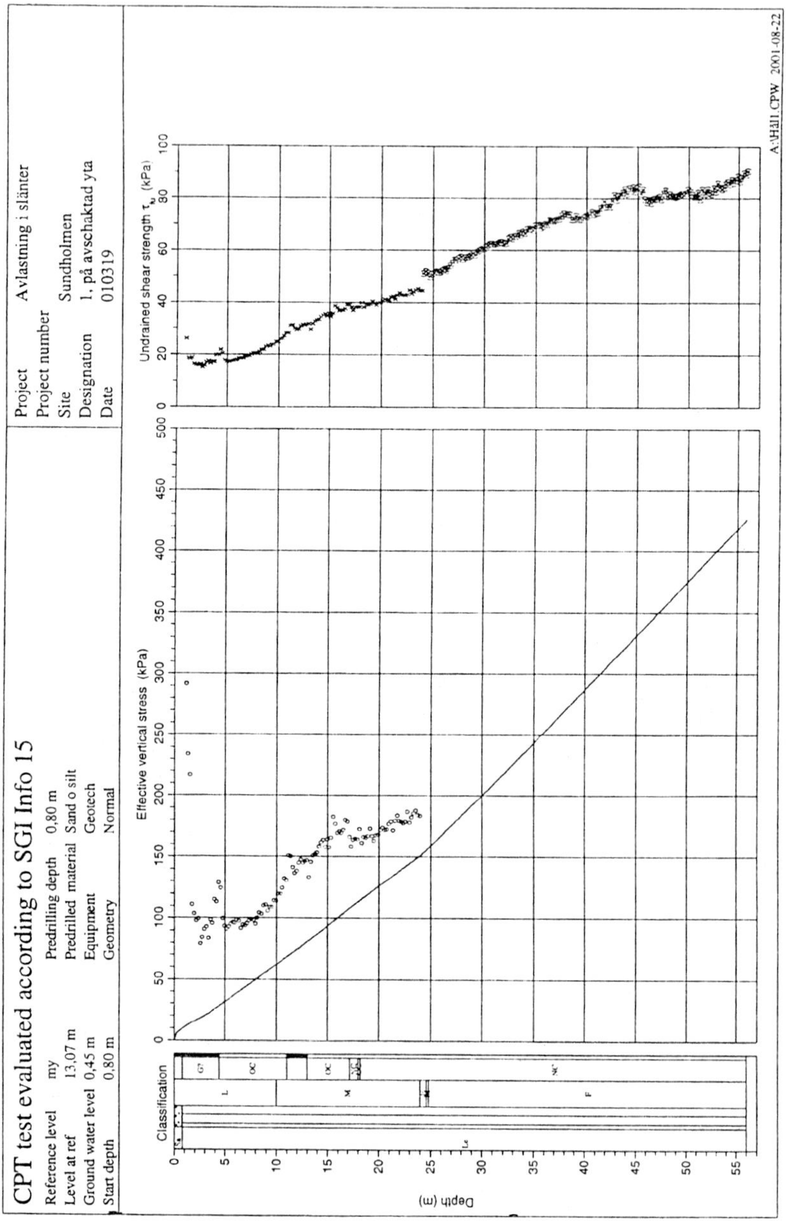

Fig. 8 *Evaluated preconsolidation pressure and undrained shear strength in a clay profile.*

undrained shear strength and preconsolidation pressure. In Sweden, the traditional way of determining the undrained shear strength is by field vane tests and a local calibration of N_{KT} can also be performed against such tests.

The Swedish experience is mainly related to normally consolidated and only slightly overconsolidated soils. Recent investigations in truly overconsolidated soils have indicated that the shear strength evaluated from CPT tests in clay and gyttja should be corrected for overconsolidation ratio (Larsson and Åhnberg 2002). The evaluation then becomes

$$c_u = \frac{q_T - \sigma_{v0}}{13.4 + 6.65\, w_L} \cdot \left(\frac{\sigma'_c}{1.3\, \sigma'_{v0}} \right)^{-0.2}$$

The preconsolidation pressure in this formula can normally be obtained with sufficient accuracy from the CPT tests. The exponent -0.2 is an empirical average for the variation in undrained shear strength with overconsolidation ratio, which can be checked and modified by results from direct simple shear tests in the laboratory.

The undrained shear strengths evaluated with the presented formulas should be directly compatible with the strength evaluated from corrected field vane tests or direct simple shear tests.

(b) Field vane tests

In Sweden, the field vane test, see Fig. 9, has been the main way of determining undrained shear strength in clay and gyttja for the last 50 years. The test is performed according to the recommendations from the Swedish Geotechnical Society (1993). These are similar to EC7 Part 3. In clay till, the equipment and procedures recommended by the Danish Geotechnical Society (DGF Feltkomité, 1993), are normally used.

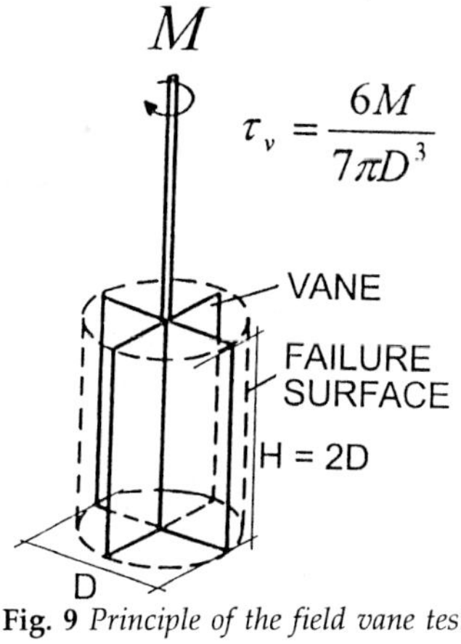

$$\tau_v = \frac{6M}{7\pi D^3}$$

Fig. 9 *Principle of the field vane test.*

According to the Swedish recommendations, vanes with a constant height to diameter ratio of 2 and diameters varying from 40 mm to 100 mm may be used. The commonly used equipment also has sets of interchangeable vanes with diameters of 55, 65 and 80 mm. The vane normally used in soft to stiff clays has a diameter of 65 mm and height of 130 mm, and is replaced when the soil becomes too stiff or very soft. However, recent investigations have shown that significant effects of size on the measured shear strength may occur when the vanes are exchanged (Åhnberg et al., 2001). These effects are considered to be related to disturbance when pushing the vane into the soil and in organic soils they may also be due to fibre effects. They are most pronounced for very small vanes. It is therefore recommended primarily to use the normal vane and to be observant concerning possible size effects when using vanes of other dimensions.

There are mainly two types of equipment in use, the original type with a casing for the vane and cased rods. This system is pushed down to 0.4 m above the test level, whereupon the vane is pushed out of the casing and rotated by the "friction-free" inner rods. After the test, the vane is retracted into the casing and the whole system is pushed to the next test level. On re-entering the casing, any soil adhering to the vane is scraped off.

The other type of equipment is a lightweight system without casing. The vane is provided with a slip coupling to the rods, whereby the rods can first be rotated until failure occurs between them and the surrounding soil. The vane is then engaged and the whole system is loaded. The torque on the vane is obtained by subtracting the rod friction from the total measured torque. After the test, the rod is rotated backwards to open the slip coupling and the system is pushed to the next level on the assumption that any soil sticking to the vane will be continuously scraped off as the vane penetrates through the underlying stiffer soil. This equipment has shown to give results comparable with those from the original equipment in homogeneous clay profiles where the strength increases continuously with depth. In other profiles with varying soil conditions and mixed layering, it is recommended to use the original equipment. An overlying crust or fill layer always has to be predrilled when using the lightweight equipment.

The results from field vane tests have to be corrected for effects of disturbance and testing rate. These effects are related to the type of soil and different empirical relations are used in different parts of the world. In Sweden, the measured strength values are corrected with respect to long empirical experience from the types of clays (and organic soils) in this country. The correction is based on the liquid limit and the undrained shear strength is evaluated from

$$c_u = \mu \tau_v$$

where τ_v is the uncorrected shear strength value and μ is the correction factor calculated from

$$\mu = \left(\frac{0.43}{w_L}\right)^{0.45} \geq 0.5 \text{ (Larsson et al., 1985)}$$

Higher correction factors than 1.2 should not be applied without support from supplementary investigations.

The evaluation of field vane tests thus requires samples to be taken at the corresponding levels in adjacent boreholes and determination of the liquid limits.

The correction factors are empirical and there may be considerable deviations in some soils. A check on the relevance should therefore be made. One such check makes use of the empirical relation between normal uncorrected shear strength values from the vane tests and the preconsolidation pressure (Hansbo, 1957)

$$\tau_v = \sigma'_c \cdot 0.45\, w_L$$

When considerably higher values than those corresponding to this relation are measured, there is an obvious risk that the shear strength value will have to be reduced by more than the general correction. If the uncorrected shear strength value is considerably lower than the correlation yields, there is a high probability that supplementary investigations will give higher strengths.

Experience has shown that lower correction factors have to be applied in overconsolidated clays. A proposal for two separate sets of correction factors for normally consolidated or only slightly overconsolidated (young) clays and overconsolidated (aged) clays respectively was thus presented by Aas et al. (1986). Later results in Sweden indicate that the field vane test results in clays with OCR > 1.3, apart from the correction with respect to the liquid limit, also should be corrected for overconsolidation ratio (Larsson and Åhnberg 2002)

$$c_u = \mu \tau_v \left(\frac{OCR}{1.3}\right)^{-0,2} \qquad OCR > 1.3$$

It should also be borne in mind that the general correction factors apply to normal clays (and gyttjas) and that they cannot be applied directly to soils with other compositions. The general correction factors have thus been found to be invalid for a number of clays containing large amounts of silt, sand, shells, intact organic matter (such as root threads and plant remains) and sulphides. In these soils, a local calibration has to be made versus direct simple shear tests or a combination of such tests and triaxial tests in the laboratory. The corrected undrained shear strength from field vane tests is normally considered to be directly compatible with results from direct simple shear tests.

Results of vane tests in clay till using the Danish equipment are normally not corrected with respect to the liquid limit. However, lower values are normally chosen as a cautious estimate of the strength from the normally widely dispersed results.

(c) Dilatometer tests

Dilatometer tests, see Fig. 10, are performed according to the recommendations issued by the Swedish Geotechnical Society (1995). These recommendations essentially conform with the recommendations given elsewhere, e.g. ASTM Subcommittee D 18.02.10 (1986), ISSMGE Committee TC16(2001).

The test results consist of the two measured parameters, pressures A and B, which, after correction for membrane stiffness and possible other factors, yield the lift-off pressure p_0 and the expansion pressure p_I.

From these pressures, the overburden pressure and the pore pressure at the test level, the dilatometer parameters material index, I_D, horizontal stress index, K_D, and dilatometer modulus, E_D, are calculated.

$$I_D = \frac{p_1 - p_0}{p_0 - u_0}$$

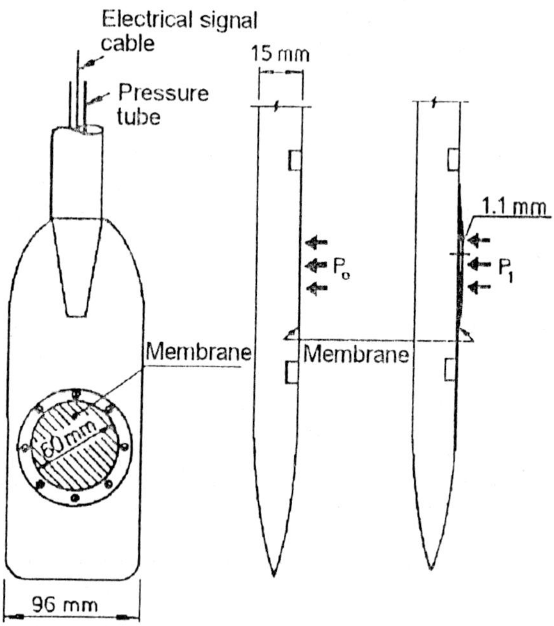

Fig. 10 *Principle of the dilatometer test.*

$$K_D = \frac{p_1 - u_0}{\sigma_{v0} - u_0}$$

$$E_D = 34.7(p_1 - p_0)$$

In the Swedish classification chart, a small correction of the material index is made for overconsolidation effects (Larsson, 1989):

$I_{D(corr)} = I_D - 0.075\ (K_D - 2.5)$ when $K_D > 2.5$ and depth < 2 m

$I_{D(corr)} = I_D - 0.035\ (K_D - 2.5)$ when $K_D > 2.5$ and depth ≥ 2 m

Like the CPT test, the dilatometer test provides a preliminary classification of the soil profile based on the measured parameters and information on the in-situ pore pressure profile, see Fig. 11.

The classification yields the corresponding "soil type behaviour" and a preliminary estimate of the density, from which the overburden pressure at the next test level can be calculated.

Dilatometer tests are normally employed only in clay soils and gyttja for special purposes. These are:

- Estimation of the coefficient of lateral earth pressure.
- Estimation of overconsolidation ratio and preconsolidation pressure in clays in which it is very difficult to obtain undisturbed samples.
- Estimation of the modulus of compressibility in rather heavily overconsolidated clays, such as clay tills, into which the dilatometer can be pushed.
- Obtaining an additional estimate of the shear strength in heavily stratified soils, in which the disturbance effects for both CPT and field vane tests are exceptionally high and it is difficult to take undisturbed samples.

The dilatometer test is also employed in soil profiles where silt overlies clay and/or gyttja, clay and/or gyttja overlies loose silt or alternating layers of clay and/or gyttja and loose coarser soils occur. Here, the main purpose of the dilatometer test is to determine the compressibility of the coarser layers.

An estimation of the in-situ horizontal stress has to be made whenever a more complex calculation starting from the in-situ stresses is to be performed or when the specimens in more advanced laboratory tests are to be reconsolidated to the in-situ stresses. Although empirical in nature, the dilatometer test is then the most rational way of obtaining an estimate of this parameter. In Swedish clays and gyttjas, the coefficient of earth pressure is evaluated from

$$K_0 = 0.24\ K_D^{0.84}$$

where K_D is the horizontal stress index evaluated from the dilatometer test.

This evaluation has generally proved to yield reasonable values.

The overconsolidation ratio is evaluated from the same horizontal

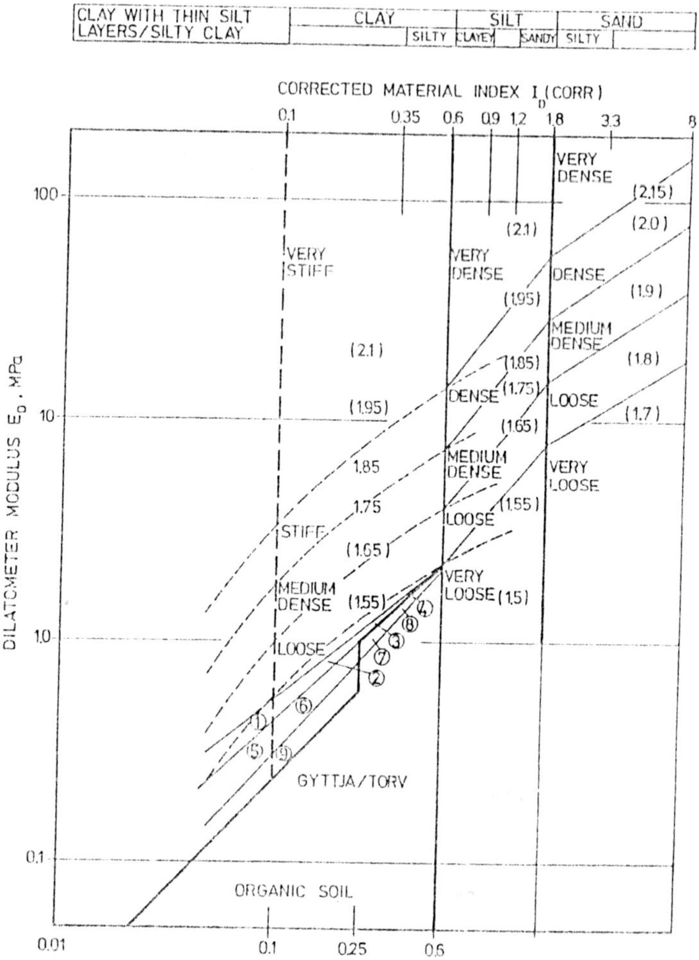

ZONE	CLASSIFICATION*	
1	CLAY WITH THIN SILT LAYERS/SILTY CLAY/(ORGANIC CLAY)	(1.45)
2	CLAY/(ORGANIC CLAY)	(1.45)
3	CLAY/(GYTTJA/PEAT)	(1.4)
4	SILTY CLAY/(GYTTJA/PEAT)	(1.4)
5	ORGANIC CLAY/(CLAY WITH THIN SILT LAYERS/SILTY CLAY)	(1.3)
6	ORGANIC CLAY/(CLAY)	(1.3)
7	GYTTJA/PEAT/(CLAY)	(1.25)
8	GYTTJA/PEAT/(SILTY CLAY)	(1.25)
9	ORGANIC CLAY/GYTTJA/PEAT	(1.15)

*FOR CLAYEY AND ORGANIC SOILS, THE NOTATIONS VERY LOOSE, LOOSE, MEDIUM STIFF, STIFF AND VERY STIFF ARE GIVEN ON THE BASIS OF THE ESTIMATED UNDRAINED SHEAR STRENGTH.

Fig. 11 *Chart for classification of Swedish soils from dilatometer test results (Larsson, 1989).*

stress index. A number of relations have been proposed in the literature. The Swedish experience is mainly based on normally consolidated and slightly overconsolidated soils, for which a good relation was obtained (Larsson and Eskilsson, 1989). Another relation was proposed for non-fissured U.K. soils by Powell and Uglow (1988). This relation has later shown to yield good results also in heavily overconsolidated Swedish soils. A combination of these relations yields the following evaluations:

$$OCR = 10^{0.16(K_D-2,5)} \qquad K_D \leq 5$$
$$OCR = 2.51 + 0.368\,(K_D - 5) \qquad 5 < K_D \leq 7.5$$
$$OCR = 0.24 K_D^{1.32} \qquad K_D > 7.5$$

The preconsolidation pressure is then calculated from the overconsolidation ratio and the effective vertical stress.

The modulus of compressibility in overconsolidated clay may be estimated from

$$M_0 = E_D(0.14 + 2.36 \log K_D) \geq 0.85\,E_D$$

Or alternatively from

$$M_0 \approx 250\,c_u \quad \text{when } I_{D(corr)} \leq 0.35$$
$$M_0 \approx 500\,c_u \quad \text{when } 0.35 < I_{D(corr)} < 0.6$$
$$M_0 \approx 1000\,c_u \quad \text{when } I_{D(corr)} \geq 0.6$$

In the latter case, the undrained shear strength may be estimated from the results of the dilatometer test.

This evaluation is normally performed only when stiff clay layers are intermixed with layers of coarser soil. The evaluated modulus corresponds to an oedometer modulus. An absolute condition for the relevance of the evaluated modulus is that the vertical effective stresses remain well below the preconsolidation pressure also after consolidation for the applied load. Moduli for normally consolidated clays, or clays in which the applied load will straddle the preconsolidation stresses, cannot be estimated from dilatometer tests.

The undrained shear strength can be calculated in two ways. Either the empirical relation

$$c_u \approx 0.22\,\sigma'_{v0}\,OCR^{0.8}$$

can be used together with the estimated values of the effective overburden pressure and overconsolidation ratio, or the more direct evaluation

$$c_u = \frac{p_l - K_0 \sigma'_{v0}}{f}$$

can be employed (Roque et al., 1988; Larsson and Eskilsson, 1989).

The first equation cannot be used for gyttja, for which the factor 0.22 does not apply.

In the second equation, both the coefficient of horizontal earth pressure, K_0, and the effective overburden pressure are evaluated from the dilatometer test, but the product is small compared to pressure p_l and thus the possible error is also limited. In Swedish soils, the factor f has been found to be 10.3 for clay and 9 for gyttja. The shear strength in clay till should not be evaluated from dilatometer tests.

(d) Pressuremeter tests

Pressuremeter tests are used in clay till for determining parameters used in the calculation of settlements and bearing capacity. Normally, the traditional methods and equipment elaborated by Menard (Baguelin et al., 1978), are employed. However, new equipment and methods have been developed for testing and interpretation with special reference to moduli at small strains in clay till (Ekdahl and Bengtsson, 1996; Larsson, 2001). These tests, which are more elaborate, are strain controlled and use more precise measurements of the expansion of the pressuremeter cell.

(e) Pore pressure measurements

The pore pressures in an area should be measured at a sufficient number of points to establish their variation both laterally and with depth. Closed piezometers are used in clay and gyttja, and the material in the piezometers and connecting push rods should be selected in order to prevent chemical reactions and possible formation of gas. The location of the piezometers is also modified to coincide with possible draining and semi-draining layers and with borders between significantly different soil layers with guidance from the results of the CPT tests. The piezometers are monitored for a sufficient period of time to enable a prediction of extreme maximum and minimum values. Open systems may be used in permeable layers below or within the fine-grained soils.

Laboratory Tests

General

Most geotechnical laboratory tests and their interpretation are standardised by the Swedish Building Standards Institution. In tests for which there is no Swedish standard, the recommendations issued by the Swedish Geotechnical Society's Laboratory Committee and the ISSMGE Geotechnical Laboratory Testing Committee ETC 5 (1998) are normally followed.

Samples

In Sweden, laboratory tests on undisturbed samples of fine-grained soils are normally performed on samples taken with the Swedish standard piston sampler (Swedish Committee on Piston Sampling (1961). This is not only a sampler and sampling procedure designed and elaborated to take samples that are as undisturbed as possible, but a complete system including

- Stiff tubes for transportation and storage of the samples
- Airtight end caps
- Thermally insulated, shock-absorbing transport boxes
- Temperature and humidity controlled storage rooms
- Sample extruders and routine testing equipment specially designed for or adapted to the sample tubes
- Testing equipment adapted to the sample tubes to facilitate transfer of the specimens from the sample tubes to the test equipment and to mount the specimens with a minimum of disturbance.

Samples of clay till are often taken with larger thin-walled tube samplers, usually 70 mm diameter, or as continuous cores by the Geobor-S triple tube wire line core sampling system taking 102 mm diameter samples (e.g. Larsson, 2001).

Disturbed samples of clay are normally only taken for classification and determination of water content and liquid limit in superficial layers.

Routine tests

Routine tests on clay samples in the laboratory comprise as a minimum the following determinations:

- Classification by inspection
- Bulk density
- Water content
- Liquid limit
- Undrained shear strength by fall-cone tests
- Remoulded shear strength and sensitivity by fall cone tests.

When required, the following determinations are also made

- Organic content
- Carbonate content
- Sulphide content
- Specific density
- Grain size distribution.

In the classification, the system adopted by the Swedish Geotechnical Society in 1982 (Karlsson and Hansbo, 1984), is used. This system is adapted to Swedish soil conditions with considerable deposits of very soft clays and organic soils and also a great variety of tills. The Swedish classification system is thus more elaborated than most others concerning soft clays, where, for example, the limits for designations such as "very soft" or "soft" are set lower and the limits for sensitivity ranges are normally higher. The variety of the organic soils in content, origin and texture of the organic substance is also considered in more detail.

The liquid limit is determined by the fall-cone test method, which is also the method preferred according to the recommendations by the ISSMGE Committee ETC5 (1998). The plastic limit is not determined on a routine basis. The reason for this is the very time-consuming procedure for determining this parameter in sensitive clays with water contents close to and often higher than the liquid limit. According to Swedish experience, most correlations based on the plasticity index can also be replaced by corresponding and equally good relations based on the liquid limit.

The determination of the undrained shear strength in clay and gyttja by the fall-cone test may be considered as an index test which shows the state of the specimen in its current condition in the laboratory. However, in Swedish practice the undrained shear strength in the field is often estimated from these tests using the following relation:

$$\tau_c = c\frac{mg}{i^2} = \text{(Hansbo, 1957)}$$

where τ_c = uncorrected shear strength value, *kPa*
c = cone factor=1.0 for cone with apex angle 30° and 0.25 for apex angle 60°.
m = mass of the cone, *g*
g = specific gravity, 9.81 m/s^2
i = cone penetration, mm.

The uncorrected shear strength value is then converted to an in-situ shear strength using the same correction factor, μ, as for the field vane test. It should be observed that this relation is elaborated for samples taken with the Swedish standard piston sampler and with the specimen kept in the sample tube during the test. For other samples and test procedures, the relation may be different. The evaluation can only be made for samples taken at shallow depths, i.e. not exceeding 10-15 metres. At greater depths, the evaluated strengths generally become too low because of the stress relief in the sample.

The remoulded shear strength, τ_R, (uncorrected) is determined by the fall-cone test and the sensitivity, S_t, is calculated from this value as $S_t=\tau_c/\tau_R$.

The organic content in clays, organic clays and clayey organic soils is determined by the colorimetric method (Graham, 1948, Larsson et al., 1987). Alternatively, any other accurate method for determination of organic carbon may be used. Determination by weight loss at ignition should not be performed, except for carbonate free soils with organic contents above about 20 %. In other soils, the potential errors are too great to enable a proper determination and classification.

The carbonate content is determined by measuring the volume of gas developing when treating the samples with hydrochloric acid. Various types of apparatus have been developed for this purpose. Probably the simplest reliable apparatus was designed by Moum (1967) in Larsson et al. (1987).

Determinations of sulphide contents are relatively rare in everyday practice, but should probably be performed more often. A fairly simple method for this purpose was developed by Svensson (1983) in Larsson et al. (1987).

The specific density is mostly determined in connection with grain size analyses, which in clays take the form of sedimentation analyses. This determination is normally only performed when there are doubts about classification or when the clay content is to be used in an empirical correlation. Determinations of specific density and grain size distribution are thus more common for clay tills than for sedimentary clays.

Oedometer tests

Oedometer tests on clay and gyttja are normally performed with constant rate of strain, CRS tests, or more correctly with constant rate of deformation. The test procedure and evaluation are specified in Swedish Standard SS 02 71 26. The tests are performed with drainage at one end and pore pressure measurement at the other. The adopted criterion is that the pore pressure at the undrained end shall not exceed 15% of the applied vertical stress (Sällfors, 1975). This entails that the normal rate of compression for clays is about 15% per day and that a test is completed in 1-2 days. Tests on very compressible organic soils often take twice as long.

The tests are used to evaluate the following:

- Preconsolidation pressure
- Modulus for stresses just above the preconsolidation pressure
- Limit pressure where the modulus starts to increase
- Rate of increase in modulus with increasing stresses
- Initial permeability of the soil
- Rate of decrease in permeability with increasing compression.

The modulus below the preconsolidation pressure and the rate of secondary compression cannot be evaluated from these tests. For clay and gyttja, they are normally estimated by empirical correlations based on soil type and undrained shear strength (or preconsolidation pressure) or soil type and water content (or void ratio) respectively.

For an accurate determination of the modulus at stresses below the preconsolidation pressure, incremental oedometer tests are used. The specimens are then loaded in steps to stresses just below the preconsolidation pressure, unloaded in steps to the current or otherwise relevant overburden pressure in the field and then reloaded. The compressive properties for stresses below the preconsolidation pressure are then evaluated from the unload-reload loop. This type of test is normally also used for the overconsolidated clay tills.

Incremental tests are also employed for determination of the coefficient of secondary consolidation. Several tests on specimens from the same sample may then have to be performed to enable an evaluation of the variation of the coefficient with the compression of the specimens. The different tests are then given different starting loads and the loads are doubled in the subsequent steps.

For cases where settlements are a potential problem, oedometer tests should be performed at a sufficient number of levels to permit reliable evaluation of the variation in preconsolidation pressure, compressibility and permeability with depth in the soil profile. The position of the drainage boundaries and possible presence of draining layers is also of the utmost significance. These factors are best judged from the results of CPT tests together with classification in the laboratory.

Apart from determining the compression characteristics and permeability properties of fine-grained soils, oedometer tests are also used as pre-tests before advanced tests to determine the shear strength or deformation properties. They can also be used to check the estimated shear strength properties from field tests. The tests are then run only long enough to enable an evaluation of the preconsolidation pressure. For checking the field test results, the preconsolidation pressure is used together with empirical relations such as

$$c_u \approx 0.22\sigma'_{v0} OCR^{0.8}$$

for clay or more elaborate relations taking into account the variation in the relation with soil type. This is normally done by relating it to the liquid limit of the soil, see Fig. 12 and 13. For clay till, the corresponding approximate relation is

$$c_u \approx 0.4\sigma'_{v0} OCR^{0.85}$$

Direct simple shear tests

Direct simple shear tests are performed in the Swedish direct simple shear apparatus. The apparatus is described in works such as Larsson (1981). Undrained direct simple shear tests are performed mainly for three purposes:

- To check shear strength values obtained by field vane and CPT tests in the field.
- To create a local model for the variation in undrained shear strength with overconsolidation.
- To investigate what increase in shear strength can be expected when the soil consolidates for applied loads.

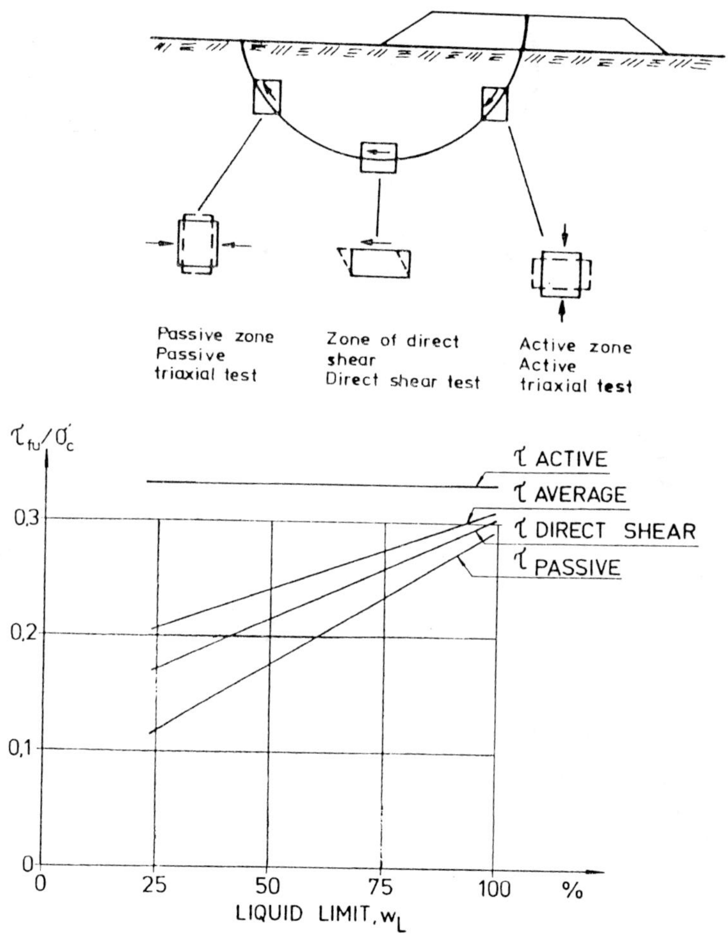

Fig. 12 *Undrained shear strength in Scandinavian inorganic clays as a function of the preconsolidation pressure and the liquid limit (Larsson et al., 1985).*

In order to check the in-situ shear strength values obtained by other methods, the specimen in the direct simple shear apparatus is first consolidated for vertical stresses close to the preconsolidation pressure. This requires a pre-test to be performed in the CRS oedometer in order to accurately determine the preconsolidation pressure. A pressure of 85-90% of the preconsolidation pressure is normally used at reconsolidation of the specimen in the direct simple shear test in order to avoid excessive compression at repeated loading. The specimen is then unloaded to the in-situ vertical stress and given ample time to adjust and possibly swell for this pressure. The volume of the specimen is then locked, and the specimen is sheared in undrained conditions. The rate of shear is normally set low and

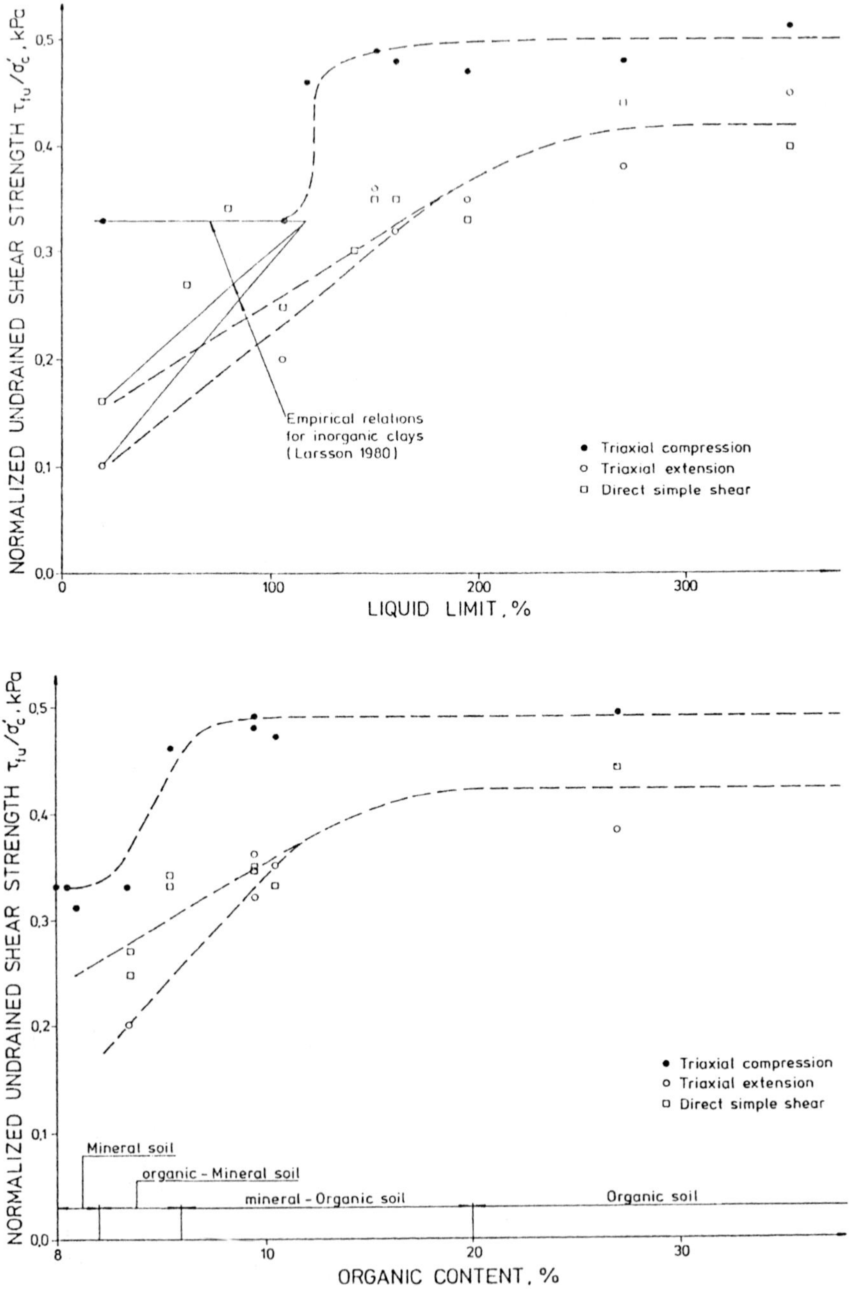

Fig. 13 *Undrained shear strength in Scandinavian organic soils as a function of the preconsolidation pressure and the liquid limit, alternatively the organic content (Larsson, 1990).*

in routine tests it corresponds to a shear strain in the specimen of 0.6% per hour. In a soil profile, tests are performed at various depths selected to cover the whole depth profile and all the constituent types of soil.

In order to create a local model for the variation in undrained shear strength with overconsolidation, a number of specimens are taken from the same sample tube or adjacent tubes at the same sampling level. All these specimens are then consolidated to 85-90% of the preconsolidation pressure. Thereafter, the individual specimens are unloaded to different vertical stresses selected in such a way that a broad range of overconsolidation ratios is covered. After the specimens have been adjusted to these stresses, the volumes are locked and the shear tests are started. The results are plotted in a diagram enabling evaluation of the parameters *a* and *b* in the formula

$$c_u \approx a\sigma'_{v0} OCR^b \text{ , (see Fig. 14)}$$

In order to determine the increase in undrained shear strength with increasing preconsolidation pressure, a number of specimens are consolidated to varying stresses equal to and higher than the existing preconsolidation pressure. The samples are then sheared in undrained

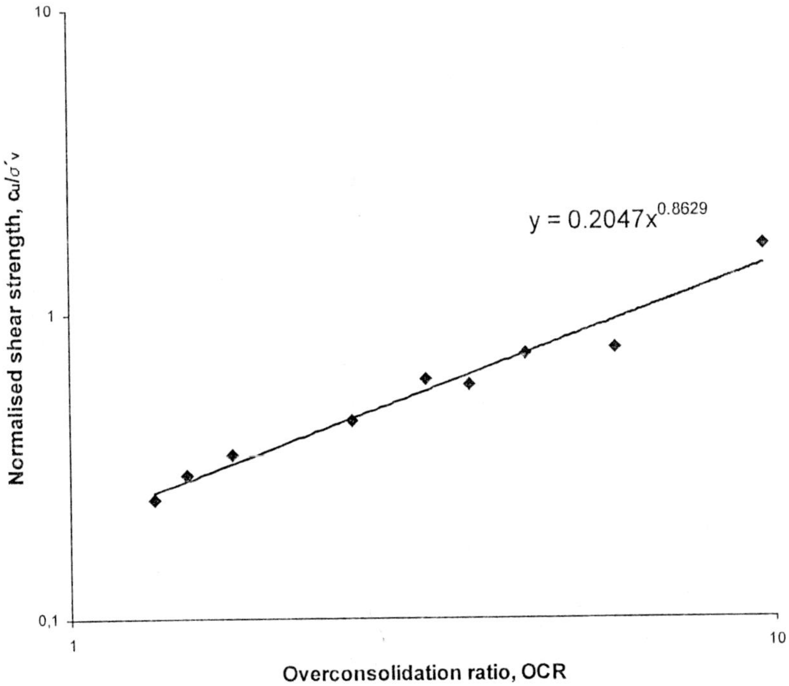

Fig. 14 *Example of an evaluation of the parameters a and b in the strength equation from Larsson and Åhnberg (2002)*

conditions and the relation between the undrained shear strength and the preconsolidation pressure is studied. For clays, this relation is normally assumed to be constant, i.e. the factor a in the above equation is constant. For organic soils, which often have very low initial preconsolidation pressures, the variation in the factor a with increasing preconsolidation pressure may be significant (Wolski et al., 1989).

Triaxial tests

Triaxial tests are normally performed on specimens reconsolidated to their in-situ stresses. This is done by first reconsolidating the specimens for 85-90% of their estimated effective preconsolidation stresses, both vertically and horizontally. This requires a priori knowledge of both the preconsolidation pressure and the coefficient of earth pressure at rest. The preconsolidation pressure is determined by oedometer tests and the coefficient of earth pressure is estimated from empirical correlations, possibly supported by results from dilatometer tests. More elaborate determinations are possible, both in the laboratory and in the field, but are rarely used in everyday practice. A check to ensure that the preconsolidation stresses have not been overestimated is made by studying the compressive strains during this first reconsolidation stage. The loading is applied in steps and any break in the stress-compression curves should be observed. The vertical and volumetric strains at the end of this first stage should normally not exceed 2% in a good quality clay specimen. In organic soils they may be somewhat higher, which can be judged from the oedometer curve.

The specimens are then unloaded to the in-situ stresses. Specimens of soft clay and organic soils are often brought back to the in-situ stresses both in terms of effective stresses and pore pressures. This normally entails lower "back-pressures" than those usually recommended in the literature, but these soils are normally fully saturated and the B-values are still almost 1.0. The resulting lower cell pressures and pore pressures allow more sensitive measuring equipment to be used, which is important in these very soft soils. This procedure also entails that the field conditions are fully simulated. The tests are normally performed at a rate of 0.6% axial deformation per hour.

At this rate, all the normal criteria for pore pressure equalisation are safely met. Because of strong rate effects on the shear strength, it may be desirable to run tests on organic soils at an even slower rate, but in practice the normal rate is most often used and a reduction factor of 0.9 is applied to the evaluated strength for these soils.

Tests on clay till are normally performed according to the same procedure but with higher back pressures in order to ensure full water saturation. The normal criteria for B-values are usually applied. These tests are performed with "friction-free" lubricated end plates.

Undrained triaxial tests on clay and organic soils are performed in cases where the anisotropy of the soil plays a significant role. Such instances occur where the loading cases are best simulated by active compression tests or passive extension tests or where the average of the potential slip surface is dominated by one of these modes, as in the analysis of a steep slope.

The active undrained shear strength in normally consolidated clays has been found to be almost constant around 0.33 σ'_c. The corresponding value for organic soils is $0.5 \cdot \sigma'_c$ (see Figs. 12 and 13). The variation with overconsolidation ratio also follows the normal pattern found in the direct simple shear tests described above. Active undrained triaxial tests are therefore normally performed only to a limited extent in order to verify the validity of these empirical relations for the actual soil. The empirical relations for the passive shear strength are more uncertain and tests have to be performed in cases where this loading mode plays a significant role. The averages of the active, passive and direct sear strengths are normally close to the strengths obtained by direct simple shear tests and also to those obtained by corrected field vane tests and CPT tests.

Drained triaxial tests are performed to determine the shear strength parameters in drained conditions. This strength is normally important only for design in clay tills and other overconsolidated clays or in parts of a larger slip surface where the overburden pressures are low. It requires most tests to be run on overconsolidated or heavily overconsolidated soils. The specimens are normally reconsolidated to the in-situ stresses in the same way as for undrained tests and the test is then performed under drained conditions. To ensure that the specimens are fully drained and that no extra strength is measured because of a suppressed tendency to dilate at failure, a rate of deformation of one tenth of the normal rate in undrained tests is generally used. This means that the time for the tests may be fairly long, sometimes more than two weeks, but failure normally develops at a relatively small strain and the tests can be stopped shortly afterwards.

The drained shear strength parameters in clay are normally estimated from empirical relations, which for Swedish clays yield $c' \approx 0.10\ c_u$(or $c' \approx 0.03\ \sigma'_c$) and $\phi' \approx 30°$. Drained triaxial tests in clay are normally performed only to check the applicability of these values.

Drained and undrained triaxial tests may also be used to determine deformation characteristics. This is normally done in order to determine the required parameters in various models used in advanced numerical calculation methods. The specimens are then often reconsolidated to the in-situ stresses, whereupon selected stress paths are followed in drained or undrained conditions. The testing program may also be designed with regard to the specific requirements on the soil model used.

PEAT

Geological Features and Composition

Peat is composed of more or less mouldered plant debris. The peat forming process has been going on ever since plant production began and is still going on. Peat is formed in places where the complete decay of plants is prevented by submergence in water, such as lakes, gently sloping ground with running surface water and permanently flooded ground. The composition of the peat material depends on the biological environment. In low nutrition ground, peat from mosses forms bogs and in high nutrition ground the peat is composed of sedge, reeds and other water-dependent plants, forming fens. For geotechnical purposes, peat is classified by high, moderate and low degrees of decomposition.

With a few exceptions, peat is a top soil. Normally, it is underlain by gyttja. Below the highest shoreline, the gyttja is in turn normally underlain by clay. Above that line, the gyttja is normally underlain by till. The thickness of a peat layer is normally a few metres, but thicknesses exceeding five metres are not uncommon.

Investigation methods and typical results

Peat soils differ from clays and gyttjas in their origin, composition and texture, which means that investigation requires other methods to be used. However, there are only a few investigation methods specially designed for investigation in peat.

Field Investigations

The thickness of a peat deposit and its variation along a line can be measured by georadars (Hartlén and Wolski, 1996). This method can also be used to determine the thickness of old road embankments on peat and its variation along the road. The method has become increasingly used for both purposes in Sweden and Finland.

CPT tests are performed through peat mainly when it overlies other soils to be investigated. The depth of the peat layer can then be registered, but no properties can be evaluated.

Dilatometer tests are used in the same circumstances as CPT tests and with similar results. Problems may occur at shallow depths in peat, since the earth pressure is insufficient to press back the membrane after the test.

Also coarser penetration tests can be used, but the borders for a transition from peat to gyttja and clay, for example, cannot then be registered.

Field vane tests can only be used in highly decomposed peat. The results are then corrected by a factor μ of 0.5. When the peat becomes significantly fibrous, the results of the field vane tests become meaningless.

Sampling

Undisturbed sampling in peat is performed with a special peat sampler (Carlsten, 1993). This consists of a very sharp wave-toothed steel edge mounted on a plastic tube with a driving head attached to its upper end. The sampler has an inner diameter of 100 mm and is driven by light percussion blows from a drill rig. Samples can be taken from the ground surface or from the bottom of an open, pre-bored hole. Samples can be taken down to at least 5 m depth.

The recovery ratio, i.e. the relation between the length of the sample and the driven length, is measured. Normally, this becomes very high, indicating that the structure, water content and void ratio of the natural soil have been retained. After extraction of the sampler to the ground surface, the cutting edge and the driving head are removed and the ends of the plastic tube are sealed. The tubes are then carefully transported to the laboratory.

In the laboratory, the samples cannot be pushed out of the tubes but have to be cut into segments and possibly split up before the specimens can be inspected and tested.

In peat with a very high degree of humification, the standard piston sampler may be used. Disturbed samples to be used for a rough classification may be taken with screw augers.

Laboratory tests

Classification and routine tests

The classification of peats is made according to the von Post system (in Carlsten, 1993 and Hartlén and Wolski, 1996), or its simplified version in the Swedish geotechnical classification system. The von Post system contains a ten-grade scale for the degree of humification, a five-grade scale for water content and four-grade scale for content of fibres and stems from sedge, content of root threads and content of wood remnants. In the geotechnical classification system, this is reduced to a three-grade scale based on degree of humification and the designation can then be made on the basis of a quick visual inspection and a squeeze test on a specimen by hand.

Apart from the visual inspection and squeeze test, the following parameters are normally determined as routine tests on peat:

- Water content, w_N
- Ash content after ignition, A_c
- Organic content ($\approx$1-A_c)
- Bulk density, γ
- Dry density, γ_d

Drying of peats for determination of water content is often performed at 80 °C.

These parameters are normally sufficient for classification, for using the most common empirical relations for properties in peat used in Sweden and for estimating the load from the peat in design. If required for correlation with other relations, the specific gravity may also be determined and the void ratio and porosity calculated.

Oedometer tests

Oedometer tests on peat are normally performed as step loaded compressometer tests. A compressometer is an oedometer where the oedometer ring is replaced by a rubber membrane to prevent loss of water and a number of suspended outer rings to keep the diameter of the specimen constant. The purpose of this arrangement is to eliminate side friction. The tests are performed on specimens with a diameter of 100 mm and height of about 45 mm.

From the oedometer tests, the following parameters are evaluated:

- the preconsolidation pressure
- the modulus for stresses just above the preconsolidation pressure
- the limit pressure where the modulus starts to increase
- the rate of increase in modulus with increasing stresses
- the rate of secondary compression
- the initial permeability of the soil
- the rate of decrease in permeability with increasing compression

The oedometer tests are performed in sufficient number to provide a sound basis for the estimation of settlements. The required number of tests depends both on the thickness of the peat layer and the variation within the layer with regard to classification and water content. The results are checked against and combined with existing empirical relations (Carlsten, 1993).

Direct simple shear tests

Direct simple shear tests are performed in the same way as tests on clay and gyttja, with the exception that the specimens are larger and have the same sizes as in compressometer tests.

Triaxial tests

Triaxial tests for determination of strength properties are normally not performed on peat, with the possible exception of highly decomposed peats. Special testing for determination of deformation parameters may be performed in certain cases.

SILTY SOILS

Geological Features and Composition

Silt has been formed both by sedimentation of material transported by meltwater (glacial silt) and by sedimentation at river outlets (postglacial silt). Silt is found mainly in valleys where former lakes and seas formed fjords and bays, but also in the neighbourhood of glaciofluvial formations such as eskers and deltas. Silt is also formed as wave-washed material.

Silt is normally found in connection with clay, forming a stratification of silt and clay. In some places, especially close to glaciofluvial deposits such as eskers and deltas, also sand layers are intermingled. In the large river valleys in central and northern Sweden, there are thick deposits of silt with possible infusions of sand and clay layers. Such deposits are normally tens of metres thick. In many places the thickness exceeds 50 m.

Investigation Methods and Typical Results

Field investigations

Field investigations in an area with silty soils are normally carried out in the following way (Knutsson et al., 1998):

- Performing CPT tests with measurement of generated pore pressures and measurement of excess pore pressure dissipation at selected levels.
- Performing supplementary dynamic penetration tests in dense silty soils and in silty soils with coarser infusions.
- Taking samples for verification of the soil classification, preferably with piston sampler.
- When soft clay layers occur, taking undisturbed samples in these layers.
- Measuring the groundwater conditions
- When a more accurate measure of the soil compressibility is required, performing dilatometer tests.

CPT TESTS

The CPT tests are performed as piezocone tests according to the standard procedure. In profiles with dense silt, or where there is a significant upper zone with negative pore pressures and possibly unsaturated soil until the free groundwater level is reached, it is recommended to use slot filters with grease (Elmgren, 1995; Larsson, 1995). This arrangement has proved to be superior to ordinary filters in retaining full saturation under these circumstances.

It is also recommended to stop the penetrations for a few minutes at a number of levels below the free groundwater level in order to study the pore pressure dissipation. In layers with coarse silt and coarser soil, a full pore pressure equalisation is reached within 5 minutes and the in-situ pore pressure can be read off. In finer silt, a relative measure of the drainage and consolidation properties is obtained at the same time.

The interpretation of the test results requires the in-situ pore pressure profile to be established. Particularly in areas with low lying free groundwater levels, the negative pore pressures in the ground above the groundwater level have to be taken into account. This negative pressure affects both the effective overburden pressure and the generated excess pore pressures used in the evaluations.

The classification starts with use of the chart in Fig. 6. Certain limits for generated excess pore pressure and cone resistance are also applied for separation of the various soil type behaviours.

Before evaluation of shear strength parameters, an assessment has to be made of the relevance of different evaluation methods with consideration to the soil behaviour during the test. When the results from the dissipation tests, the generated excess pore pressures and the classification of soil samples indicate that the soil has behaved in an undrained manner during the test, an undrained shear strength can be evaluated. The cone factor N_{KT} is normally set to 14.5 for silt.

When no significant excess pore pressures have been developed, the friction angle is evaluated in the same way as for sand. The method proposed by Marchetti (1985) is then normally applied.

In layers where partial drainage has occurred, the effective stress parameters may be estimated by a method proposed by Senneset and Janbu (1985).

The effective shear strength parameters evaluated for drained and semidrained conditions as above are mainly used for cases with relatively constant or decreasing stresses, such as the stability of natural slopes. In cases with significantly increasing stresses, such as calculation of bearing capacity of footings, the relevant friction angle is estimated from the classification of the silt. Guidelines for this have been issued by the Swedish Road Administration (1994) and the Swedish handbook for shallow foundations (Bergdahl et al., 1993) and were slightly modified by Larsson (1997) to take effects of negative pore pressures into account, see Table 1.

Table 1

Empirical values for friction angles in silt to be used in foundation design.

Relative stiffness	Friction angle,°
Very loose	26-29
Loose	29-32
Medium	32-34
Dense	34-37
Very dense	37-39

According to the same guidelines, the modulus of the silt to be used in assessment of settlements under shallow foundations is estimated as

$$E = 4.3\ q_T^{0.93} \leq 90\ \text{MPa}$$

This modulus is only valid when the penetration has occurred under essentially drained conditions. The modulus calculated in this way is a best estimate without any built-in safety factor. The scatter in the empirical data is considerable, which has to be considered in a design situation.

Dynamic probing tests

Dynamic probing tests are performed when the CPT probe cannot be pushed deep enough. In Sweden, the tests are performed according to the HfA method. The relative densities, friction angles and moduli are evaluated according to the guidelines issued by the Swedish Road Administration (1994) or the Swedish handbook for shallow foundations (Bergdahl et al., 1993). The given figures for the friction angle and modulus can be written as

$$\phi' = 26 + 2\sqrt{N_{20}} \leq 39°$$

$$E = 2.8N_{20}^{0.91} \leq 90 \text{ MPa}$$

The modulus calculated in this way is also a best estimate and the considerable scatter in the reference data has to be considered in the actual design situation.

The value N_{20} (number of blows for 0.2 m of penetration), is numerically approximately equal to the resistance number r_d in MPA, often used internationally.

Field vane tests

Field vane tests are only used when the soil profile contains layers of clay or clayey silt, which is considered to behave in an undrained manner during the test. The tests are then performed and evaluated in these layers according to the same procedure as for clays.

Dilatometer tests

In Sweden, dilatometer tests in silt are mainly performed when a better estimate of the compressibility of the soil is required. The standard test procedure for the test is followed.

When interpreting the test results, the in-situ pore pressure conditions have to be established and both positive and negative pore pressures taken into account. The compressibility of the soil classified as silt (oedometer modulus M) is then evaluated from (Marchetti, 1980).

$$M = E_D[[(0.14+0.15\ (I_D\text{-}0.6)](1\text{-}\log K_D) + 2.5 \log K_D] \geq 0.85E_D$$

In some densely layered profiles with alternating layers of finer and coarser soil, the disturbance at insertion of the dilatometer may become excessive. Test levels where this is the case can be found by studying the evaluated I_D and OCR values. Where these values are found to be extremely low, the results can be discarded in the assessment, but should be presented within parentheses in the report.

Pressuremeter tests

The results of ordinary pressuremeter tests are highly dependent on the quality of the prepared test cavity and it is very difficult to prepare good test cavities in silt below the groundwater table. Above the groundwater table, it is relatively easy to prepare the cavities and these usually stand open unsupported because of the negative pore pressure. On the other hand, the test results are also highly dependent on a negative in-situ pore pressure. Unless the pressure is stable, there is a considerable risk that the stiffness of the soil will be overestimated for sites with higher groundwater levels and increased pore pressure. There is no method of correcting this. Therefore, pressuremeter tests in silt have very limited use in Sweden.

Pore pressure measurements

In silt, positive pore pressure measurements below the free groundwater level may be measured with open standpipes using filter tips connected to inner, small-diameter, plastic tubes. In fine-grained and clayey silt, closed systems should be used if rapid changes are to be registered.

Methods for measuring negative pore pressures between 0 and -90 kPa with closed systems have been developed and are used particularly in cases with slope stability problems (Öberg, 1995). Methods exist for measuring even more negative pressures in the field, but so far they are not used in everyday practice. For estimation of in-situ negative pressures when evaluating field test results, the position of the free groundwater level and the capillarity of the soil are normally used.

Sampling

- Sampling in homogeneous silt for classification purposes may be performed with a screw auger. Sampling for tests where the structure of the soil needs to be retained normally makes use of the Swedish standard piston sampler. Extreme care has to be taken during sampling, transportation and handling, particularly of samples of loose silt, when tests on undisturbed soil are to be performed in the laboratory.

Laboratory Tests

Inspection

Testing of silt samples in the laboratory involves:

- Ocular inspection and classification of the samples
- An assessment of the degree of disturbance and relevant testing methods
- Appropriate routine tests on selected specimens
- Special tests as required

ROUTINE TESTS

The main routine tests for silt comprise

- Bulk density
- Grain size distribution
- Specific density
- Water content and degree of saturation

Determination of bulk density, water content and degree of saturation requires the samples to be taken with a piston sampler and sealed with a locked volume during transportation. The determination of specific density is often replaced by a common normal value for Swedish silts of 2.65 t/m^3 (crystalline rock of quartz and feldspar). The consistency limits are normally determined only for embedded layers of clay or in clayey silt.

SPECIAL TESTS

Permeability

The permeability in silt is often estimated on the basis of the grain size distribution. More precise measurements should be made on undisturbed specimens in permeability cells in which a back-pressure can be applied to ensure full saturation. In fully saturated clayey silts, the permeability can be measured more efficiently in CRS oedometer tests.

Capillarity and water retention capacity

The capillarity of coarse silty soils can be measured by a simple suction apparatus and that of fine-grained silts by a capillarimeter in which a gradually elevated air pressure is applied on one side of a specimen until air is blown through. The water retention capacity and the variation in negative pore pressures with degree of saturation can be measured with a special p_f apparatus.

However, there is good empirical knowledge about the relations between the grain size distribution of the soils and their water retention curves. Methods have also been elaborated to estimate the capillarity and the effective negative pressures from the mean grain size d_{50} and the full grain-size distribution curve for clay free silts and clayey silts respectively (Öberg, 1995). These methods are normally used in practice.

Frost susceptibility

The general frost susceptibility of the soil is normally classified from its grain size distribution (Swedish Road Administration, 1994). This classification should then be coupled to the local climatic data, possible insulation and access to free water in order to evaluate the possible problems associated with frost heave (Knutsson, 1994).

Compressibility and shear strength

Compressibility and shear strength in silt are normally evaluated from the field tests. In special cases and in layered soil, these determinations may have to be supplemented with laboratory tests.

Oedometer tests and direct simple shear tests are normally performed only on undisturbed specimens from embedded layers of clay or clayey silt. They are then performed in the same way as on clays.

Triaxial tests are performed on undisturbed specimens. For determination of strength properties in silt, they are normally performed as undrained or drained compression tests on samples reconsolidated to the in-situ stresses. Stress path testing in order to simulate loading paths in the field is also used, particularly in drained tests.

Cyclic undrained triaxial tests are also performed on undisturbed specimens reconsolidated to the in-situ stresses. The cyclic stress amplitude is then increased in steps in order to determine the amplitude and number of cycles where there is a risk of structural breakdown in the soil. In Sweden, these problems are related to vibrations from traffic, heavy machinery, wave loads or construction activities, rather than from earthquakes.

Triaxial tests on undisturbed silt samples are also performed in order to obtain deformation parameters required in advanced numerical calculation methods.

SAND AND GRAVEL DEPOSITS

Geological Features and Composition

During the melting of the land-ice, large amounts of coarse-grained sediments—sand, gravel, cobbles and boulders—were transported by the glacial rivers to the ice front. The soil masses were deposited outside the retreating land-ice forming long ridges—eskers —that stretched for several hundred kilometres. The eskers were formed both in water and on land. Where the meltwater met the shorelines, large deltas were formed.

The material at the bottom of the eskers is normally composed of boulders and cobbles. Above this "esker core" the material consists mainly of gravel and sand.

Large amounts of sand were also deposited in valleys in central and northern Sweden. Large deltas were also formed where the landice retreat came to a halt due to colder climate. At these areas complex deposites ofcoarse sediments and tills were formed.

During the upheaval of the land above the shorelines, the deposited soils were washed out by waves and redeposited below the shorelines. Such wave-washed material is composed of sand, gravel and cobbles. From a geotechnical point of view, it is important to realise that these deposits below the highest shore line may cover layers of clay, especially along eskers and deltas.

The thickness of sand and gravel deposits in the form of eskers and deltas is normally more than ten metres. The thickness of wave-washed deposits may be several metres along eskers and deltas. Along till deposits, the thickness seldom exceeds a few metres.

Investigation methods and typical results

Field investigations

The field investigations in sand and gravel deposits normally comprise:

- Penetration tests
- Sampling
- Groundwater measurements
- In-situ testing

Penetration test may be performed as CPT test, weight sounding test or dynamic probing test. CPT tests are preferred in sand layers where the relative density is not too high. Normal penetration depths are about 15 m, but they may be considerably greater in loose soils. Weight sounding test can be used in similar soil layers. CPT test should not be used in soils with very coarse particles since the cone cannot penetrate any coarse objects encountered, Fig. 15. The cone resistance then reflects the penetration resistance of the coarse particles and not the density of the soil.

In very dense and deep sand layers and in gravel, the dynamic probing method should be used since it normally better indicates the average density of the soil and provides a continuous resistance curve as a function of depth.

Above the groundwater table, *sampling* of sandy soils for classification and laboratory testing purposes can be performed with screw augers, open tube drive samplers and in some cases piston samplers. Below the groundwater table, open tube drive samplers with a core catcher and in some cases piston samplers can be used.

Auger samples are considered as remoulded, and drive piston samples and open tube samples are considered as disturbed. Normally, samples taken with a pushed in piston sampler are least disturbed.

Groundwater measurements are normally performed in open standpipes driven to the desired depth. The pipes are sometimes provided with a filter tip and an inner plastic tube.

In-situ testing is performed only in cases when more detailed information on the deformation characteristics is required, e.g. when the layers are believed to be considerably overconsolidated or compacted.

In sand, the dilatometer test is often used when the depth or soil stiffness is not too great to prevent insertion of the dilatometer by pushing. The evaluation methods proposed by Marchetti are then employed.

In sand layers above the groundwater table, where it is possible to prepare a stable test cavity without excessive disturbance, ordinary Menard pressuremeter tests are often performed. Below the groundwater table,

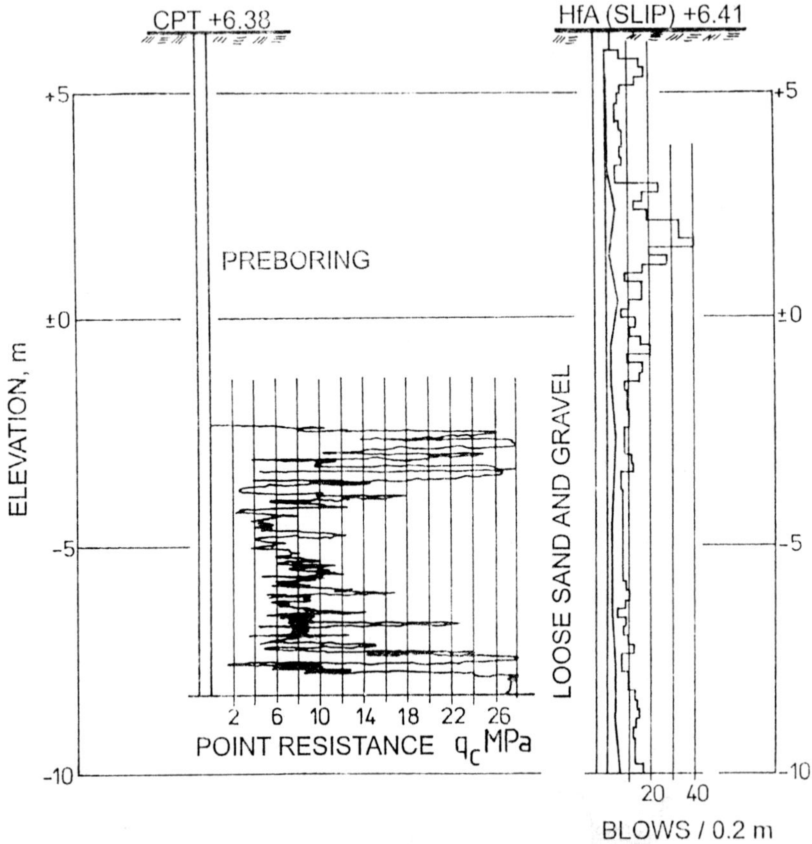

Fig. 15 *CPT test should not be used for parameter assessment in coarse material. In this case, it is better to use dynamic probing tests.*

auger borings stabilised by a bentonite slurry are used. Alternatively, a technique with the pressuremeter lowered into a slotted tube casing is employed. The latter technique is also used for tests in coarser soils than sand and the slotted tube is then inserted with a soil-rock drilling method.

In sand and gravel, the angle of internal friction is normally assessed empirically from the results of any penetration test or in-situ test. Extensive investigations at the Swedish Geotechnical Institute based on comparisons between different penetration tests and a number of large-scale plate load tests have resulted in Table 2, from which the angle of internal friction and the modulus of elasticity can be estimated (Bergdahl et al., 1984, 1986). The correlations between penetration resistances and the design parameters are primarily intended to be used in the design of shallow foundations. The values given in the table are best estimates (or mean values of the parameters) and not a "cautious estimate" as prescribed in EC7 Part 1 (ENV 1997-1, 1994), for example.

This table was elaborated before the CPT(U) test came into practice. The evaluation in silt is uncertain. This has been improved by the current use of the CPT(U) test with correction of the results for in-situ pore pressures and checking that no significant excess pore pressures are generated during the test (see Section 4.2).

Laboratory tests

The laboratory tests on sand and gravel samples normally comprise:

- Classification by ocular inspection
- Bulk density
- Grain size distribution
- Water content

Triaxial tests and oedometer tests can be performed on reconstituted samples in the laboratory. However, this is only done in special cases where the composition of the ground invalidates the use of the empirical evaluation methods or in simulation of conditions other than those prevailing in the ground.

MORAINE/TILL DEPOSITS

Geological features and composition

Till is normally a well compacted soil deposited directly on bedrock, then called bottom till, Fig. 16. In some places, till also occurs as layers or lenses in sediments. Less compacted, loose till occurs in widely dispersed areas as a surface layer on top of bottom till. It is then called surface till.

The composition of tills—matrix (gravel, sand silt and clay particles) and boulder/cobble content—exhibits a large variation in Sweden. The composition is dependent on a number of factors, the most important of which are believed to be the type of parent bedrock, transport distance from the bedrock source to the place of deposition and the way in which the till was deposited.

The most common till in Sweden is well compacted sandy-silty and sandy till. Clay till is frequent in areas with Cambro-Silurian bedrock: certain regions in southern Sweden, the islands in the Baltic sea and parts of the mountains in the north. Gravelly tills occur in places close to glacio-fluvial deposited eskers and in areas with bedrock of porphyry and quartzite.

The boulder and cobble content is highly variable in both frequency and size. Clay till normally has a relatively low content of boulders and cobbles.

Fig. 16 *Cut in bottom till. The material is a well compacted silty sandy till. This material is common in Sweden. It is difficult to excavate but offers a good subbase for foundations. Photo: L. Viberg.*

Investigation methods and typical results

Field investigations

The possibilities of performing field investigations in till layers are reduced due to the high content of coarse materials such as boulders and pebbles. However, at the same time at least the bottom till is often well compacted, which reduces the need for field investigations for normal buildings and civil engineering projects since there are seldom any problems with bearing capacity or settlements. The main geotechnical problems are often construction problems due to a high content of boulders and pebbles, a high groundwater level due to low permeability and high frost susceptibility.

A field investigation program for till deposits may have the following composition:

- Penetration tests
- Sampling
- Groundwater measurements
- In-situ testing

Penetration test is normally used to find the density of the upper layer of surface till and any redeposited till layers where the density may be less than in the bottom till.

In order to find the depth to the rock surface, percussion drilling is often performed. When the anticipated rock has been reached, drilling is normally continued for 3-5 m to ensure that the resistance was not created by a boulder.

Above the groundwater table, *sampling* for classification, laboratory testing and determination of boulder content is often performed as a test pit made by an excavator. Samples are taken by hand every 0.5-1.0 m at the bottom or in the walls of the excavation. In addition, the number and size of boulders and the excavatability of the till are estimated and recorded.

In the case of deep investigations, below the groundwater table in till deposits, casings are driven to the desired sampling depths by percussion machines and an open tube sampler ($\varnothing_i$ 34 - 90 mm) with a core catcher is used inside the casings to obtain the samples. For very important projects such as earth dams, continuous core sampling is also used.

Groundwater measurements are normally performed in open standpipes provided with a filter tip. Normally, preboring with percussion machines is necessary in order to install the standpipe.

In-situ testing is normally not performed in till deposits since preboring is required. The procedure disturbs the surrounding soil to such an extent that it is not representative for the natural ground.

The strength and deformation characteristics in till deposits can be evaluated in the same way as for sand and gravel, Table 2. However, the effect of pebbles has to be considered when evaluating the penetration resistance. In addition, the silt content has to be considered in the parameter assessment, since this may be high. For clayey till layers, refer to Section 2.2 or laboratory tests.

Laboratory tests

Laboratory tests on till materials normally comprise:

- Classification by ocular inspection
- Grain size distribution
- Water content

CONCLUDING COMMENTS

This paper summarises the normal procedures for site investigation and

Table 2

Best estimates of the angle of internal friction φ' and long-term modulus of elasticity E for quartz and feldspar sands from CPT tests, weight sounding tests and dynamic probing tests.

Relative density	*Cone resistance (q_c) [Mpa] (from CPT test)*	*Weight sounding resistance [1,5] half-turns/0.2,m*	*Dynamic Probing resistance [4,5] Blows per (0.2 m)*	*Angle of shearing resistance[2] (φ')*	*Drained Young's modulus (E)[3] [Mpa]*
Very low	0.0-2.5	0-10	0-4	29-32	<10
Low	2.5-5.0	10-30	2-8	32-35	10-20
Medium	5.0-10.0	20-50	6-14	35-37	20-30
High	10.0-20.0	40-90	10-30	37-40	30-60
Very high	>20.0	>80	>25	40-42	60-90

[1]Before determination of the relative density, the weight sounding resistance in silty soils should be divided by a factor of 1.3.

[2]Values given are valid for sands. For silty soils, a reduction of 3° should be made. For gravels, 2° may be added.

[3]Values given for the drained modulus correspond to settlements after 10 years. They are obtained on the assumption that the vertical stress distribution follows the 2:1 approximation. Furthermore, some investigations indicate that these values may be 50% higher in gravelly soils. In overconsolidated soils, the modulus may be considerably higher. When calculating settlements for ground pressures greater than 2/3 of the design bearing pressure in ultimate limit state, the modulus should be set to half of the values given in this table.

[4]The dynamic probing resistance should be the net resistance; total resistance subtracted by the number of blows necessary to overcome the skin friction along the rods.

[5]If results of weight sounding tests or dynamic probing tests alone are available, the lower value in each interval for the angle of shearing resistance and the modulus of elasticity should be selected.

parameter assessment used in Sweden today. It is important to point out that in order to produce a useful site investigation report as a basis for sound geotechnical design, it is important to use a combination of geological knowledge and field and laboratory test results.

The use of electronic equipment has increased the possibilities of taking records from different kinds of tests both in the laboratory and in the field. However, the quality of samples for testing in the laboratory is not improved, which leads to reduced possibilities of performing accurate laboratory tests for different purposes where more complicated stress-deformation relations than those relations presented in this paper need to be modelled.

It must also be stressed that the parameter assessments presented in this paper are so-called best estimate values, which means that they are often a mean value calculated from a number of estimates. Thus there is no safety margin in the assessments, such as the "cautious estimate" required in Eurocode 7. To obtain a "cautious estimate", it is necessary to divide or multiply the effect of the parameter by a factor of 1.1-1.2 depending on the criticality of the particular structure. Division is used when a lower cautious estimate is to be used for design and multiplication when an upper value is to be used, e.g. tg∅/1.2 or 1.2·tg∅.

The accuracy in the parameter assessment may be questioned, since there is a scatter in the natural soil itself, as well as in the field and laboratory test results. In addition, the scatter is related to the method of analysing, bearing capacity or settlements, for example. Hopefully, increased research on bearing capacity and settlements will improve both the investigation methods, parameter assessment and calculation methods. In the meantime, when high accuracy is needed for safety reasons, it is advisable not to rely on only one investigation method and one calculation method but to use several methods. Investigations in Sweden have indicated that an average of a number of assessments and calculations has much greater accuracy than a single estimate and the accompanying calculation.

This paper focuses on describing current established practice in soil investigations in Sweden. Methods related to ongoing research and future trends are partly included. However, results from recent attempts to include geophysical survey methods in the geotechnical investigation program are not taken up.

The paper also seeks to emphasise that there is no general way of carrying out soil investigations suitable for all types of soil conditions. On the contrary, investigation methods and procedures must be carefully selected with respect to the soil conditions, the problem at hand and the purpose of the investigation. In most cases, the investigation combines results from appropriate field and laboratory tests with knowledge from the local geology to yield a reliable model of the soil conditions.

Sweden's geology, with its wide range of soil conditions, has forced the country's engineers to develop and procure a correspondingly broad range of equipment enabling a rational assessment of the soil conditions and properties in every situation. Much of this equipment, particularly for investigations in very soft soils, has been developed in Sweden. Other equipment has been imported and introduced in Swedish practice. In some of these cases, it has been necessary to modify the equipment for local conditions and to perform local calibrations for the evaluation methods.

REFERENCES

Aas, G., Lacasse, S., Lunne, T. and Hoeg, K. (1986). Use of in Situ Tests for Foundation Design on Clay. ASCE Proceedings of In Situ'86, Blacksburg, Virginia, USA, June 23-25, 1986, pp. 1-30.

Ahlberg, P. (editor), 1979, Våtmarker. Mineralråvaror. Geologiska och geotekniska förhållanden. Fysisk riksplanering FRP Underlagsmaterial 3.79.

ASTM Subcommittee D 18.02.10, 1986, Suggested method for Performing the Flat Dilatometer Test. *ASTM Geotechnical Testing Journal,* Vol. 9, No. 2.

Bergdahl, U., Eriksson, U., 1983, Bestämning av jordegenskaper med sondering—en litteraturstudie. Statens geotekniska institut, Rapport 22, Linköping.

Bergdahl, U., 1984, Geotekniska undersökningar i Fält. Swedish Geotechnical Institute. Information 2.

Bergdahl, U., Ottosson E., 1984, "Jordegenskaper ur sonderingsresultat: en jämförelse mellan olika undersökningsmetoder i friktionsjord." Proc. NGM 84. Vol. 2, pp. 725-738, Linköping.

Bergdahl, U., Hult, G., Ottosson, E., 1984, Belastningförsök på plattor grundlagda i friktionsjord för bestämning av jords deformationsegenskaper. Försök vid SGI:s provfält Kolbyttemon, Linköping, SGI Varia 138.

Bergdahl, U., Hult, G., Ottosson, E., 1986, Belastningsförsök på plattor grundlagda i friktionsjord för bestämning av jords deformationsegenskaper. Försök vid Albysjön, Fittja. SGI Varia 168.

Bergdahl, U., Mamborg, B.S. and Ottosson, E., 1993, Plattgrundläggning (Shallow foundations). Svensk Byggtjänst, Solna. (In Swedish).

Carlsten, P., 1993, Peat - Geotechnical Properties and Up-to-Date Methods of Design and Construction. State-of-the-Art-Report. International Workshop "Advances in Understanding and Modelling the Mechanical Behaviour of Peat. Delft. Also in Swedish Geotechnical Institute, Varia No. 215, Linköping.

CEN European Prestandard, 1994, ENV 1997-1: Eurocode 7, Part 1. Geotechnical Design—General Rules.

CEN European Prestandard, 1999, Eurocode 7: Geotechnical design Part 3. Design assisted by field testing.

Dahlberg, R., 1974, Penetration testing in Sweden. Proc. European Symposium on Penetration Testing in Stockholm June 5-7, 1974, Vol. 1. National Swedish Building Research.

DGF Feltkomité (The Field Committee of the Danish Geotechnical Society), 1993, Referenceblad for vingeforsog. Referenceblad 1, Revision 1, Dansk Geoteknisk Forening, Copenhagen.

Ekdahl, U. and Bengtsson, P-E., 1996, Ny metod att räkna sättningar vid plattgrundläggning. Väg-och Vattenbyggaren, No. 3, 1996.

Elmgren, K., 1995, Slot-type pore pressure CPT-u filters—Behaviour of different filling media. International Symposium on Cone Penetration Testing, Vol. 2. Linköping.

Graham, E.R., 1948, Determination of soil organic matter by means of a photoelectric colorimeter. Soil Sci. 65.

Hansbo, S., 1975, A new approach to the determination of the shear strength of clay by the fall-cone test. Swedish Geotechnical Institute, Proceedings No. 14, Stockholm.

Hartlén, J. and Wolski, W. Eds., 1996, Embankments on Organic Soils. Elsevier.

ISSMFE Technical Committee on Penetration Testing of Soils TC16, 1989, Report on the ISSMFE Technical Committee on Penetration Testing of Soils TC16 with Reference Test Procedures; CPT—SPT—DP—WST. Swedish Geotechnical Institute, Information No. 7, Linköping.

ISSMGE European Regional Technical Committee ETC5 "Laboratory Testing", 1998, Recommendations of the ISSMGE For Geotechnical Laboratory Testing. Beuth.

ISSMGE Committee TC16, 2001, The Flat Dilatometer Test (DMT) in soil investigations. Report prepared by Marchetti, S., Monaco, P., Totani, G. and Calabrese, M. IN-SITU 2001. Int. Conf. on In-Situ Measurement of Soil Properties, Bali, Indonesia.

Karlsson, R. and Hansbo, S., 1984, Soil classification and identification. Swedish Council for Building Research., T21:1982. Second Edition, Revised in 1984, Stockholm.

Knutsson, S., 1994, Tjälningsprocessen och beräkning av tjäldjup. Kompendium. Luleå tekniska universitet, Avdelningen för geoteknik, Skrift 94:6. Luleå.

Knutsson, S., Larsson, R., Tremblay, M. and Öberg-Högsta, A-L., 1998, Siltjordars egenskaper. Swedish Geotechnical Institute, Information No. 16, Linköping.

Larsson, R., 1989, Dilatometerförsök—En in-situ metod för bestämning av lagerföljd och egenskaper i jord—Utförande och utvärdering. Swedish Geotechnical Institute, Information No. 10, Linköping.

Larsson, R., 1990, Behaviour of organic clay and gyttja. Swedish Geotechnical Institute, Report No. 38. Linköping.

Larsson, R., 1993, The CPT - test; Equipment—Testing—Evaluation. Swedish Geotechnical Institute, Information No. 15E. Linköping.

Larsson, R., 1995, Use of a thin slot as filter in piezocone tests. International Symposium on Cone Penetration Testing, Vol. 2. Linköping.

Larsson, R., 1997, Investigations and load tests in silty soils.—Results from a series of investigations in silty soils in Sweden. Swedish Geotechnical Institute, Report No. 54, Linköping.

Larsson, R., 2001, Investigations and load tests in clay till.—Results from a series of investigations in the test field at Tornhill outside Lund in southern Sweden. Swedish Geotechnical Institute, Report No. 59, Linköping.

Larsson, R., Bergdahl, U. and Eriksson, L., 1984, Evaluation of shear strength in cohesive soils with special reference to Swedish practice and experience. Swedish Geotechnical Institute, Information No. 3E, Linköping.

Larsson, R. and Eskilsson, S., 1989, Dilatometerförsök i lera. Swedish Geotechnical Institute, Varia No. 243, Linköping.

Larsson, R., Löfroth, B. and Möller, B., 1995, Processing of data from CPT tests. International Symposium on Cone Penetration Testing, Vol. 2. Linköping.

Larsson, R. and Mulabdic, M., 1991, Piezocone Tests in clay. Swedish Geotechnical Institute, Report No. 42, Linköping.

Larsson, R., Nilsson, G. and Rogbeck, J., 1987, Determination of organic content, carbonate content and sulphide content in soil. Swedish Geotechnical Institue, Report No. 27E, Linköping.

Larsson, R. and Åhnberg, H., 2002. Effects of excavations at clay slopes. Swedish Geotechnical Institute. Report in preparation, Linköping.

Marchetti, S., 1980, In-Situ Tests by Flat Dilatometer. ASCE, Journal of the Geotechnical Engineering Division, Vol. 106, No. GT3.

Marchetti, S., 1985, On the Field Determination of K_0 in Sand. Panel Discussion. 11th International Conference on Soil Mechanics and Foundation Engineering. San Francisco, Session 2A.

Orre, B., 1979, Redovisning av geotekniska utredningar. Byggforskningen T33:1979.

Powell, J.J.M. and Uglow, I.M., 1988, The Interpretation of the Marchetti Dilatometer in UK Clays. Penetration Testing in the UK. Thomas Telford, London.

Roque, R., Janbu, N. and Senneset, K., 1988, Basic Interpretation Procedures of Flat Dilatometer Tests. Proceedings of 1st International Symposium on Penetration Testing, Orlando, Florida, Vol. 1.

Senneset, K. and Janbu, N., 1985, Shear Strength Parameters obtained from

Static Cone Penetration Tests. ASTM Special Technical Publication, STP 883.

Swedish Committee on Piston Sampling, 1961, Standard Piston Sampling. Swedish Geotechnical Institute, Proceedings No. 19, Stockholm.

Swedish Council for Building Research, 1990, Soil constituents and structure. Laboratory Committee of the Swedish Geotechnical Society. BFR Report D 23:1990.

Swedish Geotechnical Society, 1993, Recommended Standard for Cone Penetration Tests. Swedish Geotechnical Society, Report No. 1:93E, Linköping.

Swedish Geotechnical Society, 1993, Recommended Standard for Field Vane Tests. Swedish Geotechnical Society, Report No. 2:93E, Linköping.

Swedish Geotechnical Society, 1995, Recommended Standard for Dilatometer Tests. Swedish Geotechnical Society, Report No. 1:95E, Linköping.

Swedish Geotechnical Society, 1996, Geoteknisk Fälthandbok, Allmänna råd och metodbeskrivningar, SGF Report 1:96, Linköping.

Swedish Road Administration, 1994, Jords hållfasthets—och deformationsegenskaper. VV Publication 1994:15, Vägverket, Borlänge.

Swedish Road Administration, 1994, Väg 94. VV Publication 1994:86, Vägverket, Borlänge.

Swedish Standard SS 02 71 26, 1991, Geotekniska provningsmetoder—Kompressionsegenskaper—ödometerförsök—CRS-försök—kohesionsjord. Swedish Building Standards Institution, Stockholm.

Sällfors, G., 1975, Preconsolidation Pressure of Soft, High-Plastic Clays. Thesis, Chalmers University of Technology. Division of Geotechnical Engineering, Göteborg.

Åhnberg, H., Larsson, R. and Berglund, C., 2001, Nygamla vingar, stora som små. Swedish Geotechnical Institute, Varia No. 509, Linköping.

Öberg, A.-L., 1997, Matrix Suction in Silt and Sand Slopes—Significance and Practical Use in Stability Analysis. Thesis, Chalmers University of Technology. Division of Geotechnical Engineering, Göteborg.

NOTATIONS AND SYMBOLS

A	Pressure reading in dilatometer test
A_c	Ash content
a	Parameter in general equation for undrained shear strength
B-value	Pore pressure response value in triaxial test
B	Pressure reading in dilatometer test
B_q	Pore pressure parameter in CPT test $(=\Delta u / (q_T - \sigma_{v0})$
b	Parameter in general equation for undrained shear strength
c	Cone constant in fall-cone test
c'	Effective strength parameter
c_u	Undrained shear strength

DPPR	Differential pore pressure ratio, pore pressure parameter in CPT test $(=\Delta u/q_T)$
d_{50}	Mean grain size
E	Modulus of elasticity
E_D	Dilatometer modulus
f	Factor in evaluation of undrained shear strength from dilatometer test
f_T	Total sleeve friction in CPT test
g	Specific gravity (9.81 m/s^2)
I_D	Material index in dilatometer test
$I_{D(corr)}$	Material index in dilatometer test corrected for overconsolidation
i	Cone penetration in fall-cone test
K_0	Coefficient of lateral earth pressure
K_D	Horizontal stress index in dilatometer test
M	Modulus of compressibility, oedometer modulus
M_0	Modulus of compressibility, oedometer modulus for vertical stresses $<\sigma'_c$
m	Mass of cone in fall cone test
N_{KT}	Cone factor at evaluation of undrained shear strength from CPT test
N_{20}	Number of blows for 0.2 m penetration in dynamic probing test type HfA
OCR	Overconsolidation ratio $(=\sigma'_c/\sigma'_{v0})$
p_0	Lift-off pressure in dilatometer test
p_1	Expansion pressure in dilatometer test
q_T	Total cone resistance in CPT test
R_f	Friction ratio in CPT test $(= f_T/ q_T)$
r_d	Resistance number in dynamic probing tests
S_t	Sensitivity
u	Total pore pressure
Δu	Generated pore pressure in CPT test $(= u-u_0)$
u_0	In-situ pore pressure
w_L	Liquid limit
w_N	Natural water content
ϕ	Friction angle, effective strength parameter
γ	Bulk density
γ_d	Dry density
μ	Correction factor for field vane- and fall-cone tests
σ'_c	Preconsolidation pressure
σ_{v0}	Total overburden pressure
σ'_{v0}	Effective overburden pressure
τ_c	Uncorrected shear strength value from fall cone test
τ_v	Uncorrected shear strength value from field vane test
τ_R	Remoulded shear strength

6

In-Situ Tests for Dynamic Properties of Soils

Vijay K. Puri[1] *and Shamsher Prakash*[2]

INTRODUCTION

Several problems in geotechnical earthquake engineering and soil dynamics require knowledge of dynamic soil properties. Problems involving dynamic loading of soils are divided into small and large strain response problems. For example a foundation supporting a well balanced machine represents a low strain situation. A structure subjected to an earthquake or blast loading or an off-shore struture subjected to storm loading exemplify a large strain problem. For solution of almost all problems dealing with dynamic loading of soil we need information on dynamic shear modulus and damping. The dynamic soil properties are affected by several factors. In fact it is the understanding of these factors which has lead to procedures for determination and rational evaluation of the dynamic soil properties for a given design or analysis problem (Prakash 1981, Prakash and Puri 1981, 1988, Woods 1991). The factors affecting the dynamic soil properties are:

1. Type of soil and soil properties such as void ratio.
2. Shear strain level.
3. Nature of applied stress e.g. static, static + dynamic, or dynamic only.
4. Confining pressure (mean effective confining pressure).
5. Particle size and shape in coarse sands.
6. Percent finer in fine grained soils.
7. Depositional history of soil.
8. Ageing.
9. Degree of cementation.
10. Mineral composition
11. Stress history.
12. Dynamic pre-strain.
13. Stress path.
14. Frequency of loading.

1. Professor of Civil Engineering, Southern Illinois University Carbondale, IL 62901, USA
2. Professor Emeritus, Civil Engineering Deptt., University of Missouri Rolla, MO 65409, USA

Shear strain level and effective confining pressure have the most significant effect on dynamic soil properties and are discussed first.

EFFECT OF SHEAR STRAIN LEVEL

The response of a soil under dynamic loading depends on the cyclic shear strains induced in the soil. At small strains, the stress strain response is generally linear and $G \sim G_{max}$, where G_{max} is the maximum value of the shear modulus G (Dobry et al., 1989). The upper limit of shear strains for this linear behaviour, known as the "elastic threshold" is about 10^{-5} in sands and 10^{-4} in normally consolidated clays with plasticity index, PI = 50. When the shear strains are smaller than the elastic threshold, the value of G is only slightly lower then G_{max}, the material damping (ξ) is low, and there is no stress-strain degradation or hardening with the number of load cycles. The upper limit of this range known as "threshold strain" or the "plastic threshold" is about 3 to 5×10^{-5} for gravel, about 10^{-4} for sands, and about 10^{-3} for clays of high plasticity. At intermediate shear strains above the plastic threshold, the stress strain response is non-linear, with much higher material damping and considerable degradation of stiffness or hardening caused by cyclic loading. When the shear strains are large of the order of 1% (10^{-2}), G is only a small fraction of G_{max}. Under such loading condition, the stress-strain response can change significantly even in as little as one or two cycles, and the cyclic stress-strain behaviour is controlled by factors other than G_{max}. These factors include mainly the degree of saturation, contractive or dilative character of the soil and its static (monotonic) and cyclic strength characteristics. For such situations G/G_{max} and ξ should be determined considering the large stain behaviour of soil and its static strength with due consideration of strain rate effects.

When response analysis of critical facilities is conducted, G_{max} is obtained from both in-situ measurements and laboratory determinations on small samples, and cyclic properties at large strains such as G/G_{max} and ξ from laboratory testing. For less critical facilities, very often some or all cyclic properties are estimated rather than obtained from measurements. This certainly is made possible by our accumulated knowledge of the factors controlling G_{max}, G/G_{max} and ξ, as well as published correlations between the cyclic strength properties and soil type, degree of saturation and effective confining pressure, index properties, penetration resistance, relative density, unconfined compressive strength, and plasticity index (Dobry et al., 1989).

EFFECT OF CONFINING PRESSURE

The dynamic shear modulus has been observed to increase with the effective confining pressure. Several investigators have developed empirical relations for G_{max} in terms of simple soil properties (Hardin and Black, 1968,1969).

$$G_{max} = 1230\,(OCR)^k \frac{(2.973-e)^2}{1+e} (\bar{\sigma}_o)^{\frac{1}{2}} \tag{1}$$

where,

G_{max} = maximum value of G in psi at $\gamma = 10^{-6}$
OCR = over-consolidation ratio
k = factor depending on PI of clay
e = void ratio
γ = shear strain

$$\bar{\sigma}_o = \frac{\sigma_1 + \sigma_2 + \sigma_3}{3}$$, average mean effective confining pressure, in psi.

Equation (1) is a general type of equation for all types of soils e.g. sand and clay. Several other similar relationships have been developed by other investigators, which show that $G_{max} \propto (\sigma_o)^{0.5}$ (Prakash, 1981, Prakash and Puri, 1988). It may therefore be reasonable to argue that G also depends on all the factors that govern G_{max}, and therefore, is a function of $(\sigma_o)^{0.5}$.

Some discussion is also needed here about the effect of frequency of loading on the values of shear modulus and damping. Hardin and Drnevich (1972) have observed that the frequency of loading has a relatively unimportant effect on small strain shear modulus G_{max} and hence the shear wave velocity in all soils. The material damping may however be significantly affected by excitation frequency. Stokoe et al. (1999) has confirmed the general observations of Hardin and Drnevich (1972) and suggested the following general trends for excitation frequencies changes in the range 1 to about 100 Hz

1. G_{max} increases by about 5 to 30% with the effect generally increasing with PI and
2. The material damping may increase by about 2 to 3 times with increase in frequency in this range.

It appears, therefore, that the frequency effects are important in determining the material damping for fine grained soil and the effect becomes more predominant with an increase in the percentage of fines.

IN-SITU METHODS

Some of the field methods used for determining the shear modulus and/or damping are

1. Cross-bore hole wave propagation test
2. Up-hole or down-hole wave propagation test
3. Surface wave propagation test
4. Vertical footing resonance test
5. Horizontal footing resonance test

6. Spectral analysis of surface waves (SASW)
7. Field cone test (Piezocone)
8. Back calculation from seismic arrays
9. Seismic cone
10. Standard penetration test

Most commonly used in-situ methods essentially utilize the principle of wave propagation through soil. Up-hole and down-hole wave propagation tests and the cross-bore hole are rather commonly used. Seismic cone and spectral analysis of surface waves (SASW) have been investigated during the last decade and found to be useful. Back calculation from seismic arrays has also been investigated in recent years. Footing vibration tests are rather uncommon now for determination of dynamic soil properties.

Cross-Bore hole Wave Propagation Test

In the cross-bore hole method (Stokoe and Woods, 1972), the velocity of wave propagation is measured from one bore hole to another. A minimum of two bore holes are required, one for generating an impulse and the other for the sensors. In Fig. 1, the impulsive rod is struck on top, causing an impulse to travel down the rod to the soil at the bottom of the hole. The shearing between the rod and the soil creates shear waves that travel

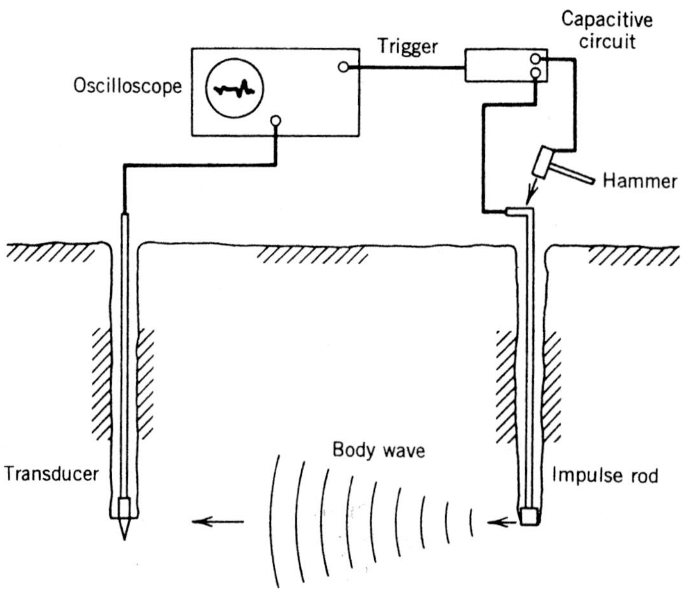

Fig. 1 *Schematic sketch showing cross-bore hole technique for wave programme velocity measurement.*

horizontally through the soil to the vertical motion sensor in the second hole; the time required for a shear wave to traverse this known distance is monitored.

There are four important considerations in conducting a cross-borehole shear test: a) the boreholes, b) the seismic source, c) the seismic receiver, and d) the recording and timing equipment.

Although a minimum of two bore holes are required, for extensive investigations and for increased accuracy, three or more bore holes should be preferred whenever possible. If bore holes are installed in a straight line, wave velocities can be calculated from the intervals of time required for passage between any two bore holes. Thus, the necessity for precisely recording the triggering time is eliminated (Stokoe and Hoar, 1978). In addition, the bore holes must be vertical for the travel distance to be measured properly. In general, any bore hole 10 m or deeper should be surveyed with an inclinometer or another logging device for determining its verticality (Woods, 1978).

Impulsive sources are more often used, although both impulsive and steady state seismic sources are in use. The major criteria for a seismic source are: a) It must be capable of generating predominantly one kind of wave. b) It must be capable of repeating desired characteristics at a predetermined energy level.

Velocity transducers (geophones) that have natural frequencies of 4 to 15 Hz are adequate for detecting (receiving) the shear waves as they arrive from the source. The receivers must be oriented in the shearing mode and should be securely coupled to the sides of the boring. The recording equipment should be able to resolve arrival times of up to 0.2 ms or 5 per cent of the travel time. Storage oscilloscopes are also often used.

Up-Hole Or Down-Hole WavePropagation Tests

Up-hole and down-hole tests are performed by using only one bore hole. In the up-hole method, the sensor is placed at the surface, and shear waves are generated at different depths within the bore hole. In the down-hole method, the excitation is applied at the surface, and one or more sensors are placed at different depths within the hole (Fig. 2). Both the up-hole and the down-hole methods give average values of wave velocities for the soil between the excitation and the sensor if one is used in the bore hole (Richart, 1977). Figure 3 is a schematic diagram of a down-hole survey with all principal elements included (Imai and Yoshimura, 1975).

Surface-Wave Propagation Test

Rayleigh waves and Love Waves can be used to determine the shear moduli of soils near the surface. The Rayleigh wave (*R*-wave) travels in a

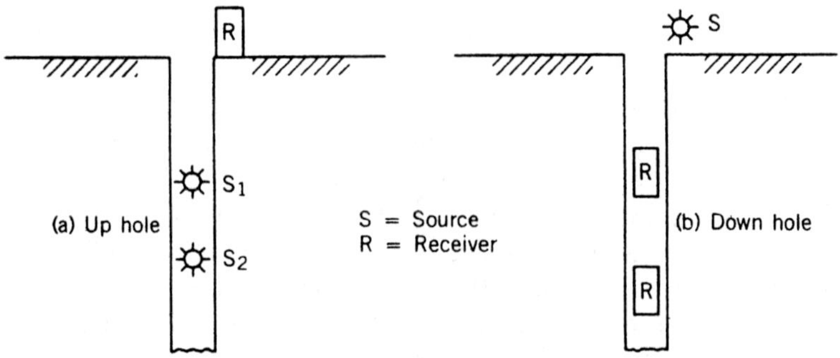

Fig. 2 *(a) Up-hole and (b) down-hole survey (Woods, 1978).*

zone close to the surface. An electro-magnetic or other harmonic vibrator can be used to generate a steady-state *R*-wave and the velocity of wave propagation can be measured. It has been observed that the Rayleigh wave velocity, V_r is about 0.919 to 0.955 of shear wave velocity V_s, depending on the value of Poisson's ratio.

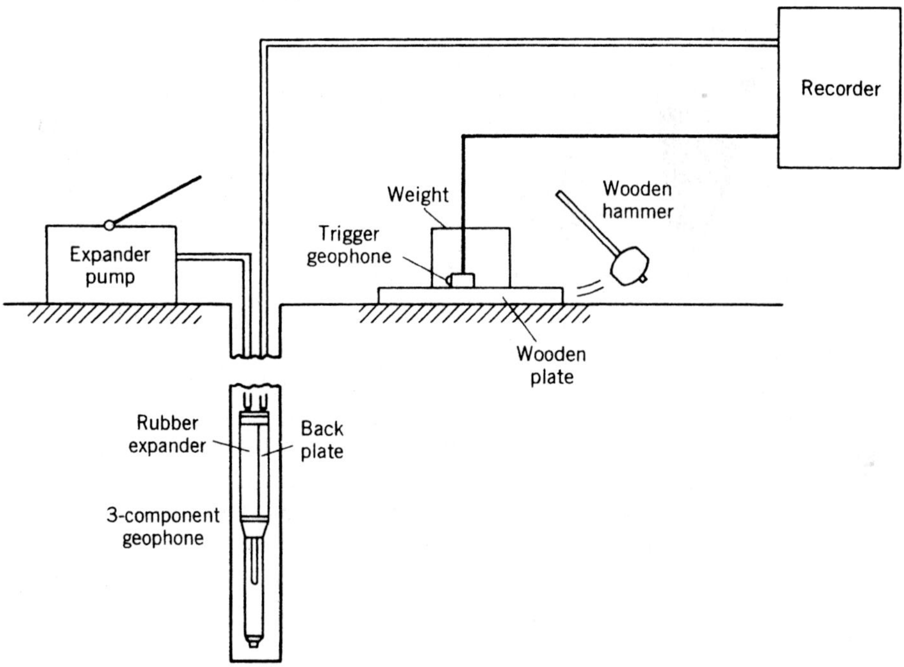

Fig. 3 *Schematic of the equipment for down-hole survey (Woods, 1978).*

Standard Penetration Test

In the standard penetration test (SPT), a standard split spoon sampler is driven with a 140 lb hammer that falls freely through a distance of 30 inches. The number of blows for 12 inches of penetration of the split spoon sampler is designated as the N value. This is $N_{measured}$. In a design problem using N values, a correction for effective overburden pressure is applied (Peck et al., 1974). Although the test is designed as a "standard" test, there are several personal errors as well as errors that are equipment based. Therefore, the use of SPT to measure any soil property has been questioned by many engineers. (Woods, 1978). The "uses and abuses" of SPT from which it is evident that although SPT is used extensively in soil investigations, there has been no documented, carefully controlled, research conducted on it. Schertmann (1975) described uses of SPT in estimating soil properties and pointed to some hope for more accurate uses of SPT in the future with improvements in the performance and understanding of the test. Recent careful studies by Kovacs (1975), Kovacs et al. (1977a, 1977b), Palacios (1977), and Schmertmann (1975, 1977) have thrown new light on the potential correlations between SPT and observed liquefaction. It has been reported that Chinese engineers also use this approach in their attempts to assess the liquefaction potential of sand deposits.

Schmertmann (1977) suggested that with SPT, either only those dynamic properties for which SPT provides a direct model of the phenomenon can be measured, if at all, or the factors, which cotrols the behaviour of SPT, similarly control the correlative dynamic behaviour. As an example of the former, pile driving is an obvious direct model; of the latter; it has been shown that the factors, which affect cyclic liquefaction, also effect SPT in a similar manner (Seed 1976). Additional advantages of using SPT to evaluate dynamic effects, according to Schmertmann (1977), are that SPT is a dynamic test for modelling a dynamic phenomenon, and it is essentially undrained for each blow in that it generates principally a shearing energy. Imai (1977) reported a correlation based on recordings at four urban locations in Japan and was able to establish the following relationship:

$$V_s = 91\ N^{0.337} \qquad (2)$$

In arriving at the above relationship, he converted the N values over 50 or under 1 from the penetrating length at the time of 50 or 1 blows into the number of blows necessary for penetration as deep as 30 cm. Prakash and Puri (1981,1984) successfully applied the above relationship in predicting dynamic soil properties at different depths.

Seismic Cone

The seismic cone has been used by several investigators (Dobry et al., 1989). This method provides information on V_s and G_{max} in addition to

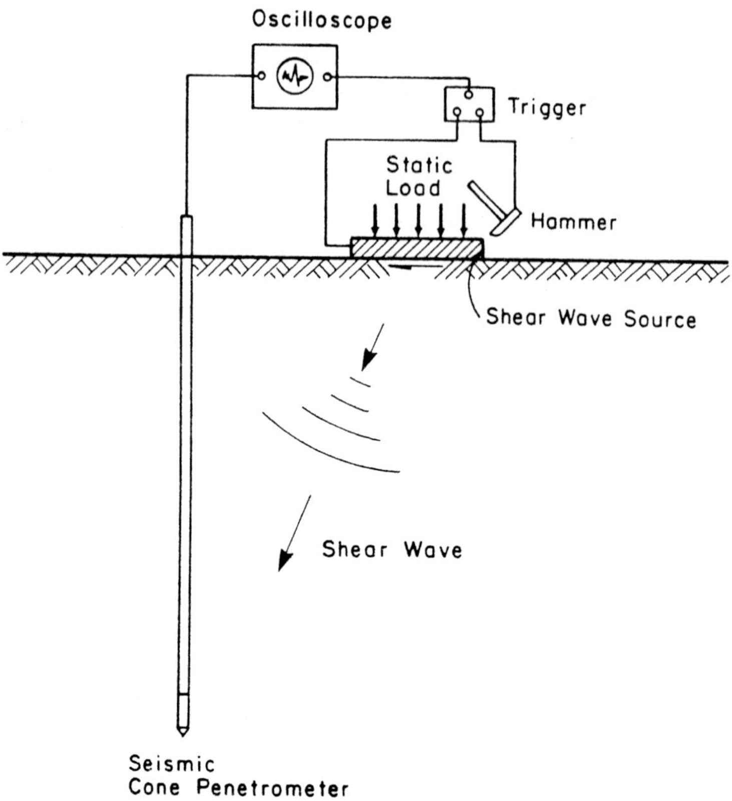

Fig. 4 *Seismic cone penetration test (Robertson et al., 1985).*

regular static cone penetration resistance. The Seismic cone test configuration is shown in Fig. 4. Basically, a cone penotrometer containing a triaxial receiver system is statically penetrated into the soil deposit. Then down hole seismic tests are conducted by exciting sources on the ground surface in the vicinity of the cone. Wave velocities and moduli are inferred from the travel times of the waves between source and receiver. Information on soil strength is obtained from the cone penetrometer results. Figure 5 shows a comparison between results from down hole seismic cone and cross-hole tests for clay at a site in Norway.

The seismic cone method offers the capability of providing information on low strain behaviour and strength characteristic at the same location. The limitations of the method include the inherent problems of cone penetrometer testing, that is, its lack of ability to test coarse-grained (gravel-type) soil deposits. Campanella et al. (1986) have shown good comparison of seismic cone data with cross-hole data.

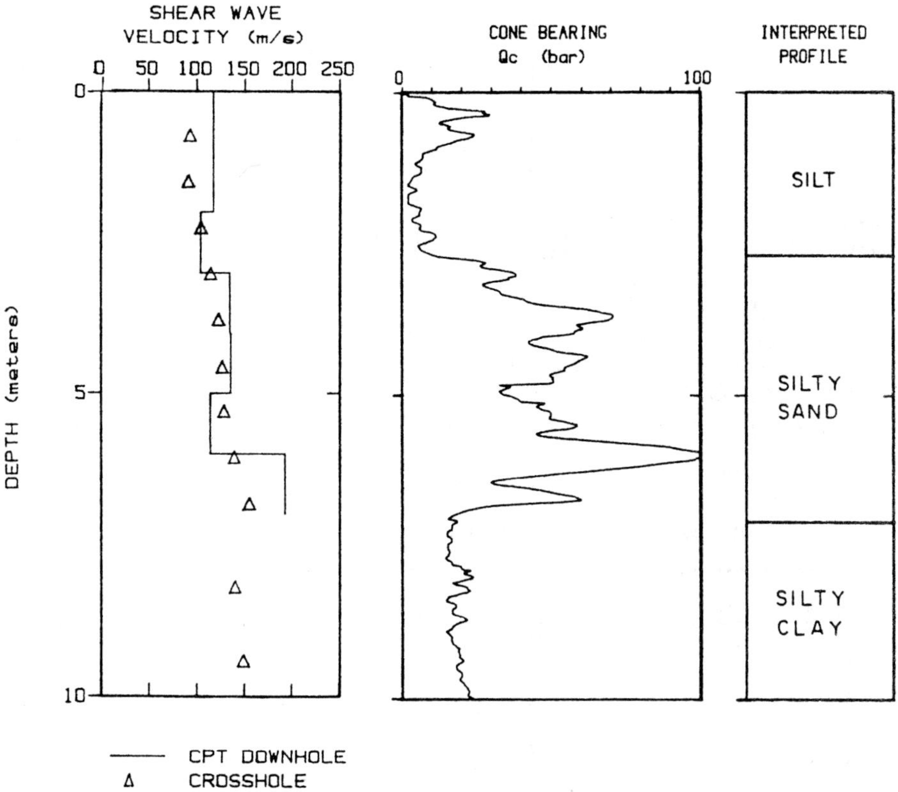

Fig. 5 *Comparison of the seismic cone data with cross-bore hole data at a site in imperical valley (Robertson et al., 1985).*

Spectral Analysis Of Surface Waves (SASW)

Stokoe and Nazarian (1985) have developed a surface wave method for monitoring G_{max} of the soil with depth without the use of boreholes. In this technique, two vertical transducers are placed on the ground surface at equal distances from an imaginary centre line as shown in Fig. 6a. A vertical impulse is then generated on the ground surface at a distance, which for the near receiver is approximately equal to the distance between receivers. Surface waves of the Rayleigh wave type are monitored as they propagate past the two transducers.

The depth of soil sampled with the SASW method is a function of the wavelength generated; low frequency waves sample greater depths. This is related to the dispersion of surface waves in actual soil deposits, and it forms the basis of the SASW method (Rix et al., 1990). Dispersion means that the propagation velocity of surface waves is representative of the

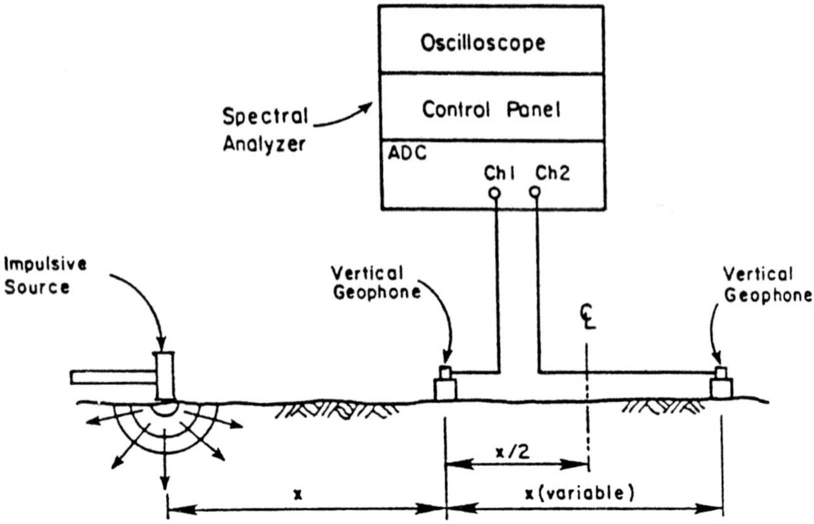

(a) General Configuration of SASW Tests

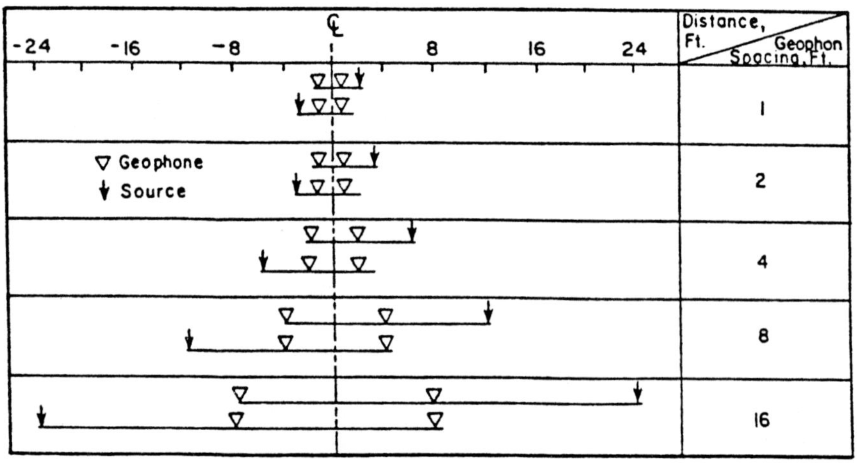

(b) Common Receivers Midpoint Geometry

Fig. 6 *Schematic of experimental setup for SASW tests (Rix et al., 1990).*

material stiffness over depths where there is a significant particle motion. For example, wave which has a wave length λ_1, less than the thickness of the top layer in the profile shown in Fig. 7 will exhibit a propagation velocity which is only dependent on the stiffness of that top layer. On the other hand, the velocity of a surface wave which has a longer wave lenght $\lambda_2 > \lambda_1$, and thus has particle motions in all of these layers will be influenced

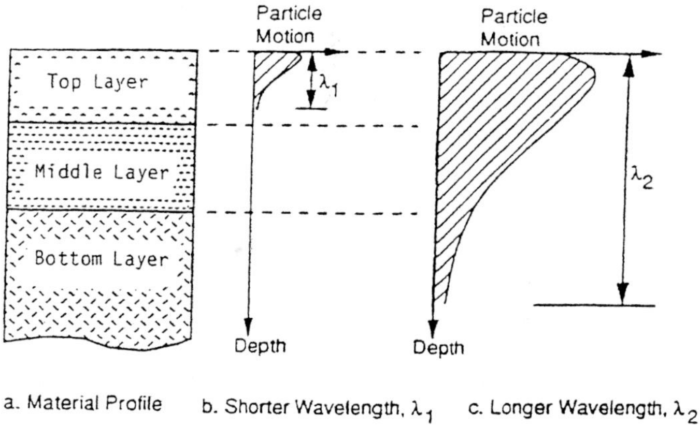

Fig. 7 *Approximate distribution of vertical particle motion with depth for two surface waves of different wave lengths (Rix et al.,* 1990*).*

by the stiffness of all layers (Fig. 7). Thus, by measuring the surface wave velocity over a wide range of wave lengths (frequencies), it is possible to assess the stiffness of the layers over a range of depths. The basic relationship between wave velocity, frequency, and wavelength is:

$$V_R = f.\ \lambda \tag{3}$$

where:

V_R = surface (Rayleigh) wave velocity
F = frequency, and
λ = wavelength

A series of receiver spacings is employed in testing one site. Initially, a close receiver spacing is used to sample the near-surface soils. Then the receiver spacings are increased about the imaginary centre line (Fig. 6b), so that deeper soils are sampled. For each receiver spacing the source is located first to the left and then to the right of the receivers. With this approach, average values of wave propagation between the receivers are collected in an attempt to minimize problems with inhomogenities between them. In this manner, a composite dispersion curve (V_R versus λ) is developed from all receiver spacings.

Once the dispersion curve has been developed for a site, it must be inverted. Inversion is the process of calculating the shear wave velocity profile from the dispersion curve. A theoretical dispersion curve is calculated from an assumed velocity profile and is then compared to the field dispersion curve. The assumed velocity profile contains a sufficiently large number of sublayers to define the in-situ variation in stiffness with depth. The theoretical curve is calculated using a modified Haskell-Thompson matrix algorithm (Thompson, 1950; Nazarian, 1984). The shear wave velocities and thickness of the sublayers in the assumed profile are adjusted by trial and error until a satisfactory match is obtained, at which time the final profile is assumed to represent the in-situ profile. The computer programs used for this SASW inversion continue to be developed, and their availability is presently (2002) quite limited.

Rix and Stokoe (1990) report good correlation between results of SASW and crosshole measurements. However, Sirles (1987, 1988) reports some discrepancies in another case study. This seems to occur in layered soils with inclined boundaries or in heterogenous soil deposits.

CALCULATION FROM BACK ARRAYS

Efforts are being directed to monitor ground motion data during earthquakes. Also soil response with depths up to 17 m has been monitored during similar events (Chang et al., 1991). The properties of soil have been back calculated at appropriate strain. G_{max} values have been determined by wave propagation test method. This method enables determination of Shear Modulus from field tests at large strains as well.

LABORATORY TESTS

The following tests are performed to determine dynamic shear modulus and material damping of soils:

1. Resonant column tests
2. Cyclic triaxial test
3. Cyclic simple shear test

Resonant column test is a low strain test while cyclic triaxial and simple shear tests are intermediate to large strain tests. Since the soil parameters cannot be determined from one test at all strain levels, the results must be reported at a common confining stress.

Laboratory tests have several advantages as they provide the opportunity to conduct parametric studies which can greatly enhance the analysis of field test data. These studies also contribute to the development of new field applications. Their major criticism is the use of small samples of material under imposed boundary conditions.

CONCLUSIONS

Field techniques are now commonly used for determination of dynamic soil properties and offer several advantages over laboratory testing. However they measure the property values at small strains only. There is a scarcity of in-situ methods to reliably measure 'G' at intermediate and large strains and at any strain. These developments would be very desirable due to (i) the difficulty of sampling and testing some soils (saturated sands, and gravels), and (ii) the uncertainty in the lab results because of sample disturbance in all soils.

REFERENCES

Campanella, R.G., Robertson, P.K. and Gillespie, D., 1986, "Seismic Cone Penetration Test", *Geotechnical Special Publication No. 6* (use of In-situ tests in Geotechnical Engineering), editor, S.P. Clemence, p. 116-130.

Chang, C.Y. Mok, C.M. Power, M.S. Tang, Y.K. Tang, H.T. and Stepp, J.C., 1991, "Development of Shear Modulus Reduction Curves Based on Lotung Downhole Ground Motion Data", *Proc. Second International Conference on Recent Advances in Geotechnical Earthquake Engineering and Soil Dynamics*, Vol. 1, March 11-15, St. Louis, Missouri, Editor, Shamsher Prakash. Paper No. 144, pp. 111-118.

Demello, V.B.F., 1971, "The Standard Penetration Test: A State-of-the-Art Report", *4th Pan Am. Conf. on Soil Mechanics and Foundation Engineering*, Puerto Rico, Vol. 1, pp. 1-86.

Dobry, R., Bhatia, S.K., Cramer, C.H., Henke, R., Higgins, C., Jamiolkowski, M., Ko, H.Y., Prakash, S., Saada, A. and Sirles, P.D., 1989,"Low and High Strain Cyclic material Properties", Chapter 3, *Proceedings, NSF/ EPRI Workshop on Dynamic Soil Properties and Soil Characterization*, Vol. 1, pp. 3-1 to 3-100.

Fletcher, G., 1965, "Standard Penetration Test: Its Uses and Abuses", *J. Soil. Mech. Found. Div.*, ASCE 91(SM4), July, pp. 67-75.

Hardin, B.O. and Black, W.L., 1968, "Vibration Modulus of Normally Consolidated Clay", *J. Soil Mech. Found. Div.*, ASCE, Vol. 94, March, pp. 358-369.

Hardin, B.O. and Black, W.L., 1969, "Closure to Vibration Modulus of Normally Consolidated Clay", *J. Soil Mech. Found. Div.*, ASCE, Vol. 95, Nov., pp. 1531-1537.

Hardin, B.O. and Drnevich, V.P., 1972, "Shear Modulus and Damping in Soils: Measurements and Parameter Effects", *J. Soil Mech. Found. Div.*, ASCE,Vol. 98, No. SM 6, June, pp. 603-624.

Imai, T. and M. Yoshimura, 1975, "The Relation of Mechanical Properties of Soils to P- and S-wave velocities for soil Ground in Japan", *Rept. RD-477*, TN-07, Urawa Research Institute. Oya Corp., Japan.

Ireland, H.O., O. Moretto and M.Vargas, 1970, "The Dynamic Penetration Test—A Standard That Is Not standardized", *Geotechnique*, Vol. 20, No. 2, p. 185.

Kovacs, W.D., 1975, Discussion of "On Dynamic Shear Moduli and Poisson's Ratios of Soil Deposits", *Soils and Found.*, Vol. 15, No. 1, March.

Kovacs, W.D., A.H. Griffith and J.C. Evans, 1977a, "An Alternate to the Cathead and Rope for the SPT", *ASTM Symp. On Dynamic Field and Laboratory Testing of Soil and Rock*, Denver, Colo., June 29.

Kovacs, W.D., J.C. Evans, and A.H. Griffith, 1977b, "Towards a More Standardized SPT", *Proc. 9th Intl. Conf. On Soil Mechanics and Foundation Engineering, Tokyo*, Japan 2, pp. 269-276.

Mohr., H.A., 1966, Discussion of "Standard Penetration Test: Its Uses and Abuses", *J. Soil Mech. Found. Div.*, ASCE, 92(SMI), Jan., pp. 196-199.

Nazarian, S., 1984, "In-situ Determination of Elastic Moduli of Soil Deposits and Pavement Systems by SASW Method", Ph.D. Dissertation, Univ. of Texas-Austin.

Palacios, A., 1977, "The Theory and Measurement of Energy Transfer during Standard Penetration Test Sampling", Ph.D. Thesis, University of Florida, August.

Peck, R.B., W.E. Hansen, and T.H. Thornburn, 1974, *Foundation Engineering*, 2nd Ed., John Wiley and Sons, New York.

Prakash, S., 1981, "Soil Dynamics", McGraw Hill Book Co., New York, (Reprinted by S.P. Foundation, Rolla, 1991).

Prakash, S. and Puri, V.K., 1981, "Dynamic Properties of Soils from In-Situ Tests", *Journal of Geotechnical Engineering Division*, American Society of Civil Engineering, Vol. 107, No. GT-7, pp. 943-963.

Prakash, S. and V.K. Puri, 1984, "Design of Compressor Foundations: Predictions and Observations", *International Conf. On Case Histories in Geotechnical Engineering*, St. Louis, MO, May 6-11.

Prakash, S. and Puri, V.K., 1988, *Foundations For Machines: Analysis and Design*, John Wiley and Son, NY.

Richart, F.E., Jr., 1977, "Dynamic Stress-Strain Relations for Soils, State-of-the-Art Report", Proc, *9th Intl. Conf. on Soil Mechanics and Foundation Engineering*, Tokyo, Japan, 2, pp. 605-612.

Rix, G.J., Bay, J.A. and Stokoe, K.H., 1990, "Use of Seismic Tests to Access In-situ the Properties of Portland Cement Concrete During Curing", Paper presented at the 1990 Annual Meeting of the Transportation Research Board.

Robertson, P.K., Campanella, R.G., Gillespie, D., and Rice, A., 1985, "Seismic CPT tp Measure In-Situ Shear Wave Velocity", *Proc. Measurement and use of Shear wave Velocity*, Ed., R.D. Woods, pp. 34-48.

Schmertmann, J.H., 1975, "Measurement of In-Situ Shear Strength", *Proc. ASCE Speciality Conference On In-Situ Measurement of Soil Properties*, Raleigh, N.C., 2, pp. 56-138.

Suchmertmann, J.H., 1977, "Use the SPT to Measure Dynamic Soil Properties?—Yes, But...", *Dynamic Geotech. Testing*, ASTM, Spec. Tech. Pub. 654, June, pp. 341-355.

Sirles, P.C., 1987, "Shear Wave Velocity and Attenuation Measurements in Liquefiable Soil Deposits in the Tru Meadows, Reno, Nevada", M.S. thesis, University Nevada, Reno.

Sirles, P.C., 1988, "Evaluation of SASW and Crosshole Velocity Measurements Before and After Dynamic Compaction Jackson Lake Dam, Wyoming", *U.S. Bureau Reclamation Report* (Unpublished).

Stokoe, K.H., II, Darendeli, M.B., Andrus, R.D., and Brown, L.T., 1999, "Dynamic Soil Properties: Laboratory, Field and Correlation Studies", *Proceedings, Second International Conference on Earthquake* Geotechnical Engineering, Seco e Pinto, Editor, A.A. Balkema, Publishers/Rotterdam & Brookfield, Netherland, Vol. 3, pp. 811-845.

Stokoe, K.H. and R.D. Woods, 1972, "In-Situ Shear Wave Velocity by Cross-Hole Method". *J. Soil Mech. Found. Div.*, ASCE, 98 (SM5), pp. 443-460.

Stokoe, K.H. and R.J. Hoar, 1978, "Variables Affecting In-Situ Seismic Measurements", *Proc. ASCE Spec. Conference on Earthquake Engineering and Soil Dynamics,* Pasadena, Calif., Vol. 2, pp. 919-939, June.

Thompson, W.T., 1950, "Transmission of Elastic waves through a Stratified Solid Medium", *Journal of Applied Physics,* Vol. 21, pp. 89-93.

Woods, R.D., 1978, "Measurement of Dynamic Soil Properties—State-of-the-Art", *Proc. ASCE Spec. Conf. on Earthquake Engineering and Soil Dynamics,* Pasadena, Calif., June. Vol. 1, pp. 91-180.

Woods, R.D., 1991, "Field and Laboratory Determination of Soil Properties at Low and High Strains", *Proc., Second International Conference on Recent Advances in Geotechnical Earthquake Engineering and Soil Dynamics,* Saint Louis, MO, Editor, Shamsher Prakash, March 11-15, pp. 1727-1742.

7

AMAP'sols Static Penetration in Dense Sandy Gravel, Sandstone and Hard Claystone *(CPT Revel in Gravel)*

G. Sanglerat[1], *M. Petit-Maire*[2], *F. Bardot*[3] *and P. Savasta*[4]

INTRODUCTION

For many years, common knowledge dictated that static penetration could not be applied to investigate gravel, hard layers or soft rock ledges. This was concluded from refusals met in these layers at shallow depth and from the damage caused to expensive electric cones.

Certain geotechnical engineers therefore limited the use of the static penetrometer to soft layers. Although, some others extended its use to sand deposits (Begemann, 1965; Schmertmann, 1967), all excluded it for the study of gravel (Mitchell and Brandon, 1998).

However, since 1992, it is now possible to conduct CPT in gravel using the AMAP' sols rig (Sanglerat, 1994; Sanglerat et al., 1995).

The name AMAP'sols means « *Ateliers Mobiles d'Auscultation par Pénétration des Sols* » (Mobile Soil Testing Unit by Penetration).

The results obtained are described below.

AMAP'SOLS STATIC-DYNAMIC PENETROMETER

Description

The development of the static-dynamic penetration was presented by Sanglerat in 1979.

In 1992, in South Central France area, geotechnical engineers decided to develop a new type of static-dynamic penetrometer that would expand

[1]*182 bis avenue Felir Faure 69003 Lyon, France.*
[2]*d' AMAP' sols grue Irevert 42.570 Saint-Héandm, France.*
[3]*26 vue Jules Renard 69007, Lyon, France.*
[4]*SETSOL, 181 montée des Aires 13880 Velaux, France.*

the applicability of this method of soil exploration (Sanglerat 1995, Sanglerat et al. 1995).

Important improvements were brought about which are summarized below:

Rig :

The penetrometer is mounted on a 6x6 Mercedes truck, of 260 kN (58,500 lb) (see Fig. 1).

The levelling of the truck, to insure true vertical penetration, is done automatically by five hydraulic jacks controlled by electronic sensors. Four of the jacks bear on a retractable caterpillar track of the truck, used as an auxiliary propulsion mechanism (see Fig. 2).

The cat-like tracks ensure mobility on soft soil terrains where conventional trucks would bog down. The tracks are a patented device of Van den Berg of the Netherlands who has many years experience with the static penetrometer.

Static mode

Totally automatic operations, penetration at 2 cm/sec with continuous numeric recording every 2 cm on memory board with simultaneous drawing of the cone resistance diagrams in real time. This allows continuous test quality control.

The data and interpretation are transmitted by modem to the office. The engineering interpretation of these test results can be used to evaluate the soil parameter necessary for soil bearing capacities and settlements computations (Sanglerat et al. 1984, Robertson 1995, Lunne et al. 1998).

Dynamic mode

The conventional way of driving a probe by a free falling hammer was

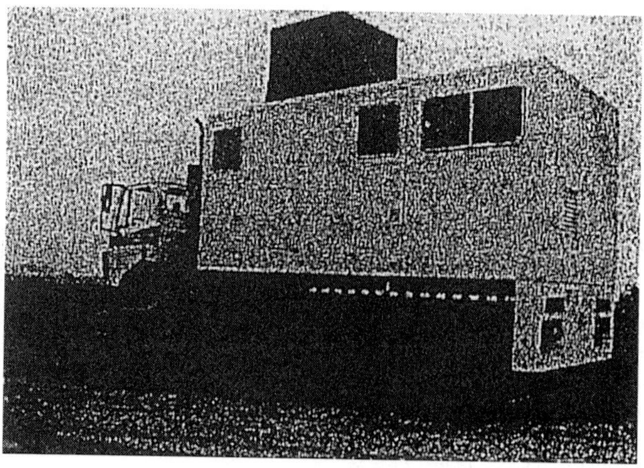

Fig. 1 *The AMAP'sols vehicle.*

Fig. 2 *Hydraulic stabilisers on caterpillar track.*

replaced by a very powerful, fast-action hydraulic hammer (see Fig. 3). By adjusting energy it is now capable of going through dense layers and penetrate into weathered bedrock.

Characteristics

The penetrometer rig allows static and dynamic penetration of soils. It can push conventional electrical or mechanical tips from 10 to 50 cm^2.

Usually, static penetration is done either with a 50 cm^2 mechanical tip equipped with a skin friction sleeve of 250 mm in length (Fig. 4), or with a 12 cm^2 tip.

In the static mode, the following measurements are made with the 50 cm^2 tip:

q_c : cone resistance (up to 30 MPa or 435 psi)

f_s : skin friction, which permits calculation of the friction ratio $FR = \dfrac{f_s}{q_c}$

Q_{st} : total penetration resistance (up to 220 kN, but usually limited to 180 kN)

When refusal is met at 30 MPa (435 psi) with the 50 cm^2 tip in a hard soil layer, static penetration may be continued with a 12 cm^2 tip not equipped with a friction sleeve. The static cone resistance of the 12 cm^2 tip can reach 148 MPa (2146 psi) in cohesionless soils (this measurement is identified as q_{c12} to differentiate it from the q_{c50}). When refusal is met, dynamic penetration is used (see Fig. 3).

It is of our opinion, that when using the AMAP' sols rig, only the static penetration data is useful for the purpose of data interpretation. Since the dynamic penetration is used when q_c > 140 MPa, it is only used to break

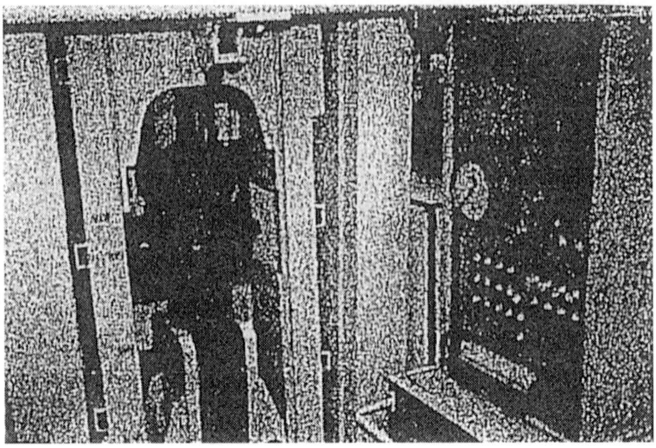

Fig. 3 *View of the hydraulic hammer and measuring equipment for the static-dynamic penetration of AMAP'sols.*

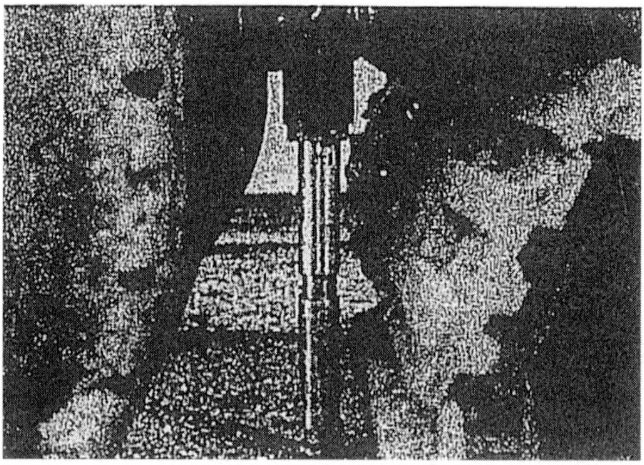

Fig. 4 *AMAP'sols 50 cm² cone with lateral friction sleeve.*

through the hard soil or weathered rock layer in search of a potentially lower weaker layer.

If and when underlying weaker layers are detected the penetration test is then reverted to the static testing mode.

Hence, dynamic penetration is used only to get through cemented soil layers.

Furthermore, in the dynamic penetration mode, a static test is performed, every 25 cm to check the total resistance Q_{st}. Each time that Q_{st} is lower than 170 or 160 kN, the static penetration mode is employed. A sound alarm is triggered every time any of the load limits are reached for each of the tubing configurations.

Dynamic penetration is therefore only used to get through very hard layers, where q_c is usually higher than 130 or 140 MPa using the small cone tip.

In the event of a sudden drop of tip resistance, the dynamic driving mechanism automatically stops immediately at the upper boundary of the less resistant layer. This stops further penetration so that measurements can be made in the softer layer.

The softer underlying layer can then be once again analysed in the static mode over its entire depth. This procedure was not always possible with the older model of static-dynamic penetrometers. These would go into soft layers without yielding any test results for several tens of centimetres.

An audible alarm warns the operator whenever a maximum load is reached for the different tubes used to transmit the stress to the measuring tip.

Other Uses

The penetrometer accepts different cone tips, such as the piezocone and the envirocone of Van den Berg (Lunne et al., 1995).

In each of these two cases, the computer programs used are those defined by Van den Berg.

Depending on its manner of use, the piezocone can measure q_c, f_s as well as pore water pressures.

With the envirocone, the following measurements, besides q_c, are made possible :

Soil conductivity

H^- and O^{++} concentration

Redox potential

Temperatures

Ph

Porewater pressure

This range of application brings about a considerable improvement in the resources available to study environmental and waste management problems, thanks to the quality of the data obtained (Mitchell and Brandon 1998, Robertson 1995, Robertson and Wride 1998).

PENETRATION INTO LYON SANDSTONE

For comparison purpose, initial testing of the AMAP'sols was made on well-known Lyon soils.

Generally, static penetration with a Gouda mechanical or electric cone is inappropriate because refusal is usually met at depth of 3 to 4 metres.

On the other hand, static dynamic penetration of the old type met refusal in cemented or dense alluvial deposits, or in the upper boundary of sandstone, about 20 metres below the surface.

First results obtained with the AMAP'sols static dynamic penetrometer in 1993 reached depths of 35 metres, thus achieving penetration of some 15 metres into the sandstone. This is a remarkable improvement in performance (Sanglerat, 1994).

At another site in Lyon in 1994, dense alluvial layers consisting of gravel with varying sand contents were penetrated to 21 metres and an additional 24 metres of penetration was achieved into the underlying sandstone (cf. Fig. 5). This established a record never reached by any other type of penetrometer, either static or dynamic.

The diagram gives important information on the characteristics of the sandstone.

The test method is considerably less expensive than the traditional rock coring methods and laboratory testing of samples.

LE HAVRE HARBOR COMPARATIVE TESTS

The operation of the AMAP'sols penetrometer 12 cm² cone tip, in its static mode, is different than that of the classical electric cone of 10 cm². Therefore it was necessary to prove the reliability of the cone tip calculated values by comparing them with the 10 cm² electrical and 50 cm² mechanical cone tips.

The comparative tests were made easy due to the fact that any one of the three cone types could be used for a soil foundation study in Le Havre (France).

These interesting results were obtained in the sedimentary deposits at the delta of the Seine river, in the Le Havre harbor (Sanglerat et al., 1995).

These results prove the validity of the method, using the small AMAP'sols cone (12 cm²) because the penetration diagrams of both penetrometer types were identical.

HIGHWAY BRIDGE NEAR CHAMONIX

A steel driven pile supported three spans highway bridge experienced differential settlements during construction. Due to insufficient resistance, pile lengths were increased from design lengths of 25 m to an as-built length of 58 m.

By the end of construction, in spite of the added pile length, the left top abutment rotated away from the deck. The pre-construction soil borings with pressuremeter tests were subsequently complemented by four AMAP'sols penetrometer tests. These penetrated 43 to 50 m deeper than the original soil tests, to a depth of 82 m. The final test results gave a good stratigraphy of both compressible and pile bearing soil layers (Sanglerat et al., 1995).

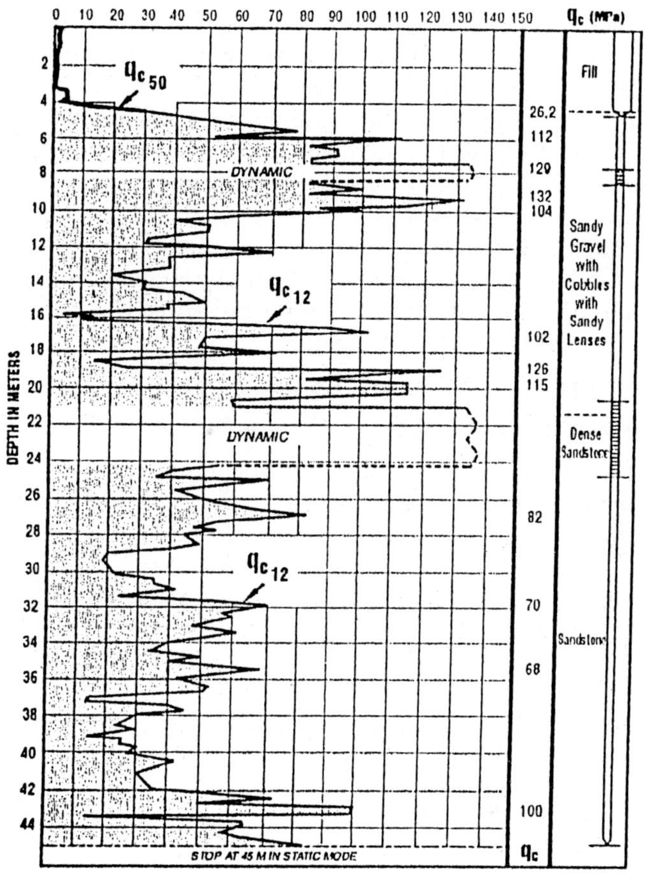

Fig. 5 *Penetration record in dense sandstone.*

Based on the information then collected, the bridge was closed for a period of 4 months while improvements were made to the left embankment. These consisted of removing 6 m of soil fill, replacing it with a light weight expanded polystyrene fill material and the installation of 8 vertical drains extending into the artesian aquifer.

DYNAMIC PENETRATION

It has been known for a long time that dynamic penetration into saturated cohesive soils should be avoided (Sanglerat, 1979). In other soil conditions however, this type of penetration is acceptable even though it is difficult to translate the dynamic penetration data into penetration resistance values for conventional dynamic equations type, such as obtained by the Dutch formula (a less than perfect formula but widely used).

This interpretation need not to be made with the data from the static—dynamic AMAP'sols penetrometer. Thanks to its great capacity and when used with the 12 cm^2 cone tip, soils with static resistances of up to 140 or 148 MPa may be penetrated. The dynamic mode therefore is only used to get through extremely dense or cemented soils. Once traversed, penetration can revert to the static mode as soon as the static resistance drops back down below 130 or 140 MPa.

Consequently whenever the static penetration is high (over 130 or 140 MPa) it is no longer necessary to evaluate the dynamic resistance or to translate the data into allowable bearing pressure. This is an important feature of the AMAP'sols penetrometer.

STATIC PENETRATION IN VERY DENSE COHESIONLESS SOILS

One of the frequent objections made to the penetrometer test, especially for static penetration and sometimes for dynamic penetration, is that these tests are blocked by dense or cemented layers and that no information can be gathered about underlying soils. Often this objection is justifiable but no longer applies to the static dynamic penetrometer of AMAP'sols.

It was shown that this device can penetrate to large depths in the Lyon (France) dense sandstone and sometimes in the case of very high cohesive soils.

Let's look at the penetrometer diagrams for six sites in the Rhône-Alpes Region of France:

—Sassenage (Isère) (Fig. 6)
—Lyon—La Guillotière (Fig. 7)
—Golf de Villette d'Anthon (Isère) (Fig. 8)
—Highway bridge at Jarrié (Isère) (Fig. 9)
—Echirolles (Isère) (Fig. 10)
—Lyon—Porte Dauphine (Fig. 11)

The diagrams on these figures show very high resistance values can often be encountered in sandy, gravely dense soils. Only the AMAP'sols penetrometer is able to measure and record them.

STATIC PENETRATION IN HARD CLAYSTONE

Static penetration can be used to investigate compressible cohesive soils (soft clays, peat, silts). As a matter of fact, it is the best way to characterize these soils by in-situ methods, especially below the water table.

On the other hand, until 1992, the drawback of the static penetrometers was that they met refusal on hard layers, even thin ones, which possibly overlayed other weaker soil zones.

Earlier, all static penetrometers met refusal in stiff cohesive soils due to either a high point resistance or, more often, because of a high total lateral friction resistance.

If in the case of cohesionless soils dynamic penetration is, at times, necessary to get through indurated layers, or layers containing large boulders, in very stiff cohesive soils, such as over-consolidated clayey silts (particularly when located below the water table), dynamic penetration does not always progress with significant results when compared with the results obtained by static mode. Hence, only the power of the static penetrometer allows the engineer to further investigate these soils. Until now, all types of static penetrometers met refusal in very stiff cohesive soils.

Moreover, because of the large capacity of the static AMAP'sols penetrometer, using at first a large diameter cone, then with a smaller cone, the total lateral friction is greatly reduced as penetration progresses and greater depths of investigation become attainable.

The following is a recent example which occurred in the Mediterranean region (France).

The soil profile is as follows:

—0 - 2m—Fill material

—2 - 8m—Recent alluvials containing soft clay and peat lenses

—> 8m—Hard silty clay (claystone) with some weathered upper boundary layer

In order to construct a car-park of 48 x 77 m in plan dimension to a depth of 17 m, the engineers had expected to construct a diaphragm wall

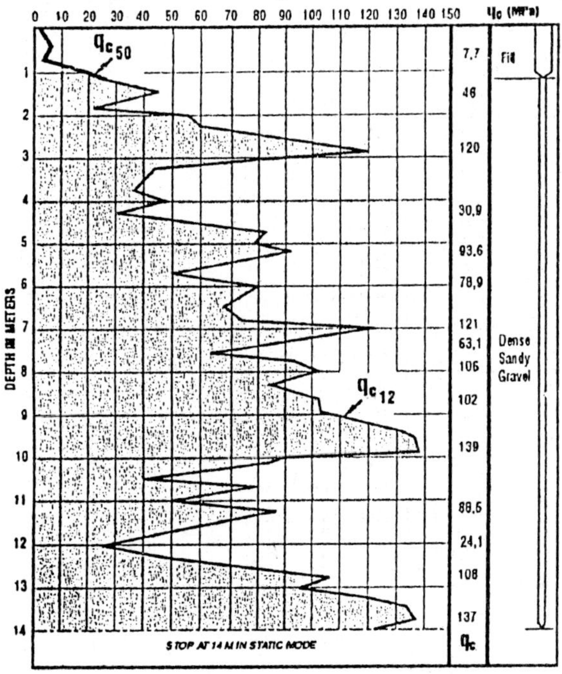

Fig. 6 *Static penetration in dense sandy gravel.*

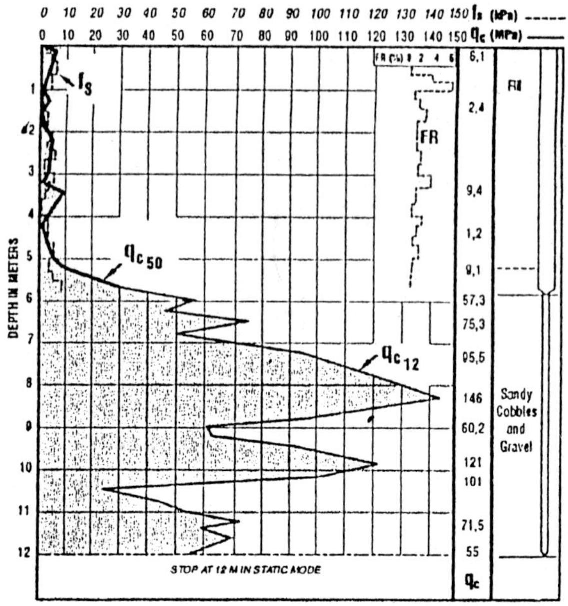

Fig. 7 *Penetration in dense sandy gravel with cobbles.*

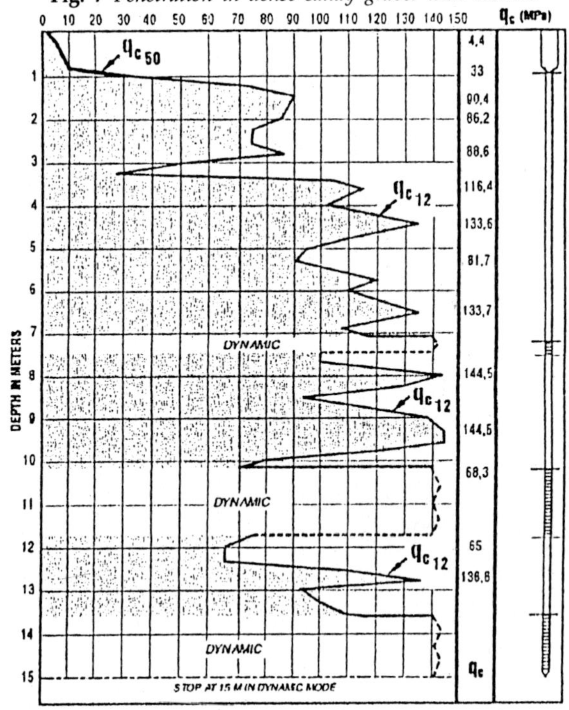

Fig. 8 *Penetration in very dense sandy gravel.*

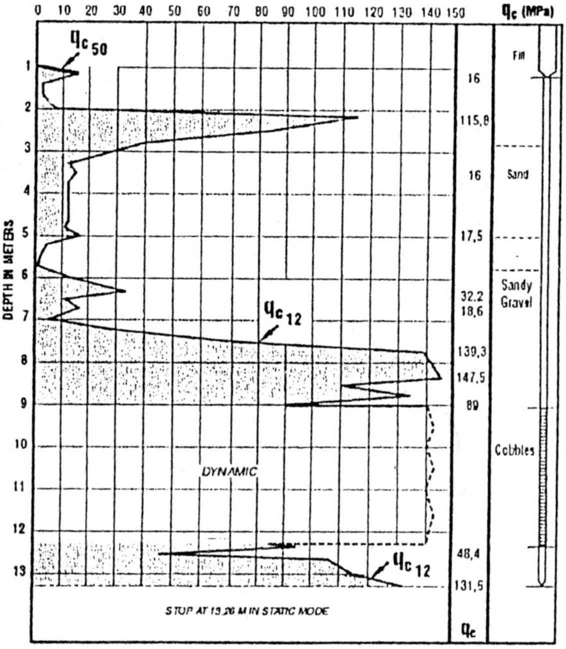

Fig. 9 *Penetration in very dense sandy gravel with cobbles.*

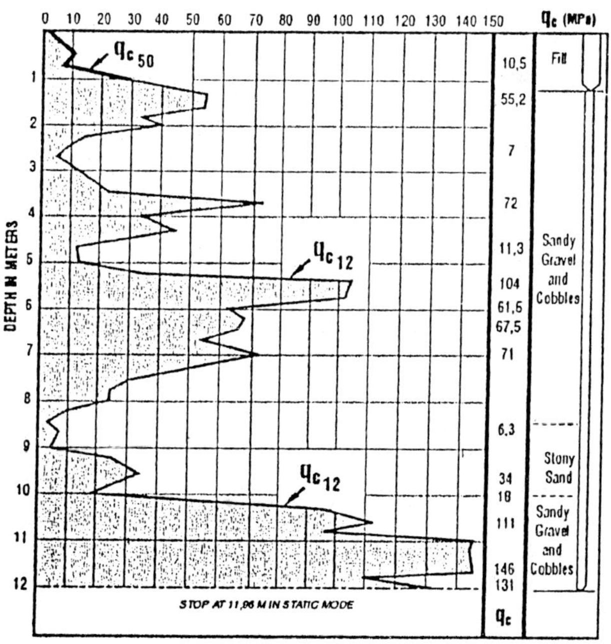

Fig. 10 *Penetration in very dense sandy gravel with cobbles.*

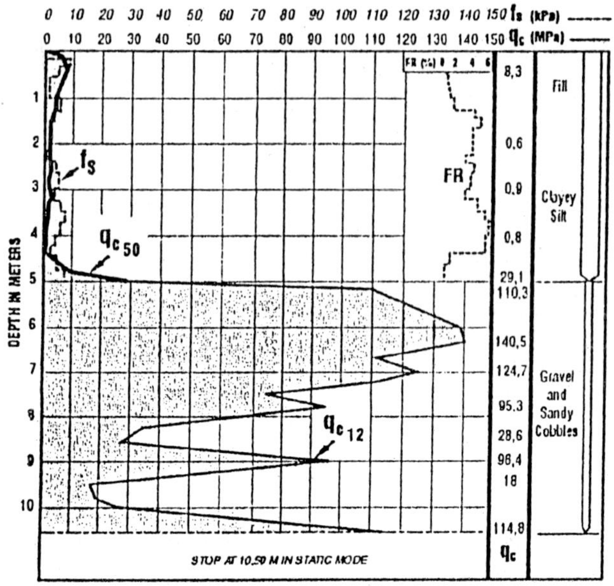

Fig. 11 *Static penetration in very dense gravel and sandy cobbles.*

of 60 cm in thickness with three levels of pre-stressed anchors (cf. Fig. 12) and horizontal bracing at the four corners at an angle of 45°.

The initial soil investigation consisted of few borings with pressuremeter tests to shallow depths. However the job specifications indicated that the soil conditions were well known in the area based on the previous construction of several tunnels. The soil properties were given in Table 1 which is reproduced above.

Two important existing buildings, dating from the 16th century were located near the proposed excavation, on the east and north sides.

After excavation and when the diaphragm wall was completed, greater wall deflections were measured than expected. Cracks even appeared at certain locations and settlements of the order of 30 to 60 mm were noted at the N-E corner of the parking, resulting in damage to the old adjacent buildings.

In-situ tests were then performed which included the Menard pressuremeter and two types of static penetrometers, one electric and the other mechanical. It was not possible to define the top of the hard layer with the pressuremeter and the two static penetrometers met refusal at 8 and 12 m. However it was necessary to define the boundary between the weathered and intact claystone in the area where the surface settlements had occurred, and where the wall deflections were the greatest. A key point was to define which of the soil layers were exerting pressures on the wall. Four AMAP'sols penetrometer tests were performed to depths of from 15.70 m to 20 m; 20, 20 m and 21, 10 m without difficulty.

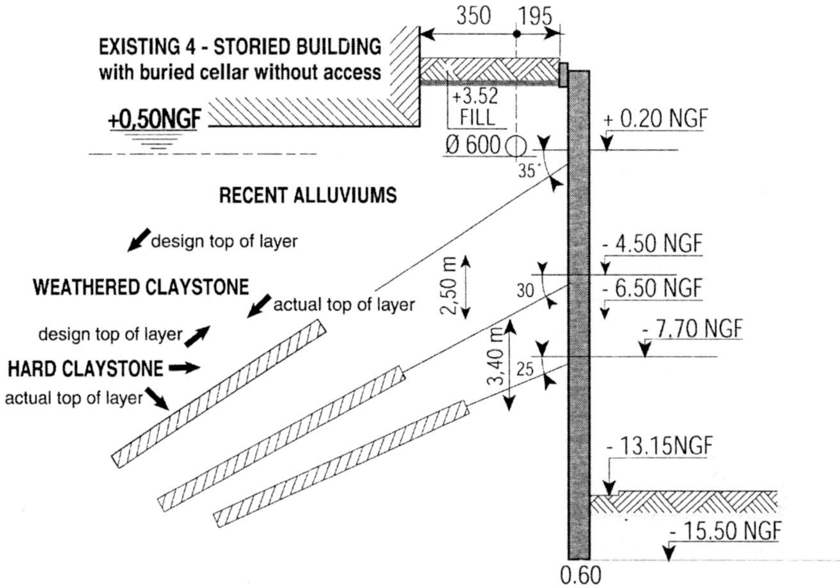

Fig. 12 *Cross-section of concrete slurry wall showing 3 levels of pre-stressed anchors.*

Figure 13 presents the AMAP'sols test results. It shows that down to a 10 m depth, recent alluvial soils were present. They are very compressible and the results indicated that the claystone started considerably lower than expected, since it appeared at depth of 7.50 m.

Furthermore, it was noticed that between 10 m and 13. 50 m the point resistance q_{c50} and the lateral friction f_s increase at a uniform rate but not at high values. The claystone expected at 10 m depth was in fact encountered at 13.50 m.

Table 1

Description	density		undrained		drained	
	γ	γ′	φ°	c (kPa)	φ′	c′(kPa)
Fill	2	1	30	0	30	0
Alluvials	1.8	0.8	25	15	30	0
Weathered claystone	2.1	1.1	20	80	25	40
Hard claystone	2.1	1.1	25	200	30	130

The penetrometer data is summarized in Table 2.

Figure 12 shows a cross-section with the design and actual location of the top of the layers. It can be noticed that due to significant errors, lateral forces are considerably greater than expected. Therefore the lateral

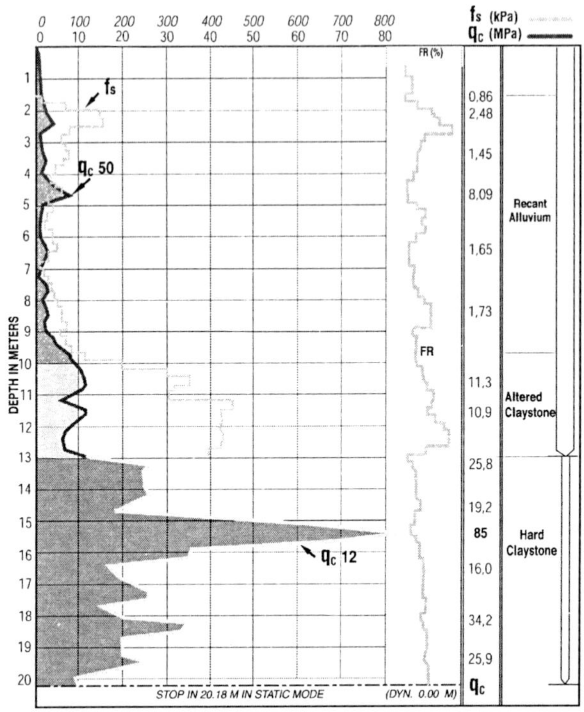

Fig. 13 *Static penetration in very dense gravel and sandy cobbles.*

wall deflections were excessive and engendered corresponding surface settlements.

Table 2

Weathered claystone
$4 < q_{c50} < 9$ MPa
$100 < f_s < 400$ kPa
$2 < FR < 5\%$
Hard claystone
$10 < q_{c12} < 85$ MPa

CONTROL OF SANDY GRAVEL FILLS

The static penetrometer is an excellent tool to control the density, therefore the quality of a sandy gravel fill, either on land or below water (in lakes or submarine conditions). Table 3 summarizes the cone resistance criteria which define the density of fills (Sanglerat, 1979).

Table 3
Correlation between consistency and q_c

	Fill	q_c(MPa)
Type	Consistency	
Clean gravel	Loose	1.5 to 2.5
Sandy gravel	Loose	3 to 4.5
Sandy gravel	Medium dense	8 to 12
Sandy gravel	Very dense	>20

Of course, when sandy gravel fills are dense, the cone tips, especially electric ones, can be seriously damaged.

However, this drawback is avoided by using the AMAP'sols penetrometer since the tip is very rigid and can withstand large thrusts.

The following example illustrates the powerful adaptability of the AMAP'sols device.

Tests were performed through a compacted gravel fill existing below a warehouse reinforced concrete floor slab. The underlying natural soil consisted of soft clay. Tests were necessary to determine the cause of slab failure due to differential settlements. Three typical penetration diagrams of some twenty tests performed are presented on Fig. 14. Measurements of q_c, cone tip resistance, were recorded every 2 cm of penetration starting just below the slab, through cored holes in the concrete slab.

The diagrams show that:

1. The thickness of the fill varying from 50 cm to 110 cm did not meet the 1 metre criteria design;
2. The upper layer of fill was well compacted at all test locations (q_c> 20 MPa), but the density varied downward from location to location.

VERIFICATION OF DENSIFIED STONE COLUMNS

French standards (*D.T.U. 13.2—Deep Foundations*) state that densified soil columns may be checked and accepted using the static penetrometer whenever

$$q_c > 10 \text{ MPa}$$

Too often, however, with small cones of 10 and 15 cm², the usual penetrometers may be deflected outside the densified column in some gravels. When this occurs, q_c values drop as the point enters the undisturbed adjacent soil.

The results then become irrelevant and may even lead to erroneous conclusions.

The advantage of the AMAP'sols penetrometer, with its 50 cm² point and great rigidity and power, is that complete penetration of densified

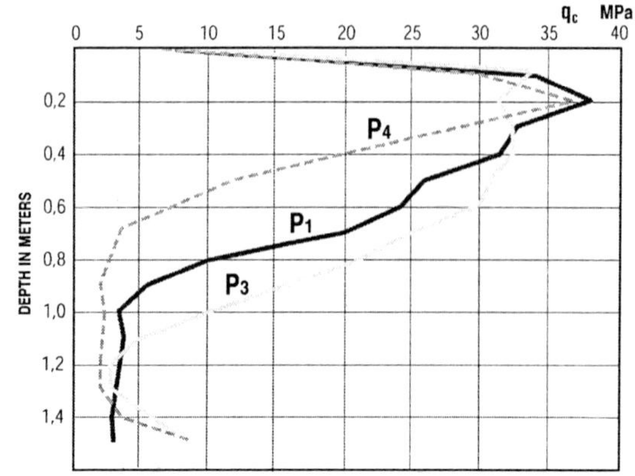

Fig. 14 *Fill density check under concrete floor.*

columns is achieved. Hence, the cone is not deviated and cannot be stopped by large gravel particles.

Figure 15 (b) shows the excellent result obtained in Paris.

COST OF AMAP'SOLS TESTS

It is obvious that thanks to the wide range of tips available to explore sub-surface soils, the AMAP'sols static-dynamic penetrometer provides an efficient and fast method to perform soil investigations. Moving it around a site is made easy thanks to the system of tracks and automatic use of its jacks (cf. Fig. 2).

Moreover, its ability to use both the piezocone and the envirocone yields additional important information in the struggle against pollution (Lunne et al., 1998).

Practical experience over the last ten years of operation of the AMAP'sols penetrometer shows that soil investigations made in this manner are less expensive than any other methods.

In France, the unit rate per metre of penetration of the typical AMAP'sols investigation is half as expensive as that of a good quality pressuremeter sounding and its cost is only about 30 to 33% that of a soil boring with sampling and drilling parameters.

There remains little doubt that demand for this penetrometer will increase greatly, not only in France, but in numerous other countries in 2000 and beyond.

No other penetrometer compares to its performance.

PENETRATION IN DUTCH GRAVEL

On the occasion of the 12th European Congress of Soil Mechanics and

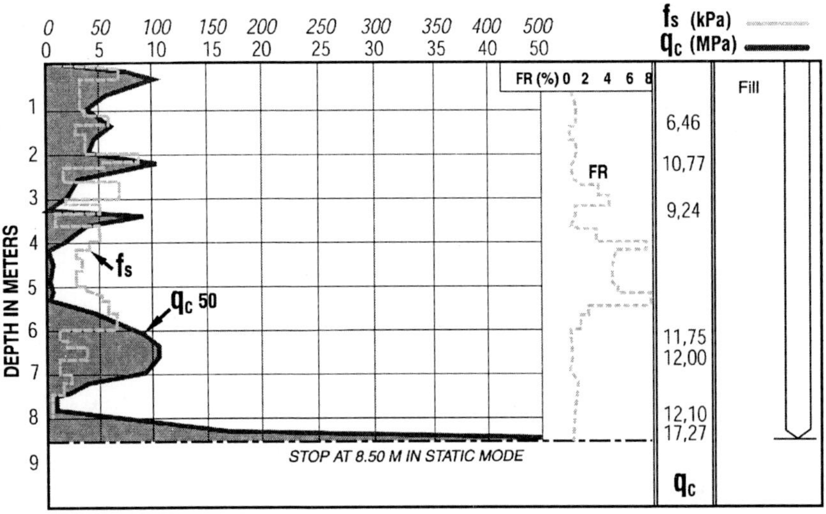

Fig. 15a *Untreated soil.*

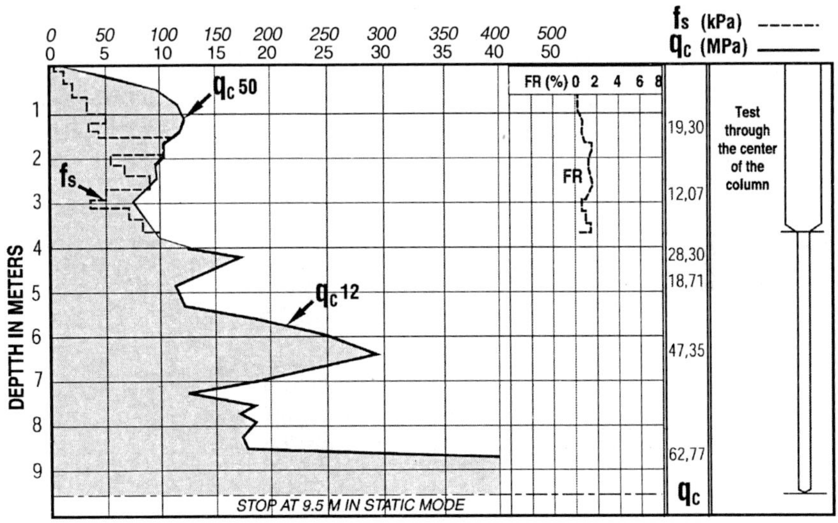

Fig. 15b *Static penetration in densified stone column.*

Foundation Engineering held in Amsterdam from 7 through 10 June 1999, the AMAP'sols penetrometer was displayed.

In the same period, the company A.P. van den Berg of Heerenveen was facing a difficult challenge in the east of the Netherlands, in the city of Nijmegen. The city is situated on the Rhine, near the border with Germany.

To avoid freeze/thaw issues at the near surface of the soccer field, the NEC football club had horizontal heating tubes embedded 25 cm below the soccer field installed. The required heat had to be provided by heat exchangers located at 28 m below ground, and a heat pump system. In the summer the same system would be used for air conditioning of the building or for returning the heat to the soil in order to re-establish the global thermal balance in the long term (heliogeothermics).

The advantage of this environmentally cautious system are:

—no more soccer field freeze/thaw problems
—reduction of the use of fossil fuels
—reduction of CO_2–emission
—at least 50% reduction of energy costs
—less maintenance.

As in general the soil layers in the Netherlands are soft, the static penetrometer is used to install the heat exchangers at the desired depth.

First, steel casing tubes Ø 56/40 mm with a lost tip Ø 64 mm are pushed into the soil. In the tube string a polyethylene heat exchanger Ø 30 mm is lowered. Then the casing tubes are removed, leaving the heat exchanger in place, tied to the lost tip.

But in Nijmegen, the soil consists of predominantly thick sand layers, with randomly distributed low resistant loess, clay and peat layer or dense gravel layers, with cimentacions zones. The thickness of these random layers varies from 1 to 4 m. The groundwater level is at 9 m below ground and fluctuates.

Two Dutch static cone penetrometer rigs were used to push the casing tubes.

However, in some locations, refusal was met on dense gravel layers at an approximate depth of 16 m below ground.

But it was necessary to continue to a depth of at least 28 m in order to obtain a certain temperature difference for the water of the heat exchangers.

Since the heat exchanger design depth location was 28 m, it became necessary to find a new installation method. A.P. van den Berg, for an important part involved in the construction of the new AMAP'sols static penetrometer in 1992, knew it to be much more powerful than all other existing static cone penetrometers, and suggested to try this French penetrometer to solve their problem.

The AMAP'sols penetrometer was able to exceed the design depth of 28 m, to 40 m depth, using the static or dynamic method alternatively as necessary.

In less than 4 weeks, the new AMAP'sols penetrometer succeeded in pushing 44 tubes of 40 m length in dense gravel. Conventional CPT rigs achieve depth of only 16 m.

A mechanical AMAP'sols static penetration test was conducted to evaluate the characteristics of the different sedimentary layers. It is presented in Fig. 16. Based on this CPT test results the following can be concluded:

1. The 50 cm² mechanical cone (ϕ 80 mm), giving q_{c50}, meets refusal at 15.5 m at the top of dense Rhine gravel that often blocked casing tubes at about this depth.
2. The 12 cm² mechanical cone (ϕ 39 mm), giving q_{c12}, reaches 40 m depth without difficulty using only static pushing.

The gravel of the Rhine has the following penetration characteristics between 15.5 m and 18.5 m:

$$18 < q_{c12} < 46 \text{ MPa}$$

Therefore it is not surprising that, at first, it was not possible to continue pushing the casing tubes with lost tip ϕ 64 mm to a depth of 40 metres with local penetrometers, given the force on the tip and the total skin friction.

This experience in Nijmegen shows that the latter were not used in their traditional field of application, for which they keep all their qualities.

In June 2001 FUGRO was involved for a tunnel project in Roermond in the South of the Netherlands. In the soil investigation carried out, consisting of static CPT and boreholes, the soil conditions comprise gravel layers, more or less infilled with sand, very dense. The general soil profile in this region is given in Table 4.

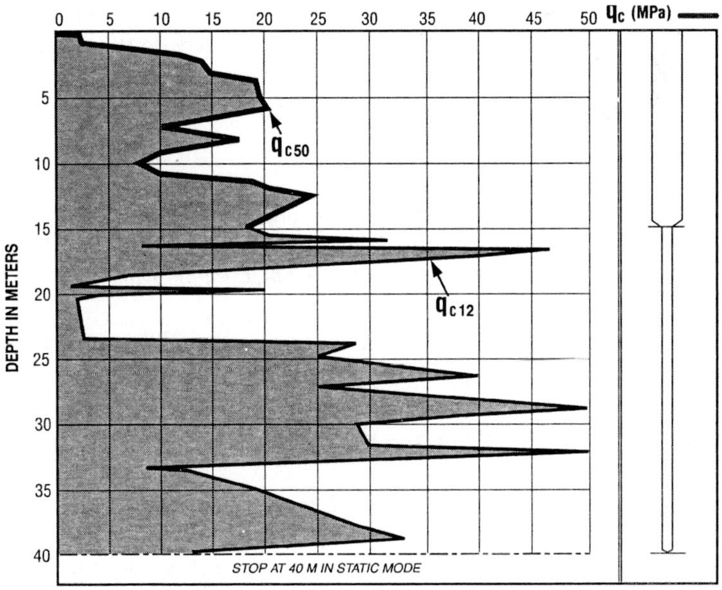

Fig. 16 *Static penetration in very dense rhine sandy gravel in Nijmegen.*

Table 4
General soil profile Roermond

Level (ref. sea level)	Soil description
+ 20.7 up to + 28.0	Surface level
+ 20.7 up to 20.8 until + 14.0	Sand and loam
+ 14.0 until + 1.0	Gravel, infilled with sand, very dense
+ 1.0 until – 2.0	Sand
–2.0 until – 20.0	Gravel and sand

The required penetration should be 50 m. The standard static penetrometer met refusal at a depth between 16 and 20 m (sea level + 7m).

Therefore FUGRO proposed to the client the AMAP'sols system to complete the investigation to greater depth.

Twenty AMAP'sols CPT easily reached depth between 45 and 50 m. These CPT could show the different characteristics of several layers through which the tunnel had to be perfore.

CONCLUSIONS

This paper presents data from France and the Netherlands which lead to the following conclusions:

- The AMAP'sols CPT rig can be used in all types of soils and weathered rock.
- The accuracy of the test reading is similar to conventional CPT rigs.
- The static-dynamic method of soil investigation, with the AMAP'sols CPT rig, is considerably more cost effective than any other traditional method of investigation soils.

REFERENCES

Begemann, H.D.S., 1965, *The friction jacket cone as an aid determining the soil profile* 6th CFMSF—Montreal—Vol. 1, pp. 17-20.

Lunne, T., Robertson, P.K., Powell, J.J.M., 1998, CPT *in geotechnical practice* New York, Blackie Academic, 312 p.

Mitchell, J.K., Brandon, T.L., 1998, *Analysis and use of CPT in earthquake and environmental engineering*, ISC'98, pp. 69-97.

Robertson, P.K., 1995, *Application of CPT to evaluate liquefaction potential* CPT'95—Linköping (Sweden)—Vol. 3, pp. 57-79. Swedish Geotechnical Society.

Robertson, P.K., Wride, C.E., 1998, *Evaluation cyclic liquefaction potential using the CPT* Canadian Geotech—J 35—pp. 442-459.

Sanglerat, G., 1979, *The penetrometer and soil exploration.* Second enlarged edition, 488 p., Amsterdam, New York : Elsevier.

Sanglerat, G., Olivari, G., Cambou, B., 1984, *Practical Problems in S.M.F.E.* Vol. I. *Physical Characteristics of Soils, Plasticity, Settlement Calculations, Interpretation of in-situ Tests.* 283 p. Amsterdam, Elsevier.

Sanglerat, G., 1994, *It took 2000 years of penetration testing to arrive at the AMAP'sols static-dynamic penetrometer.* Symposium on Development in Geotechnical Engineering, Bangkok, pp. 101–105.

Sanglerat, G., Petit-Maire, M., Bardot, F., Savasta, P., 1995, *Additional test results of the AMAP'sols static dynamic penetrometer* CPT'95—Linköping (Sweden) Vol. 2, pp. 85-91.

Schmertmann, J.H., 1967, *Static cone penetrometer for soil exploitation.* Civil Engineering—Vol. 37, No. 6, pp. 71-73.

Nota Bene :

This paper has been publised in French by « *Revue Française de Géotechnique*» —Paris 1999, n° 87, pp. 43-54.

8

In-Situ Characterization of Creep Parameters by Instrumentations

J.L. Justo[1], *P. Durand*[2] *and E. Justo*[3]

INTRODUCTION

Creep is a slow deformation of soil caused by constant effective stress. In many rockfill and earth embankments, creep is responsible for the majority of post-constructive displacements which in turn may produce cracking of the impervious element (v. Macedo-Gómez et al., 2000). Justo (1973) has found positive correlations between the maximum crest settlement and the start of cracking or piping in embankment dams (v. also Thomas, 1979). A careful study of dam characteristics collected by Justo (1973) and others has made the establishment of limits of Table 1 possible.

Figure 1 shows the influence of maximum post-constructive settlement and abutment slope upon transverse cracking. The lines indicated in the graph separate uncracked dams (below) from cracked dams (above). Most of the dams used to elaborate the graph are cited by Justo (1973). Only the few that define the lines have been included in this figure.

Positive correlations have also been found between post-constructive settlement and the amount of leakage in concrete face rockfill dams (Justo and González, 1986).

Once the construction of the embankment has been completed, a dam should have sufficient camber at the top to allow post-constructive settlement without loss of freeboard. The importance of this point is supported by the fact that in 15% of embankment dams with inadequate performance, the cause was overtopping (U.S. Committee on Large Dams, 1975). Due to this fact, it is preferable to add an excess of camber during construction, as later correction may be difficult and expensive. On the other hand, allowing for this excess might imply a great deal of extra cost.

[1]Professor of Ground Engineering, [2]Associate Professor of Ground Engineering, [3]Associate Professor of Continuum Mechanics Department of Continuum Mechanics, University of Seville, Spain

Table 1
Influence of post-constructive settlement at crest on cracking

Settlement (cm)	*Kind of cracking*
< 5	No cracking
≥ 5	Transverse cracking of dry compacted dams may appear
> 10	Reinforced concrete facing without perimetral joint may crack
≥ 13	Longitudinal cracking between core and shell may appear
> 16	Longitudinal cracking of dry compacted core may appear
> 18	Hydraulic fracturing may appear
≤ 21	No cracks in asphaltic concrete diaphragm
≥ 22	Transverse cracking of wet compacted core, or longitudinal cracking between wet compacted core and shell may appear
≥ 35	Asphaltic concrete facing may crack (self-healed for settlement of 35 cm)
≥ 37	Reinforced concrete facing with perimetral joint may crack
> 40	Longitudinal cracking of wet compacted core may appear
> 100	No uncracked dam
≥ 140	Serious cracking of asphaltic concrete facing
> 200	All dams exhibit transverse cracking
≥ 270	Cracking leading to failure of dam

In a road embankment, post-constructive settlements may produce distortions in the grade, cracks on the road surface and steps between the road and structures.

The need for a method to calculate creep in embankments is more than merely a matter of academic interest, as creep settlements up to 1 m have been measured in some high modern embankments (Justo, 1973; Justo and González, 1986).

Post-constructive settlements are not the only consequences of creep. In using elastic methods for building face dams, the modulus applied for water load is 1.6 to 4.0 times greater than it is during construction (Cooke, 1984; Justo et al., 1988; Mori and Pinto, 1988) which may be partly a result of creep. Finally, if creep is not included, calculated stresses produced by water load in the concrete face of dams may show opposite sign to what measured stresses indicate (Central Board of Irrigation and Power, 1992; Justo et al., 1995) apart from other major discrepancies. When creep is not taken into account there will be an underprediction of settlement even if elasto-plastic methods are utilized (Justo, 1979).

Several authors have proposed linear relationships between creep strain and log of time in soils (Buisman, 1936; Sowers et al., 1965; Singh and Mitchell, 1968; Parking, 1991; Central Board of Irrigation and Power, 1992). This semilogarithmic law implies that strain will tend to infinity when time tends to infinity. However, at least in some cases, it seems that settlement finally ceases (v. Holtz and Hilf, 1961; Marsal et al., 1965; Singh and

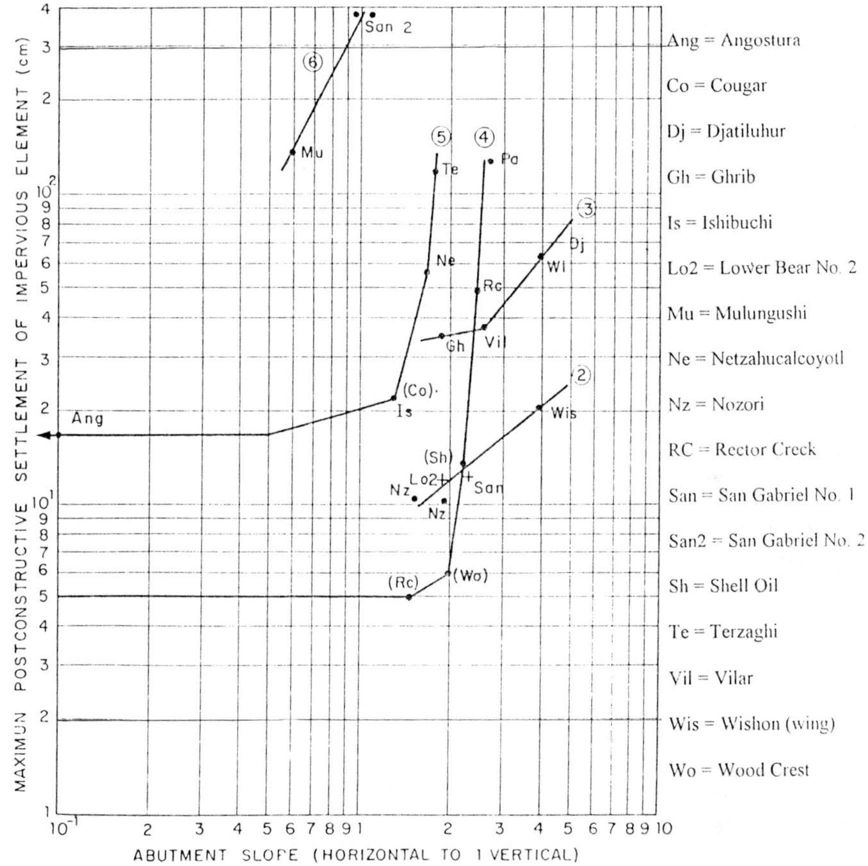

Fig. 1 *Transverse cracking of impervious element produced by deformations:*
(+)Without transverse cracks. (·) With transverse cracks
(1)No cracks
(2)Cracking of reinforced concrete facing without perimetral joint
(3)Cracking of reinforced concrete facing with perimetral joint and of asphaltic concrete facing (self-healed)
(4)Cracking of earth dry compacted core
(5)Cores compacted not wetter than +1% of optimum may crack
(6)Serious cracking

Mitchell, 1968; Justo, 1968; Pinto et al., 1985). Results are also dependent on the uncertain time at which creep may start. In addition, for very high pressure from 4 to 10 MPa (Vesic and Clough, 1968) or in some rockfill dams, such as the Australian Wilmot Dam (Parkin, 1991), the semilog law does not apply at all.

A FE method based upon the semilog law has shown good accordance with vertical settlement, but is not effective for predicting displacement and stresses in upstream face and downstream slope (Central Board of Irrigation and Power, 1992).

In rockfill dams, as creep is activated by the rockfill's weight, several authors (Boughton, 1970; Justo et al., 1995) have represented creep as a percentage of extra weight ranging from 16 to 36%. Although there are no theoretical grounds for using this method, the results obtained are valid. However a disadvantage is that it is not possible to determine this percentage beforehand.

For all these reasons, checking if viscoelastic rheological models could fit the measured strains in rockfill might prove to be illuminating. Extensive measurements of settlements in rockfill dams occurring under virtually one-dimensional conditions have shown that some of these models are effective in predicting settlement. For the sake of concision, only a small part of these published results will be included here.

THE APPLICATION OF RHEOLOGICAL MODELS TO CREEP ESTIMATE IN SOILS

Theory

Gibson and Lo (1961) applied the rheological model of Fig. 2(a) known as the standard linear solid (a linear spring in series with a Kelvin model) to the study of clay consolidation with secondary compression. They claim that there is a good agreement with experimental results in normally consolidated clays. Christie (1964) has shown that the results of this theory coincide with the equations presented in 1939 by Merchant.

In constrast, Christensen and Wu (1964) applied a system formed by a spring and a Maxwell model in parallel (Figure 2b) to the same problem. They claim that there is an accordance with the experimental results.

Figure 2 shows the differential equations corresponding to the two models, as well as the expressions of both the initial or elastic modulus, E_o, and the final modulus, E_f. It is well known that the two models of this figure lead to identical results for creep and relaxation if the solutions are expressed as functions of the fundamental parameters indicated below.

In a creep test (σ_o = constant):

$$\varepsilon = \frac{\sigma_o}{E_o}\left[R_c - \left(R_c - 1\right)\exp\left(-t / T_{cr}\right)\right] \tag{1}$$

where

R_c = creep ratio = E_o/E_f

T_{cr} = creep time, or time necessary to reach a 63.2% (1-e^{-1}) of the total creep strain

In a relaxation test (ε_o = constant):

$$\sigma = E_o\,\varepsilon_o = \left[\frac{1}{R_c} + \left(1 - \frac{1}{R_c}\right) exp\left(-t/T_r\right)\right] \tag{2}$$

where

T_r = relaxation time, or time necessary to reach 63.2% ($1\text{-}e^{-1}$) of the total stress relaxation

The expressions for the relaxation time as a function of the rheological parameters of the two models are indicated in Fig. 2.

In the model of Fig. 1(b), relaxation and creep times coincide, but in the Fig. 2(a) model both are related by the expression:

$$T_{cr} = T_r R_c \tag{3}$$

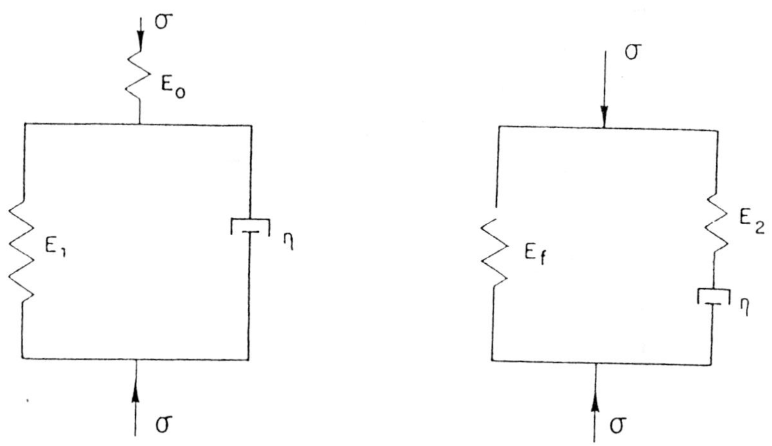

$$\sigma + T_r \frac{\partial \sigma}{\partial t} = E_f \varepsilon + E_o T_r \frac{\partial \varepsilon}{\partial t}$$

$$\frac{1}{E_f} = \frac{1}{E_o} + \frac{1}{E_1}$$

$$T_r = \eta \frac{E_o - E_f}{E_o^2}$$

a)

$$\sigma + T_r \frac{\partial \sigma}{\partial t} = E_f\left(\varepsilon + T_r \frac{\partial \varepsilon}{\partial t}\right)$$

$$E_o = E_f + E_l$$

$$T_r = \frac{\eta}{E_o - E_f}$$

b)

Fig. 2 *Differential equations for viscoelastic rheological models*
a) Standard linear solid (elastic—Kelvin model)
b) Kelvin—Maxwell model (elastic—Maxwell model)

The models in Fig. 2 can suffer one-dimensional or three-dimensional deformation. In the former case, E_o should be substituted by the oedometric modulus E_{oed}.

Zienkiewicz et al. (1968) have applied the model of Fig. 2(a) to three-dimensional problems, although they do not employ the parameters R_c and T_r. Three-dimensional applications of the Fig. 2 models to rockfill dams will be presented by the authors in a forthcoming paper.

Application to an embankment

In a creep test the time factor is defined as:

$$T = t/T_{cr} \tag{4}$$

Substituting into expression (1):

$$\varepsilon = \frac{\sigma_o}{E_o}\left[R_c - (R_c - 1)e^{-T}\right] \tag{5}$$

The height of the embankment increases with the time factor according to Fig. 3. For the moment, let us assume that the compression is one-dimensional, and the stress at a point is the product of the unit weight by the height of fill above it.

The settlement of a point, s_c, at height z_s at the end of construction has been found by Justo and Durand (2000):

$$\frac{s_c E_{oed}}{\gamma} = R_c Z_I z_I - (R_c - 1)\left[b - e^{-T_1}\, I_3(0, T_I)\right] \tag{6}$$

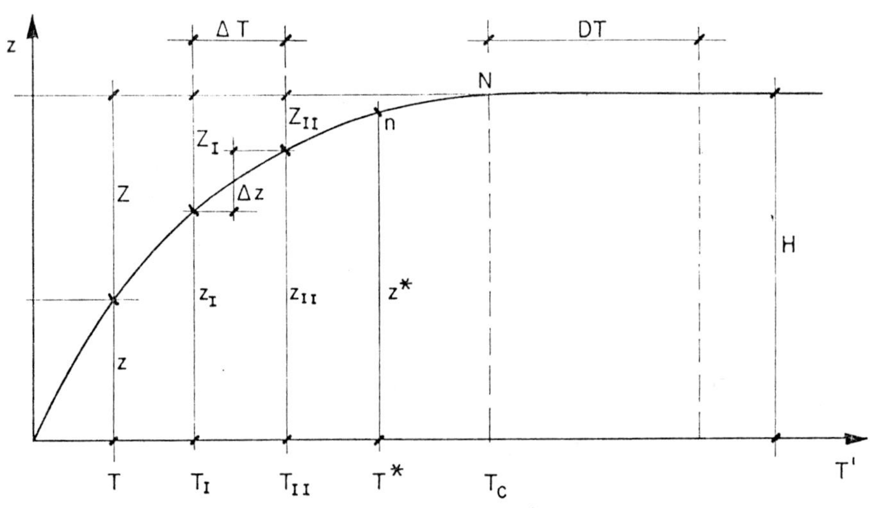

Fig. 3 *Height of embankment versus time factor.*

where

$$b = e^{-T_c}\left[z_I I_1(T_I, T_c) + I_3(0, T_I)\right] \tag{7}$$

$$I_1(T_I, T_c) = \int_{T_1}^{T_c} e^T \frac{dz}{dT}\, dT = \sum_{i=I}^{N}\left(e^{T_i}\, \Delta z_i\right) \tag{8}$$

$$\Delta z_i = \frac{z_{i+1} - z_{i-1}}{2} \qquad z_{I-1} = z_I \qquad z_{N+1} = z_N \tag{9}$$

$$I_3(0, T_I) = \int_0^{T_1} z e^T \frac{dz}{dT}\, dT = \sum_{i=1}^{I}\left(z_i e^{T_i}\, \Delta z_i\right) \qquad z_{I+1} = z_I \tag{10}$$

The 'strain' of a layer during construction may be measured from the settlements of two plates placed at the bottom (I) and top (II) of the layer (Figure 3). The compression of the layer can only be determined from the time when the upper layer is placed. The compression of the layer Δs from time T_{II} up to T^* (Fig. 3) has been calculated by Justo and Durand (2000):

$$\frac{\Delta s E_{oed}}{\gamma} = R_c(z^* - z_{II})\,\Delta z - (R_c - 1)\left[e^{-T^*}\Delta z I_1(T_{II}, T^*) + a\Delta z\left(e^{-T^*} - e^{-T_{II}}\right)\right] \tag{11}$$

where

$$a = \frac{I_3(T_I, T_{II}) - z_I I_1(T_I, T_{II})}{\Delta z} \tag{12}$$

$$I_3(T_I, T_{II}) = \int_{T_1}^{T_{II}} z e^T \frac{dz}{dT}\, dT = \sum_{i=I}^{i=II}\left(z_i e^{T_i}\, \Delta z_i\right) \tag{13}$$

$$I_1(T_I, T_{II}) = \int_{T_1}^{T_{II}} e^T \frac{dz}{dT}\, dT = \sum_{i=I}^{i=II}\left(e^{T_i}\, \Delta z_i\right) \tag{14}$$

$$\Delta z_i = \frac{z_{i+1} - z_{i-1}}{2} \qquad z_{I-1} = z_I \qquad z_{II+1} = z_{II} \tag{15}$$

$$I_1(T_{II}, T^*) = \int_{T_{II}}^{T^*} e^T \frac{dz}{dT}\, dT = \sum_{i=II}^{N}\left(e^{T_i}\, \Delta z_i\right) \qquad z_{II-1} = z_{II} \qquad z^{*+1} = z^* \tag{16}$$

The 'strain' of a finite layer during construction is defined as:

$$\varepsilon = \varepsilon\left(z_I, z_{II}\right) = \frac{\Delta s}{\Delta z} \tag{17}$$

From Eqns. (6) and (11):

$$\frac{\varepsilon E_{oed}}{\gamma} = R_c\left(z^* - z_{II}\right) - \left(R_c - 1\right)\left[e^{-T^*} I_1\left(T_{II}, T^*\right) + a\left(e^{-T^*} - e^{-T_{II}}\right)\right] \tag{18}$$

The 'strain' of a finite layer (z_I to z_{II} in Fig. 3) after construction will be:

$$\Delta\varepsilon = \frac{\Delta s\left(z_I, z_{II}\right)}{\Delta z} = \frac{\Delta s\left(z_{II}\right) - \Delta s\left(z_I\right)}{\Delta z} \tag{19}$$

$$\frac{\Delta\varepsilon E_{oed}}{\gamma\left(R_c - 1\right)} = e^{-T_c}\left(1 - e^{-DT}\right)\left[I_1\left(T_{II}, T_c\right) + a\right] \tag{20}$$

where a is obtained from equation (12) and $I_1(T_{II}, T_c)$ from equations (14) and (15), exchanging I by II and II by c or N.

The settlement of a point at height z_I after construction, Δs, has been obtained by Justo and Durand (2000) and is:

$$\frac{\Delta s E_{oed}}{\gamma\left(R_c - 1\right)} = b\left(1 - e^{-DT}\right) \tag{21}$$

where b is obtained from Eqn. (7):
If the height of the embankment varies linearly with time, then the expressions (6), (17), (20) and (21) become simplified, as indicated below (v. Appendix I).

Settlement, s_c, of a point at height z_I at the end of construction:

$$\frac{s_c E_{oed}}{\gamma} = R_c Z_I z_I + \left(R_c - 1\right)\left(\frac{H}{T_c}\right)^2\left(e^{-T_I} - e^{-T_c}\right) \tag{22}$$

'Strain', ε, of a finite layer during construction:

$$\frac{\varepsilon E_{oed}}{\gamma} = R_c\left(z^* - z_{II}\right) - \left(R_c - 1\right)\frac{H}{T_c}\,\frac{e^{-T^*}\left(e^{T_I} - e^{T_{II}}\right) + 1 - e^{-\Delta T}}{\Delta T} \tag{23}$$

'Strain', $\Delta\varepsilon$, of a finite layer after construction:

$$\frac{\Delta\varepsilon E_{oed}}{\gamma\left(R_c - 1\right)} = \frac{H}{T_c} e^{-T_c}\left(1 - e^{-DT}\right)\left(e^{T_c} - \frac{e^{T_{II}} - e^{T_I}}{\Delta T}\right) \tag{24}$$

Settlement, Δs, of a point at height z_I after construction:

$$\frac{\Delta s E_{\text{oed}}}{\gamma(R_c - 1)} = \left(\frac{H}{T_c}\right)^2 \left[T_I - e^{-(T_c - T_I)} + e^{-T_c}\right]\left(1 - e^{-DT}\right) \tag{25}$$

If the point is at the top of the embankment, then:

$$T_I = T_c \qquad z_I = H \tag{26}$$

and:

$$\frac{\Delta s E_{\text{oed}}}{\gamma(R_c - 1)} = \left(\frac{H}{T_c}\right)^2 \left(T_c - 1 + e^{-T_c}\right)\left(1 - e^{-DT}\right) \tag{27}$$

EXPERIMENTAL VERIFICATION

Instrumentation

Martin Gonzalo is an upstream face rockfill dam (Justo et al., 1988), that suffered a rupture in its membrane during filling of the reservoir. Six batteries of settlement plates (A to F) were placed (Fig. 4) and measurements made during construction and up to six years after construction completion.

Figure 5 shows the strain of the different layers for settlement plates set B for times after the end of fill placement. The accumulated rainfall was drawn in the same graph, showing a clear correlation between the creep and the rainfall rates. If collapse is defined as the compression of a material under load when wetted, then collapse is an important component of creep. It is not possible at present to relate collapse inside a mass of rockfill with the rate of rainfall, so we have included it as part of creep. In this way, calculated creep could, at most, simulate the average trend, although not the sharp steps of true strain.

So as to evaluate the strain of rockfill in a simple way, it has been assumed that in the central part of the dam there are only vertical displacements, and vertical stress corresponds to overburden pressure. In this way, the elastic vertical strain will be:

$$\varepsilon = \frac{\gamma Z}{E_{\text{oed}}} \tag{28}$$

It has been proven in Martin Gonzalo Dam that the vertical strain calculated by Eqn. (28) is similar to the vertical strain obtained from an elastic three-dimensional finite element calculation (Fig. 6). This is a consequence of quasi-oedometric conditions existing near the centre of the dam.

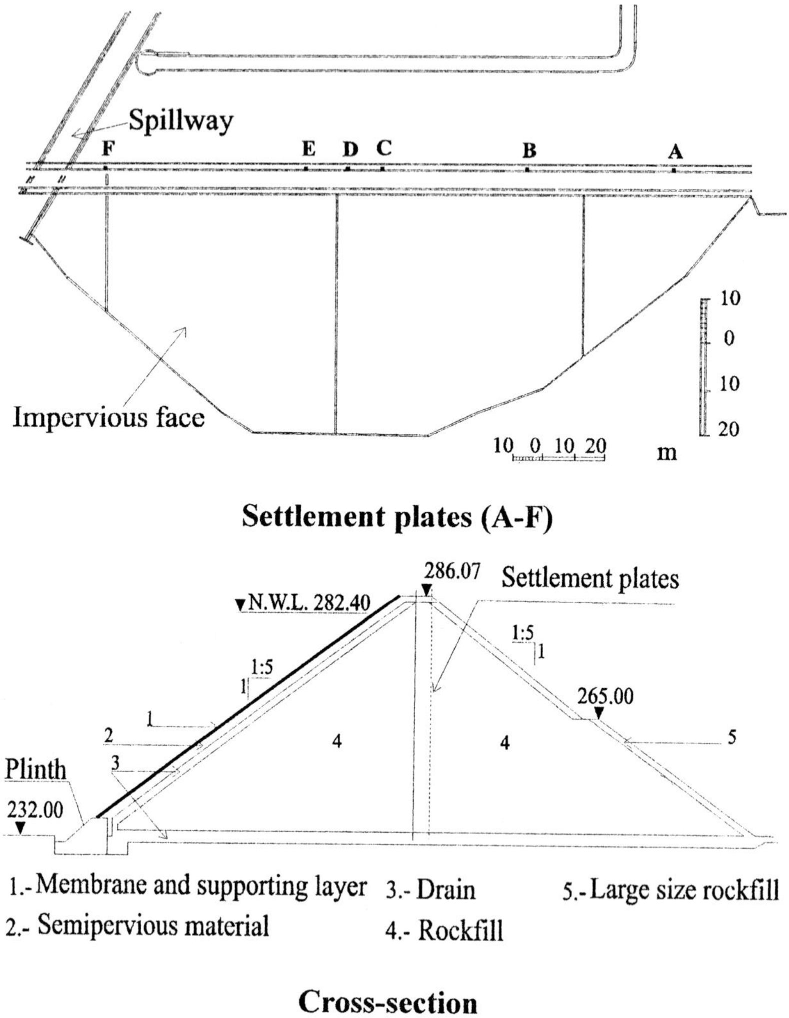

Fig. 4 *Martin gonzalo rockfill dam. Lay-out of settlement plates.*

Determining the parameters

The four parameters that define the three-dimensional model are the Poisson's ratio, ν, the initial or elastic modulus, E_o, the creep ratio, R_c, and the creep time, T_{rc}.

As indicated in section 3.1 quasi-oedometric conditions exist in important zones of the dam. Under these conditions Poisson's ratio follows the equation:

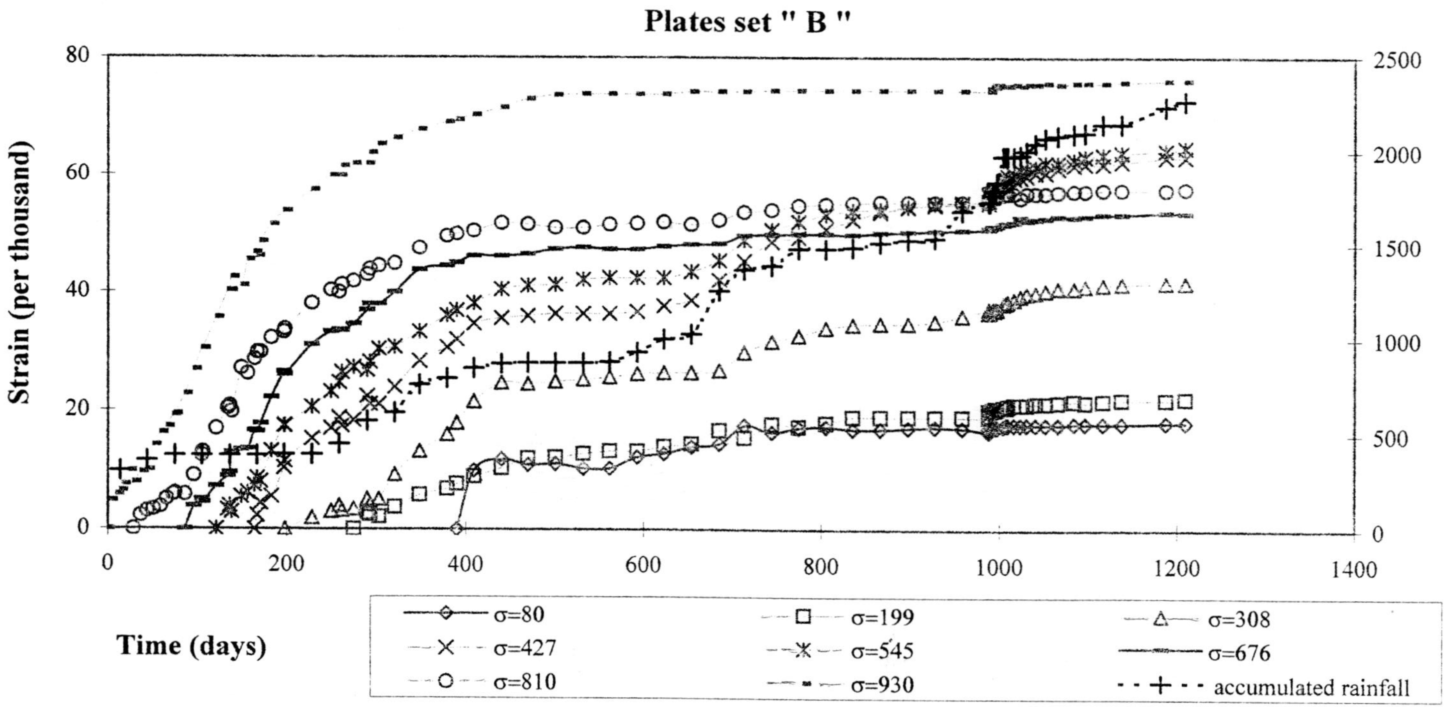

Fig. 5 *Strain of the different layers at settlement plates set B, and rainfall after filling.*
σ = *total stress (kPa) in the upper plate of layer.*

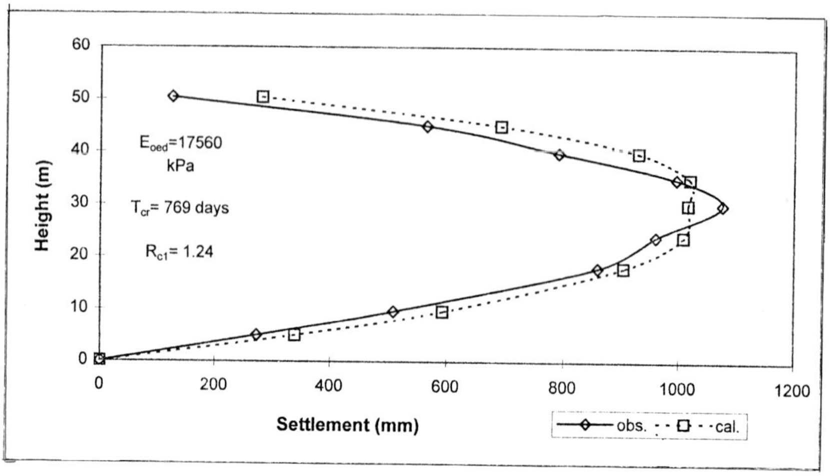

Fig. 6 *Comparison between observed and calculated settlements at the end of construction, assuming one-dimensional deformation. Plates set D.*

$$v = \frac{K_o}{1 + K_o} \tag{29}$$

where K_o (coefficient of pressure at rest) may be obtained from Jaky's formula:

$$K_o = 1\text{-}sin\ \phi' \tag{30}$$

For $\phi' = 36°$ (obtained from triaxial test) $\nu = 0.29$.

Justo and Durand (2000) indicate how the other three parameters (E_o, R_c and T_{rc}) may also be obtained from laboratory tests.

The laboratory determination of the parameters should only be used for preliminary calculations, as scale effects and differences in compaction might exist. In addition, as indicated in Sect. 3.1, creep deformations are dependent upon rainfall, which is difficult to reproduce in the laboratory. For all these reasons the calculation parameters should be corrected through in-situ testing and monitoring of settlements during and after construction. It must be taken into account that a waiting period is sometimes specified before a road pavement is laid to ensure that later settlement will not damage the rolling surface. Settlement measurements taken during this period might be used to adjust the creep parameters.

Justo and Durand (2000) indicate that the elastic modulus E_o may be obtained in field tests or in the embankment during construction by 60 cm in diameter plate loading tests.

If settlements are measured at different heights inside the rockfill, the 'strain' of a layer will permit the calculation of creep parameters for that layer. The ratio of strains after construction at any two different times is (see Eqn. 20):

$$\frac{\Delta\varepsilon(T_c, T_2)}{\Delta\varepsilon(T_c, T_1)} = \frac{1 - e^{-\Delta t_2/T_{cr}}}{1 - e^{-\Delta t_1/T_{cr}}} \tag{31}$$

where Δt is the time increment after construction.

The solution of Eqn. (31) will permit an adjustment for the creep time T_{cr}.

Once T_{cr} has been found, the creep ratio, R_c, can be obtained dividing the expressions for the 'strains' during construction (18) and after construction (20).

For $T^* = T_c$ (v. Fig. 3):

$$\frac{1}{R_c} = 1 - \frac{Z_{II}}{e^{-T_c}\left(I, I_1(T_{II}, T_c) + a\right)\left(1 + \dfrac{1-e^{-DT}}{\Delta\varepsilon}\right) - ae^{-T_{II}}} \tag{32}$$

where the subindex c indicates the end of construction.

If the height of the embankment varies linearly with time, then the expressions (23) and (24) should be divided. The corresponding expression is:

$$\frac{1}{R_c} = 1 - \frac{T_c - T_{II}}{\dfrac{\varepsilon_c}{\Delta\varepsilon}\left(1-e^{-DT}\right)\left(1-e^{-T_c}\dfrac{e^{T_{II}} - e^{T_I}}{\Delta T}\right) + \dfrac{1-e^{-T_c}\left(e^{T_{II}} - e^{T_I}\right) - e^{\Delta T}}{\Delta T}} \tag{33}$$

Once T_{cr} and R_c have been found, Eqn. (18) may be used to correct the value of E_o obtained from plate loading tests. For $T^* = T_c$:

$$E_{oed} = \frac{\gamma}{\varepsilon_c}\left(R_c Z_{II} - (R_c - 1)\left[e^{-T_c} I_1 I,(T_{II}, T_c) + a\left(e^{-T_c} - e^{-T_{II}}\right)\right]\right) \tag{34}$$

And the elastic modulus will be:

$$E_o = E_{oed}\frac{1 - \nu - 2\nu^2}{1 - \nu} \tag{35}$$

If the settlements are measured at only one height, the average creep parameters for the embankment can only be obtained up to that height. Creep time may be determined using the ratio of equation (21) for two different times:

$$\frac{\Delta s(z_I, T_2)}{\Delta s(z_I, T_1)} = \frac{1 - e^{-\Delta t_2/T_{cr}}}{1 - e^{-\Delta t_1/T_{cr}}} \tag{36}$$

From Eqns. (6) and (21):

$$\frac{1}{R_c} = 1 - \frac{Z_I z_I}{b\left(1 + \left(1 - e^{-DT} / \Delta s\right) s_c\right) - e^{T_1} I_3\left(0, T_I\right)} \tag{37}$$

From Eqn. (6):

$$E_{oed} = \frac{\gamma}{s_c}\left(R_c Z_I z_I - \left(R_c - 1\right)\left[b - e^{-T_1} I_3\left(0, T_I\right)\right]\right) \tag{38}$$

If the height of the embankment varies linearly with time, then the expressions (34), (37) and (38) become:

For $T^* = T_c$:

$$E_{oed} E_o = \frac{\gamma}{\varepsilon_c}\left(R_c Z_{II} - \left(R_c - 1\right)\frac{H}{T_c}\frac{e^{-T_c}\left(e^{T_I} - e^{T_{II}}\right) + 1 - e^{-\Delta T}}{\Delta T}\right) \tag{39}$$

$$\frac{1}{R_c} = 1 - \frac{\left(T_c - T_I\right) T_I}{\frac{s_c}{\Delta s}\left[T_I - e^{-(T_c - T_I)} + e^{-T_c}\right]\left(1 - e^{-DT}\right) - e^{T_I} + e^{-T_c}} \tag{40}$$

$$E_{oed} E_o = \frac{\gamma}{s_c}\left(R_c Z_I z_I - \left(R_c - 1\right)\left(\frac{H}{T_c}\right)^2 \left(e^{-T_c} - e^{-T_I}\right)\right) \tag{41}$$

Application to Martin Gonzalo Dam

The T_{cr}, R_c and E_{oed} parameters for every layer have been found from field records, as indicated in Sec. 3.2. The presence of collapse after rainfall produces differences in the T_{cr} and R_c values obtained at different times after construction. There is also scattering of the three parameters from layer to layer due to the variability of rockfill. However, as there is no clear trend in the variation of either T_c or T_r with respect to time, median values have been taken.

The variation of T_r with respect to pressure is complex, but the median value has also been selected for each set of plates.

On the other hand E_{oed} obtained from Eqn. (34) is constant during most of the pressure range (except for very low pressure values) indicating linear behaviour, and again the median value for each set of plates has been selected.

R_c is stress dependent, and the following median values have been selected (v. Justo and Durand, 2000):

For $\sigma < 745$ kPa $\quad R_{c2} = 2.88$

For $\sigma = 745$ kPa $\quad R_{c1} = 1.34$

If the embankment has a creep ratio R_{c1} up to height h (reached at time factor, T_h) and R_{c2} in the remainder, the non dimensional settlement at the end of construction is calculated in the following way:

For $z_I < h$ $\quad$ by Eqn. (6) with $R_c = R_{c1}$

For $z_I > h$ $\quad$ by the following equation:

$$\frac{s_c E_{oed}}{\gamma} = Z_I\left[h\left(R_{c1}-R_{c2}\right)+R_{c2}z_I\right]+h\left[e^{-T_c} I_1\left(T_h, T_c\right)-\right.$$
$$\left.e^{-T_I} I_1\left(T_h, T_I\right)\right]\left(R_{c2}-R_{c1}\right)-\left(e^{-T_c}-e^{-T_I}\right)\left[\left(R_{c1}-1\right)I_3\left(0, T_h\right)+\right. \tag{42}$$
$$\left.\left(R_{c2}-1\right)I_3\left(T_h, T_I\right)\right]-\left(R_{c2}-1\right)z_I\, e^{-T_c} I_1\left(T_I, T_c\right)$$

where $I_1(T_h,T_c)$ and $I_1(T_h,T_I)$ are obtained with equations similar to (8), $I_3(0,T_h)$ and $I_3(T_h,T_I)$ with equations similar to (10).

And settlement after construction:

For $z_I \leq h$ $\quad$ by equation (21) with $R_c = R_{c1}$

For $z_I > h$ $\quad$ by the following equation:

$$\frac{E_o \Delta s(z_I)}{\gamma} = e^{-T_c}\left(1-e^{-DT}\right)\left(\left[R_{c2}-1\right]\left[z_I I_1\left(T_I, T_c\right)+I_3\left(0, T_I\right)\right]-\right.$$
$$\left.\left[R_{c2}-R_{c1}\right]\left[hI_1\left(T_h, T_c\right)+I_3\left(0, T_h\right]\right)\right) \tag{43}$$

where the integrals $I_1(T_I,T_c)$ and $I_3(0,T_I)$ are obtained from equations (8) and (10), and $I_1(T_h,T_c)$ and $I_3(0,T_h)$ from equations similar to (8) and (10) respectively.

This discontinuity appears only in soft rockfill, where there is a critical pressure σ_c above which the creep ratio strongly decreases.

The median value of the oedometric modulus E_{oed} for the whole dam, obtained from equation (34), is 17,109 kPa. With $\nu = 0.29$ it corresponds to a modulus of elasticity of 13,056 kPa. The average elasticity modulus obtained during construction in 60 cm diameter plate loading tests was 13,240 kPa, very close to the value obtained from settlement records interpreted following the viscoelastic theory.

Figure 6 shows measured and calculated settlements at the end of construction obtained by equations (6) and (42) at plates set D. The agreement is good. Figure 7 shows measured and calculated settlements at different plates (set D) after construction obtained with equations (21) and (43). The agreement is acceptable, taking into account the heterogeneity of rockfill, although the measured settlement is more irregular due to rainfall-induced collapse.

This agreement for each layer of the dam, using single values of the parameters for the whole plates set, suggests that the viscoelastic model is consistent with the one-dimensional compression of coarse granular materials. The same result has been obtained in other plates sets (v. Justo and Durand, 2000) and other dams (v. Table 2).

Table 2 shows the average values of the creep ratio and creep time from several materials and dams. The average creep ratio may range from 1.07 to 2.88. The creep time ranges from 125 (quick creep) to 1667 days (slow creep). The critical pressure σ_c existing in the Martin Gonzalo rockfill and Yeguas transition does not exist in other similar materials.

Once the parameters, ν, E_o, R_c and T_{cr} have been estimated assuming oedometric conditions near the centre of the embankment, they may be introduced in the three-dimensional finite elements calculation indicated at the end of Sec. 2.1. Figure 8 shows measured and calculated settlements at the end of construction at settlement plates set B. As indicated in Sec. 3.3, the elasticity modulus chosen (13056 kPa) corresponds to the median

Table 2

Average creep parameters R_c and T_{cr} at several dams.

Material	*Dam*	T_{cr} *days*	σ *(kPa)*	R_c
Rockfill	Beliche (Portugal)		All	1.28
	Martin Gonzalo (Spain)	800	$<\sigma_c$	2.88
	Martin Gonzalo (Spain)	800	$>\sigma_c$	1.34
Transition	Beliche (Portugal)	1667	All	1.20
	Canales (Spain)	833	All	1.35
	Yeguas (Spain)	143	<700	1.55
	Yeguas (Spain)	143	>700	1.07
Core	Canales (Spain)		All	1.44
	Beliche (Portugal)	1064	All	1.15
	Yeguas (Spain)	125	<700	1.30
	Yeguas (Spain)	125	>700	1.07

σ_c ranges from 600 to 750 kPa

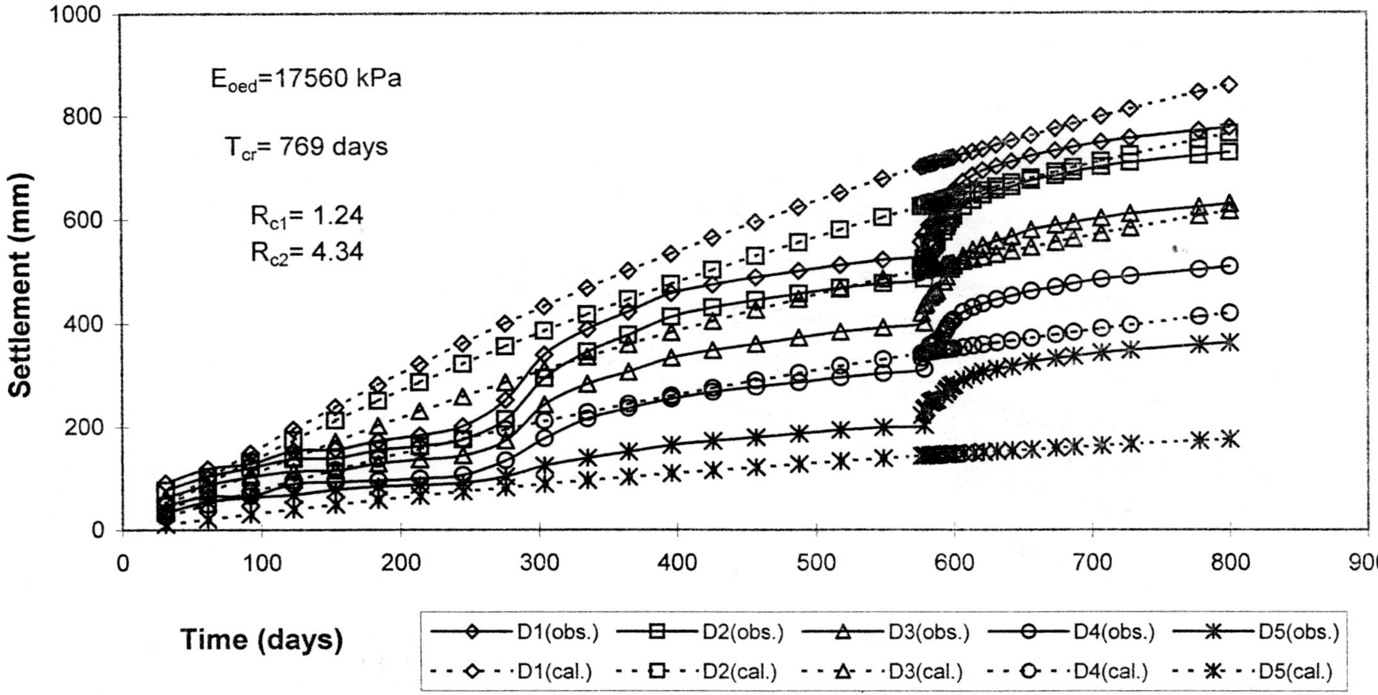

Fig. 7a *Comparison between observed and calculated settlements after construction, assuming one-dimensional deformation. Plates 1 (upper) to 5 (set D).*

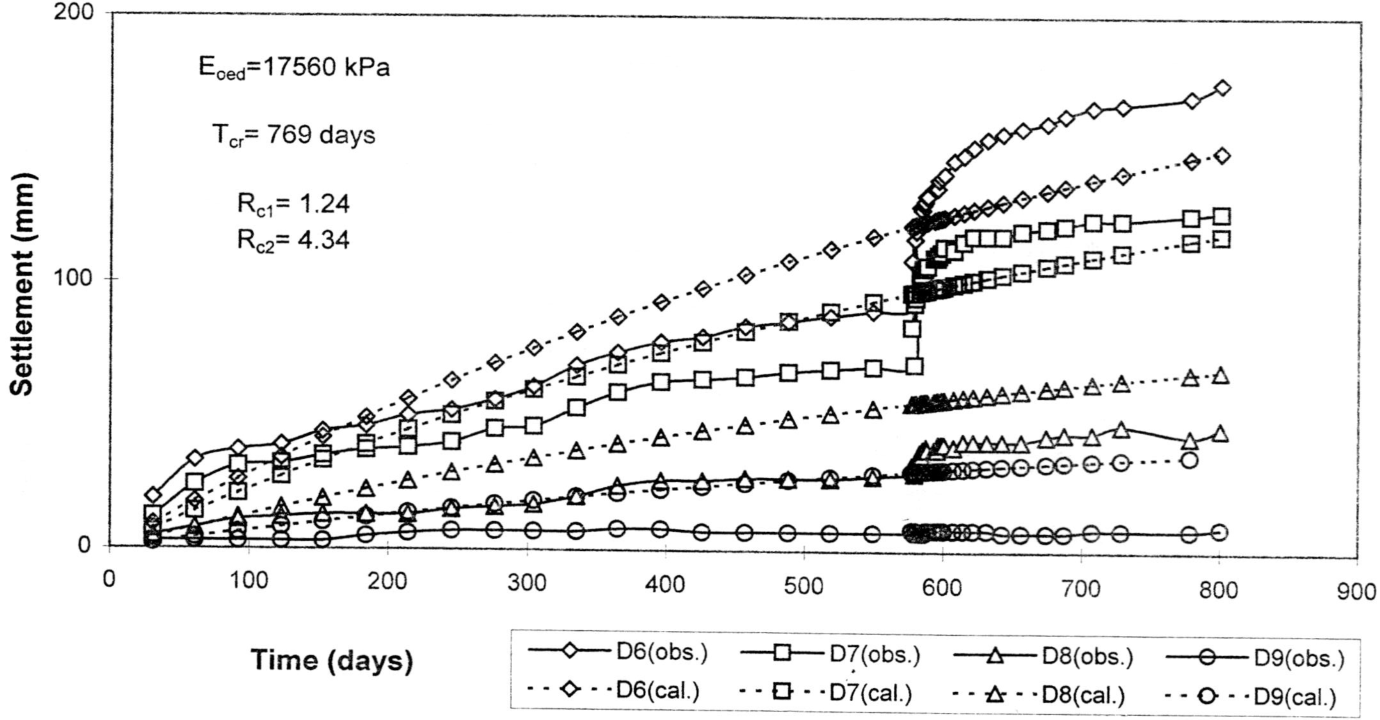

Fig. 7b *Comparison between observed and calculated settlements assuming one-dimensional deformation. Plates 6 to 9 (lower) in set D.*

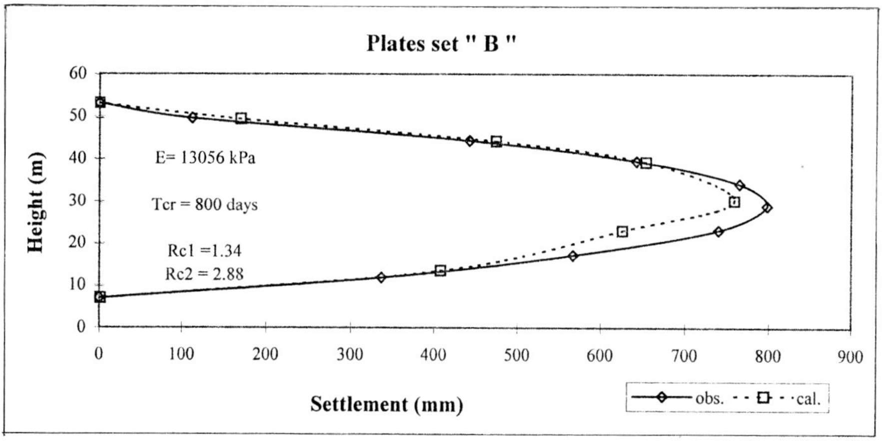

Fig. 8 *Comparison between observed and calculated settlements at the end of construction, assuming three-dimensional deformation. Plates set "B".*

value (in the different plates) of the oedometric modulus (17109 kPa) obtained from the instrumentation by the method indicated in this paper. As shown in the same section, it is nearly equal to the average elasticity modulus obtained during construction in 60 cm diameter plate loading tests. The agreement is also good.

Figure 9 compares the measured and calculated settlements at plates set B at different dates after construction. The agreement is excellent taking into account the heterogeneity of rockfill.

CONCLUSIONS

It has been shown that the use of viscoelastic rheological models to quantify the creep of granular materials leads to a good agreement between measured and calculated settlements taking into account the heterogeneity of rockfill. The two models of Fig. 2 lead to identical results if the solutions are expressed as a function of some fundamental parameters: elasticity modulus, E_o, creep ratio, R_c and creep time T_{cr}.

A preliminary estimate of these parameters may be made from laboratory tests, but they must be corrected by *in-situ* testing and field settlement measurements during and after construction. This is due to scale effects, the fact that in the laboratory the field density and particle breakage are difficult to reproduce, and most importantly the influence of collapse and weathering, especially in soft rockfill.

In three-dimensional calculations, Poisson's ratio must be added to the other three parameters indicated above. It may be estimated by the equation:

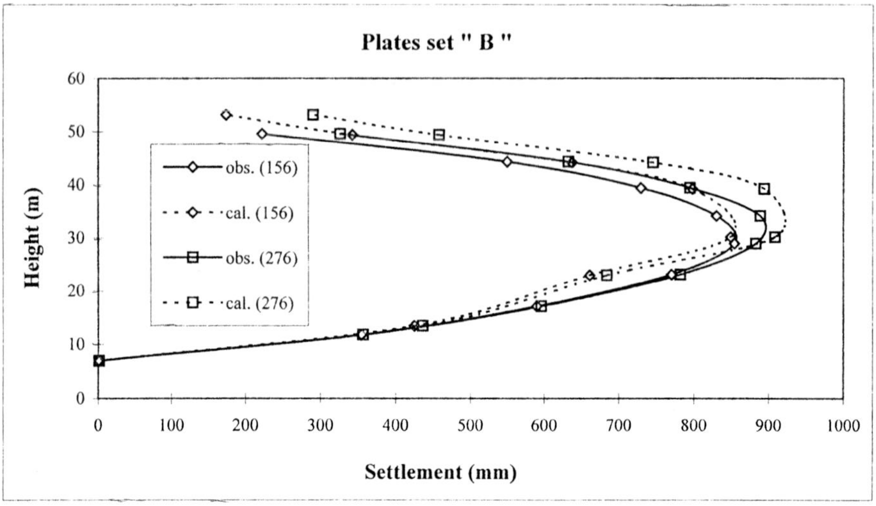

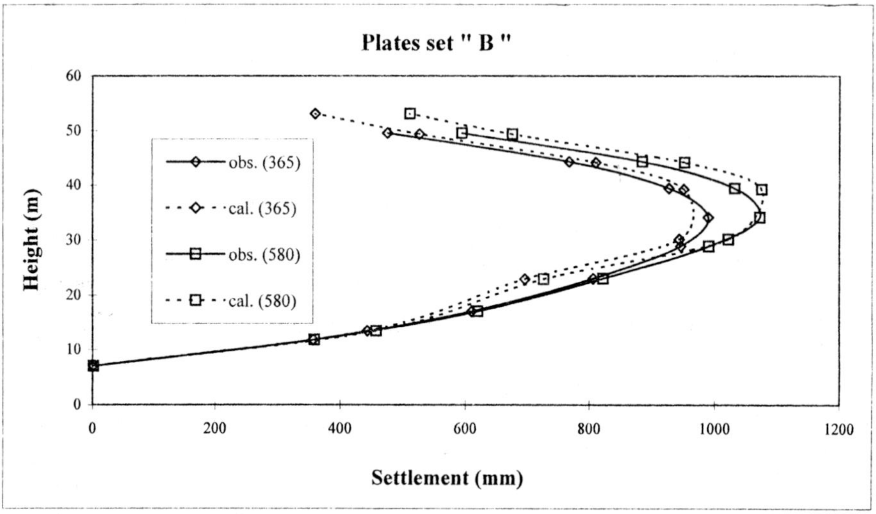

Fig. 9 *Comparison between observed and calculated settlements after construction, assuming three-dimensional deformation. Plates set "B". Inside brackets days elapsed after construction completion.*

$$\nu = \frac{1 - \sin \phi'}{2 - \sin \phi'} \tag{44}$$

The elasticity modulus may be estimated from 60 cm diameter plate loading tests, and may be corrected from Eqns. (34) or (38) and (35) after construction.

The creep ratio T_{cr} can be estimated from settlement measurements after construction through Eqns. (31) or (36).

The creep ratio, $R_{c'}$, may be estimated from settlement measurements upon construction completion, by the Eqns. (32) or (37).

Owing to the heterogeneity of the fill, and to the steps in the settlement-time curve produced by collapse, the parameters must be averaged.

A summary of the creep parameters in several materials and dams is presented in Table 2. Further results may allow preliminary estimates of the creep parameters.

The proposed method allows calculation of the wait-time needed before paving a road embankment or constructing the impervious face of a dam, so that the remaining settlement won't be significant enough to cause cracking.

REFERENCES

Boughton, N.O., 1970, Elastic analysis for behaviour of rockfill, *Journal of Soil Mechanics*. ASCE, 96:SM5: 1715-1733.

Buisman, K., 1936, Results of long duration settlement tests. *Proceedings 1st International Conference on Soil Mechanics & Foundation Engineering*, 1:103-106.

Central Board of Irrigation and Power, 1992, *Rockfill Dams*. Balkema, Rotterdam.

Christensen, R.W. and T.H. Wu., 1964, Analysis of clay deformation as a rate process. *Journal of Soil Mechanics*, ASCE, 90:125-157.

Christie, I.F., 1964, 'A re-appraisal of Merchant's contribution to the theory of consolidation', *Geotechnique*, 14, 309-320.

Cooke, J.B., 1984, Progress in rockfill dams. *Journal of Geotechnical Engineering*, ASCE, 110:10:1383-1413.

Gibson, R.E. and K.Y. Lo, 1961, A theory of consolidation for soils exhibiting secondary compression, *Norwegian Geotechnical Institute*. Publication No. 41.

Holtz W.G., and J.W. Hilf., 1961, Settlement of soil foundations due to saturation, *Proceedings. 5th International Conference on Soil Mechanics & Foundation Engineering*, 1:673-679.

Justo, J.L., 1968, *Deformación de las Presas de Escollera*, Fundación Juan March, Madrid.

Justo, J.L., 1973, The cracking of earth and rockfill dams. *The Transactions of the Eleventh International Congress on Large Dams*, 4:921-945.

Justo, J.L., 1979, Design parameters for special soil conditions. General Report. *Proceedings 7th European Conference on Soil Mechanics & Foundation Engineering*, Brighton, 5:127-158.

Justo, J.L., and A. Gonzalez, 1986, *Revista de Obras Públicas*, Madrid, 173-194.

Justo J.L., and P. Durand, 2000, Settlement-time behaviour of granular embankments. *International Journal of Numerical Analytical Methods in Geomechanics*, 24(3): 281-303.

Justo, J.L., F. Segovia and A. Jaramillo, 1995, Three-dimensional joint elements applied to concrete-faced dams. *International Journal for Numerical & Analytical Methods in Geomechanics*, 19:615-636.

Justo, J.L., P. Cañete, J.L. Manzanares, J. Del Campo and P. De Porcellinis, 1988, *Transactions 16th International Congress on Large Dams*, San Francisco, 2:815-837.

Macedo-Gómez, G., J. Castro-Abonce, L. Montañez-Cartaxo, 2000, Behaviour of Aguamilpa Dam. *Proceedings of the International Symposium on Concrete Faced Rockfill Dams* (J. Barry Cooke Volume), 117-151.

Marsal, R.J., E. Moreno, A. Núñez, R. Cuéllar, Y. R. Moreno, 1965, *Research on the Behaviour of Granular Materials and Rockfill Samples*, Comisión Federal de Electricidad, México D.F.

Mori, R.T. and N.L.S. Pinto, 1988, Analysis of deformations in concrete face rockfill dams. *Transactions 16th International Congress on Large Dams*, San Francisco, 4:27-34.

Parkin, A.K., 1991, Creep of rockfill, in *Advances in Rockfill Structures*, Kluwer Academic Publishers, Boston, 221-238.

Pinto, N.L. de S., P.L. Marques, and E. Filho, 1985, Foz the Areia Dam-Design, Construction and Behaviour, in *Concrete Face Rockfill Dams-Design, Construction and Performance*, ASCE, 173-207.

Singh, A. and J.K. Mitchell, 1968, General stress-strain-time function for soils, *Journal of Soil Mechanics*, ASCE, 94: 31-46.

Sowers, G.F., R.C. Williams and T.S. Wallace, 1965, Compressibility of broken rock and the settlement of rockfills. *Proceedings. 6th International Conference on Soil Mechanics & Foundation Engineering*, 2:561-565.

Thomas, H.H., 1976, *The Engineering of Large Dams*. John Wiley & Sons, London, Part 2: p. 409.

U.S. Committee on Large Dams, 1975, *Lessons from Dam Incidents*, ASCE, New York.

Vesic, A.S. and G.W. Clough, 1968, Behaviour of granular materials under high stresses, *Journal of Soil Mechanics*, ASCE, 94: 661-688.

Zienkiewicz, O.C., M. Watson, and I.P. King, 1968, A numerical method of visco-elastic stress analysis. *International Journal of Mechanical Sciences*, 10:807-827.

Appendix 1

Settlements when the embankment height increases linearly with time

In this case referring to Fig. 3:

For $T \leq T_c$ $$z = \frac{H}{T_c} T \qquad \text{(Ap. 1.1)}$$

For $T \geq T_c$ $$z = H \qquad \text{(Ap. 1.2)}$$

Equation (8) becomes:

$$I_1(T_I, T_c) = \frac{H}{T_c} \int_{T_1}^{T_c} e^T \, dT = \frac{H}{T_c}\left(e^{T_c} - e^{T_I}\right) \qquad \text{(Ap. 1.3)}$$

And Eqn. (10):

$$I_3(0, T_I) = \left(\frac{H}{T_c}\right)^2 \int_0^{T_1} Te^T \, dT$$

Integrating by parts:

$$I_3(0, T_I) = \left(\frac{H}{T_c}\right)^2 \left(T_I e^{T_1} - e^{T_1} + 1\right) \qquad \text{(Ap. 1.4)}$$

Equation (7) gives:

$$b = \left(\frac{H}{T_c}\right)^2 e^{-T_c}\left(T_I e^{T_c} - e^{T_I} + 1\right) \qquad \text{(Ap. 1.5)}$$

Substituting into (6):

$$\frac{s_c E_{oed}}{\gamma} = R_c Z_I z_I + (R_c - 1)\left(\frac{H}{T_c}\right)^2 \left(e^{-T_I} - e^{T_c}\right) \qquad \text{(Ap. 1.6)}$$

$$I_1(T_{II}, T^*) = \frac{H}{T_c}\left(e^{T^*} - e^{T_{II}}\right) \qquad \text{(Ap. 1.7)}$$

$$I_3(T_I, T_{II}) = \left(\frac{H}{T_c}\right)^2 \int_{T_I}^{T_{II}} Te^T \, dT = \left(\frac{H}{T_c}\right)^2 \left(T_{II} e^{T_{II}} - T_I e^{T_I} - e^{T_{II}} + e^{T_I}\right)$$

(Ap. 1.8)

$$I_1\left(T_I, T_{II}\right) = \frac{H}{T_c}\left(e^{T_{II}} - e^{T_I}\right) \qquad \text{(Ap. 1.9)}$$

Substituting (Ap. 1.8) and (Ap. 1.9) into (12):

$$a = \frac{H^2}{T_c^2\, \Delta z}\left(\Delta T e^{T_{II}} - e^{T_{II}} + e^{T_I}\right) \qquad \text{(Ap. 1.10)}$$

Substituting (Ap. 1.7) and (Ap. 1.10) into (18):

$$\frac{\varepsilon E_{oed}}{\gamma} = R_c\left(z^* - z_{II}\right) - \left(R_c - 1\right)\frac{H}{T_c}\frac{e^{-T^*}\left(e^{T_I} - e^{T_{II}}\right) + 1 - e^{-\Delta T}}{\Delta T} \qquad \text{(Ap. 1.11)}$$

$$I_1\left(T_{II}, T_c\right) = \frac{H}{T_c}\left(e^{T_c} - e^{T_{II}}\right) \qquad \text{(Ap. 1.12)}$$

Substituting (Ap. 1.7) for $T^* = T_c$ and (Ap. 1.10) into (20):

$$\frac{\Delta\varepsilon E_{oed}}{\gamma\left(R_c - 1\right)} = e^{-T_c}\left(1 - e^{-DT}\right)\frac{H}{T_c}\left(e^{T_c} - \frac{e^{T_{II}} - e^{T_I}}{\Delta T}\right) \qquad \text{(Ap. 1.13)}$$

Finally, substituting (Ap. 1.5) into (21):

$$\frac{\Delta s E_{oed}}{\gamma\left(R_c - 1\right)} = \left(\frac{H}{T_c}\right)^2 e^{-T_c}\left(T_I\, e^{T_c} - e^{T_I} + 1\right)\left(1 - e^{-DT}\right) \qquad \text{(Ap. 1.14)}$$

9

Soft Ground Improvement by Vertical Drains

Buddhima Indraratna[1], *I.W. Redana*[2] *and C. Bamunawita*[3]

INTRODUCTION

As a result of rapid increase in population and associated development activities taking place in the world, particularly in Southeast Asian countries during the last few decades, most soft clay land has been utilized for construction. The land required for industrial, commercial and residential purposes is found mainly in low-lying areas, where construction activities are now concentrated. Low-lying marshy land comprises highly compressible weak organic and peaty soils of varying thickness. These soft clay deposits have low bearing capacities as well as excessive settlement characteristics. When such low-lying areas are selected for development work, it is essential to raise the ground with fill in order to keep the surface area above the ground water table or flood level. If a fill is placed without proper compaction, a differential settlement may develop due to the heterogeneity of the fill and the compressibility of underlying weak sub-soils.

A common practice to overcome distress in structures built on filled land is to support them on special foundations, which could accommodate differential settlement to a greater degree, or to support them on pile foundations. When strong bearing stratum, is found at large depths between 15-25 m, foundation costs become prohibitively high and not commensurate with the cost of the superstructure, specially in the case of small-scale structures and low rise buildings subject to low to moderate loads. Also, construction problems arise during the construction of railways and highway embankments on compressible soft clays. A more economical solution would be to improve the engineering properties of underlying soft soil rather than use of piled foundations. Preloading is one of the more successful ground improvement techniques that can be used in the low-

[1]Professor of Civil Engineering, University of Wollongong, NSW 2522, Australia.
[2]Senior Lecturer, Dept. of Civil Engineering, Udayana University, Denpasar, Bali, Indonesia.
[3]Research Engineer, Faculty of Engineering, University of Wollongong, NSW 2522, Australia.

lying areas. This technique involves loading of the ground to induce a greater part or whole of the ultimate settlement that the ground is expected to experience under the loads imposed by the development activity. In preloading, a surcharge equal to or greater than expected foundation loading is applied on the ground surface, which allows rapid consolidation under the applied load.

One significant drawback in this technique is that the preload should be applied for a sufficient period of time. Currently, this may not be possible with busy construction schedules. Having vertical drains can reduce the preloading period significantly, because this technique accelerates the consolidation process by reducing the drainage path in the radial direction (Fig. 1). Vertical drains are applicable for moderately to highly compressible soils, which are usually normally consolidated or lightly over consolidated, and for stabilising a deep layer of a soft soil having a low permeability. These soils include organic and inorganic silts and clay with low to moderate sensitivity, varved cohesive soils and decomposed peat.

HISTORY AND DEVELOPMENT OF VERTICAL DRAINS

In the past three decades, sand compaction piles and prefabricated vertical drains have been used extensively for the purpose of soft ground

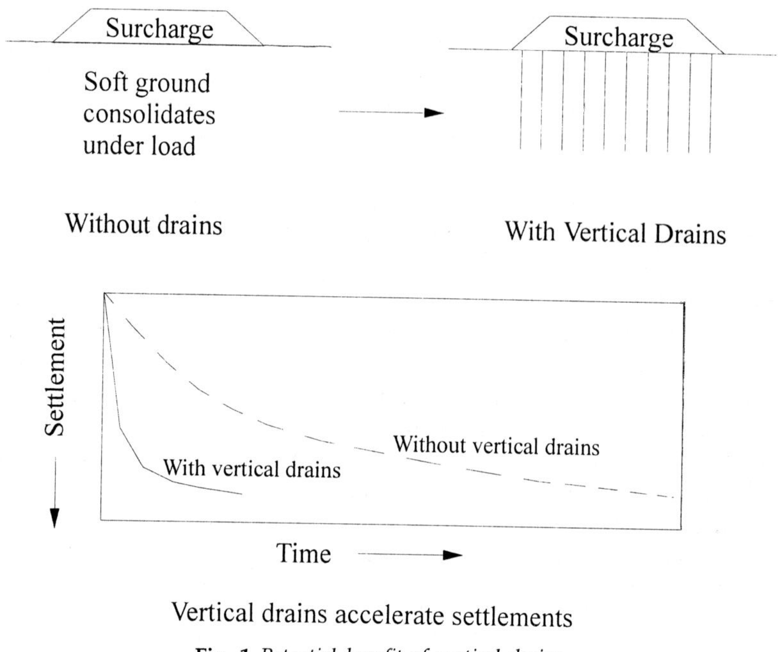

Fig. 1 *Potential benefit of vertical drains.*

improvement. Sand drains were first used in practice in and around the 1920s. The California Division of Highways has conducted the first comprehensive laboratory and field tests in 1933. In Japan, during 1940s the behaviour of vertical sand drains was not well understood, because the bearing capacity of foundations after installation of sand drains was considered to provide sufficient reinforcement, hence, the full load was placed on the foundation too quickly resulting in frequent foundation failure (Aboshi, 1992). In 1956, the method of sand compaction piles was developed in which the vertical sand drains were compacted during installation (Aboshi, 1992). After sand was poured into the pipe (casing), it was withdrawn partway and again driven down to compact the sand column and enlarge its diameter. Although no effect of densification was expected in clayey soils, these sand columns behaved as granular piles in soft ground, thereby carrying a greater load. At the same time, they also worked as vertical drains to accelerate consolidation of clayey ground, by rapid pore pressure dissipation.

Since 1940, pre-fabricated band shaped drains and Kjellman cardboard wick drains have been introduced in ground improvement. Subsequently, several types of pre-fabricated band drains were developed such as: Geodrain (Sweden), Alidrain (England), Mebradrain (Netherlands) etc. Basically, the prefabricated band drains are composed of a plastic core with a longitudinal channel wick functioning as a drain, and a sleeve of paper or fibrous material as a filter protecting the core. The size of the band drains is usually 3.5 mm by 100 mm, or in that order. Typical types of prefabricated vertical drain types are shown in Fig. 2.

INSTALLATION OF PRE-FABRICATED VERTICAL DRAINS

The vertical drains can be installed using two different methods, which are either dynamic or static. In the dynamic method of installation, the mandrel is driven into the ground using either a vibrating hammer or a conventional drop hammer. In the static procedure, the mandrel is pushed into the soil by means of a static force. Static pushing is preferable for driving the mandrel into the ground; whereas the dynamic methods seem to create a higher excess pore water pressure and a greater disturbance of the surrounding soil during installation. The installation rig is shown in Fig. 3, where the vertical drain is protected using a mandrel, which is driven into the soil using installation equipment. The installation procedure results in shear strain accompanied by the increase in total stress and pore water pressure, apart from the displacement of the soil surrounding the mandrel. The degree of disturbance depends on several factors as described below.

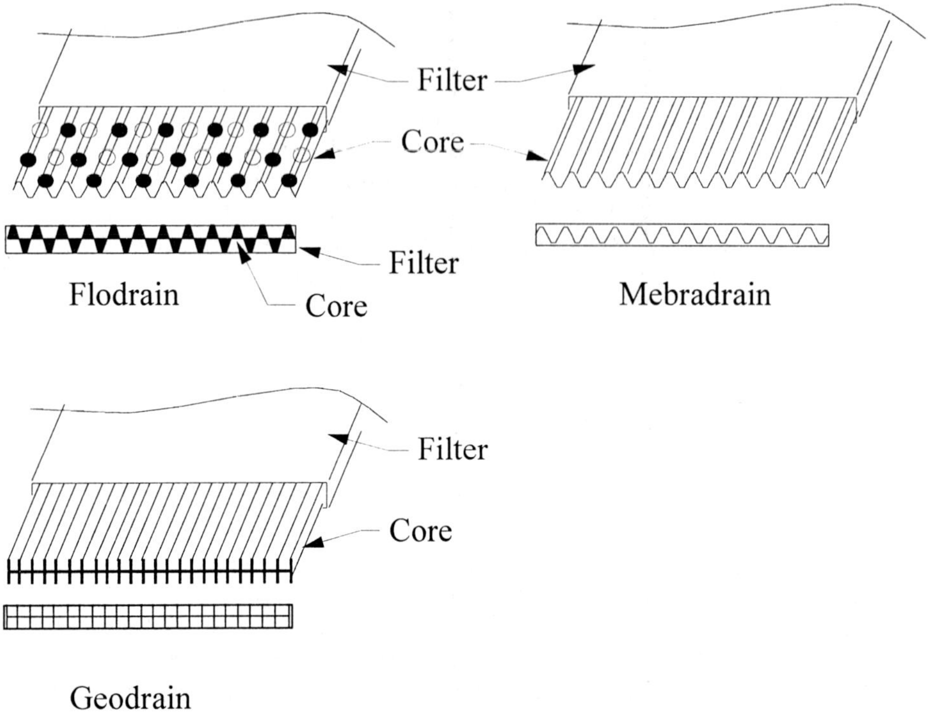

Fig. 2 *Typical types of vertical drains. (Redana, 1999; Rixner et al., 1986)*

Mandrel size and shape

Generally, disturbance increases with the total mandrel cross-sectional area. The mandrel size should be as close as possible to that of the drain to minimize displacement. While working on the effect of mandrel driven drains on soft clays, (Akagi, 1977, 1981), observed that when a closed-end mandrel is driven into a saturated clay, the clay will suffer large excess pore water pressure, associated with ground heave and lateral displacement. The strength and coefficient of consolidation of the surrounding soil can then decrease considerably. However, the excess pore pressure dissipated rapidly, followed by the process of consolidation after installation of the mandrel or before the fill is placed. Bergado et al. (1991) reported from a case study of a Bangkok clay embankment stabilized with vertical drains that a faster rate of settlement took place in the area where the drains were installed using a mandrel with a smaller cross-section area rather than a larger mandrel. This verifies that a smaller smear zone was developed in the former.

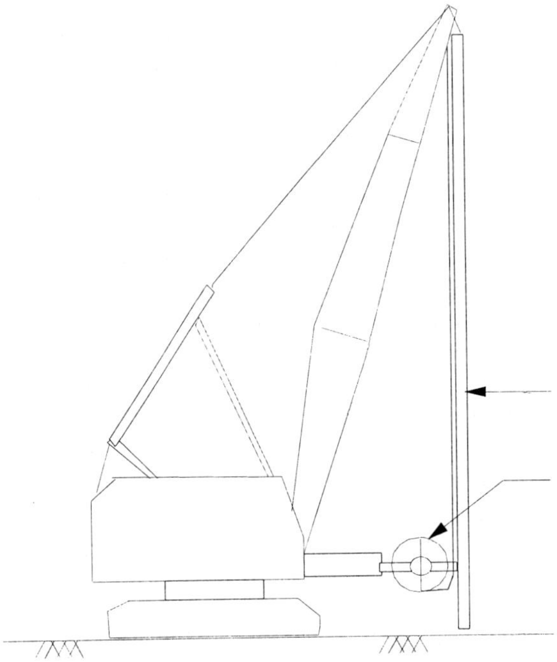

Fig. 3 *Typical installation equipment of vertical drains.*

Soil macro fabric

For soil with pronounced macrofabric, the ratio of horizontal permeability to vertical permeability (k_h/k_v) can be very high, whereas the k_h/k_v ratio becomes unity within the disturbed (smear) zone. The vertical drains are particularly efficient where the clay layers contain many thin horizontal sand or silt lenses (microlayers). However, if these microlayers are continuous in the horizontal direction, the installation of vertical drains may not be effective, since rapid drainage of the pore water out of the soil layers may occur irrespective of whether the drains are installed or not.

FACTORS INFLUENCING THE VERTICAL DRAIN EFFICIENCY

Smear zone

The installation of vertical drains by means of a mandrel causes significant remoulding of the subsoil especially in the immediate vicinity of the mandrel. Thus, a zone of smear will be developed with reduced permeability and increased compressibility. Barron (1948) stated that if the drain wells are installed by driving cased holes and then back filling as the casing is withdrawn, the driving and pulling of the casing would distort and remould the adjacent soil. In varved soils, the finer and more impervious layers will be dragged down and smeared over the more pervious layers, resulting in a zone of reduced permeability in the soil adjacent to the well periphery. As remoulding retards the consolidation process of the subsoil, it has to be considered in any theoretical solution. Barron (1948) suggested the concept of reduced permeability, which is equivalent to lowering the overall value of the coefficient of consolidation. Hansbo (1979) also introduced a zone of smear in the vicinity of the drain with a reduced value of permeability.

The combined effect of permeability and compressibility within the smear zone brought a different behaviour from the undisturbed soil, hence, the prediction of the behaviour of the soil stabilised with vertical drains cannot be made accurately if the effect of smear is ignored. Both Barron (1948) and Hansbo (1981) modelled the smear zone by dividing the soil cylinder dewatered by the central drain into two zones. The smear zone is the region in the immediate vicinity of the drain, which is disturbed, and the other zone is the intact or undisturbed region outside the smear zone. Onoue et al. (1991) introduced a three zone hypothesis defined by: plastic smear zone in the immediate vicinity of the drain where the soil is highly remoulded during the process of installation of the drain, plastic zone where the permeability is reduced moderately, and the undisturbed zone where the soil is not at all affected by the process of drain installation. This three-zone analysis was suggested after extensive laboratory testing by Ting et al. (1990). However, due to the complex variation of permeability in the radial direction from the drain, the solution of the three-zone approach becomes increasingly difficult. For practical purposes, the two-zone approach is generally sufficient.

Jamiolkowski et al.(1981) proposed that the diameter of the smear zone (d_s) and the cross-sectional area of mandrel can be related as follows:

$$d_s = \frac{(5\,\text{to}\,6)\,d_m}{2} \tag{1}$$

where d_m is the diameter of the circle with an area is equal to the cross-sectional area of the mandrel. Based on the results of Holtz and Holm (1973) and Akagi (1979) Hansbo (1987) proposed another relationship as follows:

$$d_s = 2d_m \tag{2}$$

Indraratna & Redana (1998) proposed that the estimated smear zone is about 3-4 times the cross-sectional area of the mandrel. The proposed relationship was verified using the specially designed large-scale consolidometer shown in Fig. 4 (Indraratna & Redana, 1995). The schematic

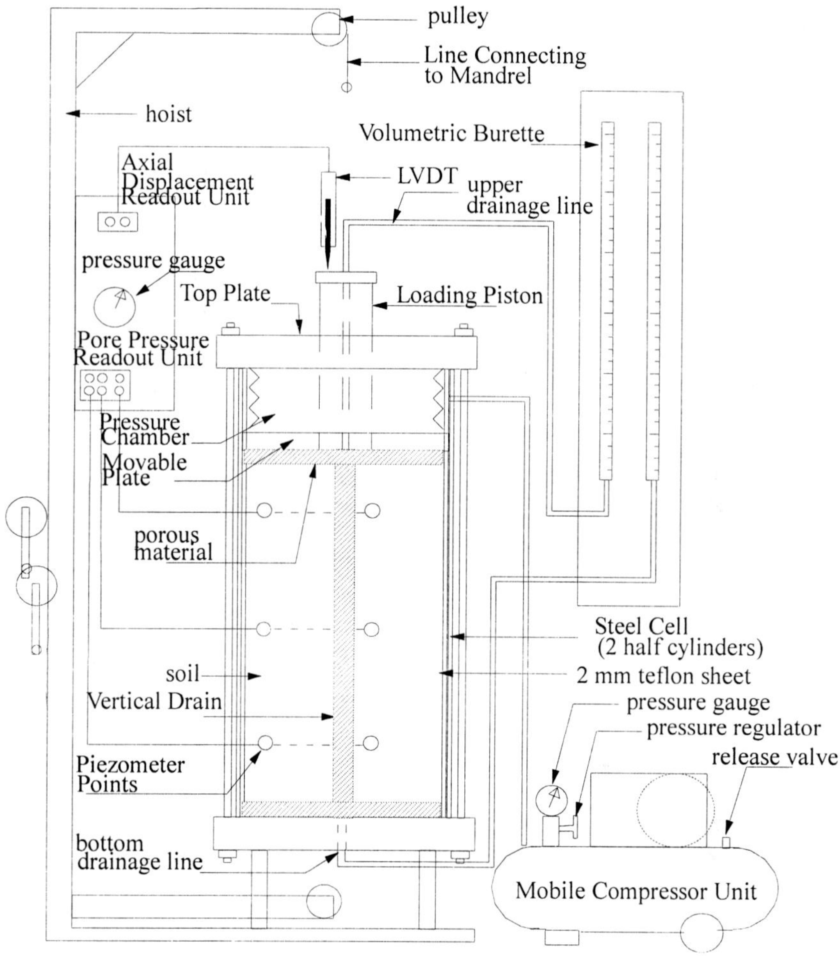

Fig. 4 *Large-scale consolidation apparatus (Indraratna & Redana, 1995).*

section of the consolidometer and the location of the recovered specimen are shown in Figs. 5(a) and 5(b). Figure 6 shows the variation of k_h/k_v ratio along the radial distance from the central distance. According to Hansbo (1987) & Bergado et al. (1991), in the smear zone, the k_h'/k_v' ratio was found to be close to unity, which is in agreement with the results of the study of Indraratna & Redana (1998).

Well resistance

The resistance to the water flowing in the vertical drains is known as the well resistance. The well resistance increases with the increase in the length

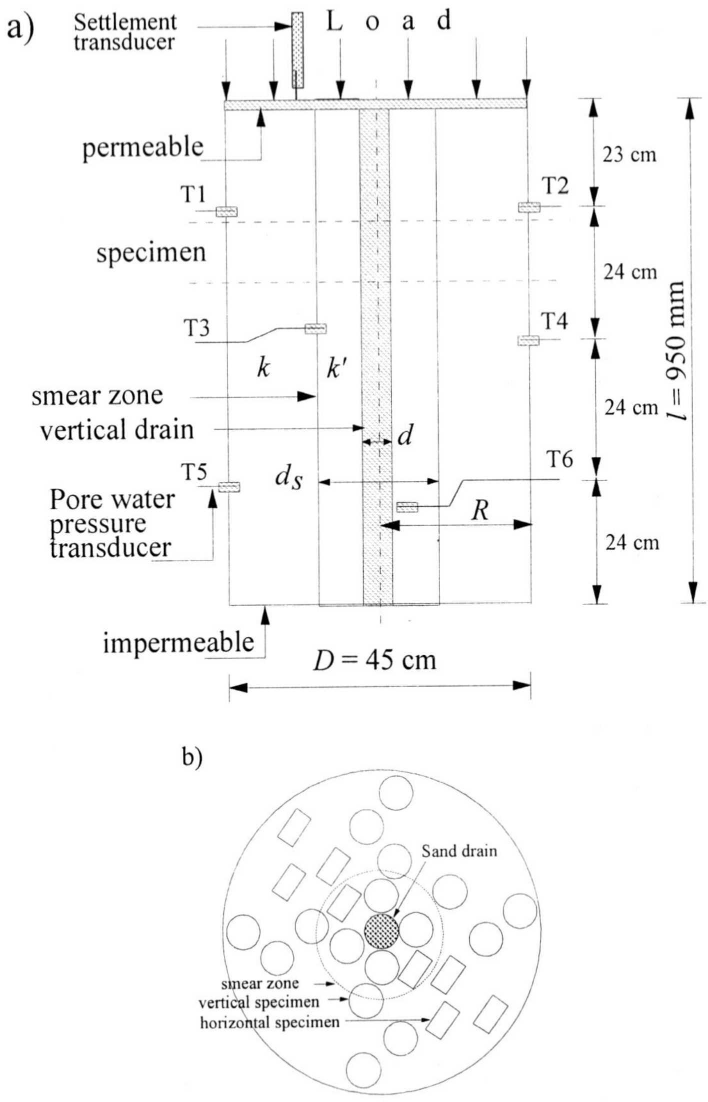

Fig. 5 *a) Schematic section of the test equipment showing the central drain and associated smear and b) locations of small specimens obtained to determine the consolidation and permeability characteristics (Indraratna & Redana, 1998).*

of the drain and reduces the consolidation rate. Well resistance retards the pore pressure dissipation, hence, retards the settlement. The three main factors which increase well resistance, are the deterioration of the drain filter (reduction of drain cross-section), passing of fine soil particles through the filter (reduction of drain cross-section) and folding of the drain because

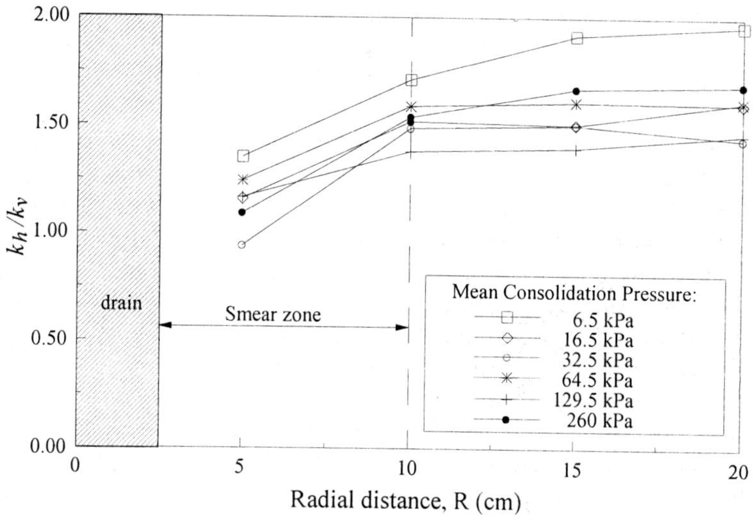

Fig. 6 *Ratio of k_h/k_v along the radial distance from the central drain. (Indraratna & Redana, 1988).*

of large settlement or lateral movement. However, these aspects are still difficult to quantify. Hansbo (1979, 1981) presented a closed form solution, which includes the effect of well resistance on drain performance.

DEVELOPMENT OF VERTICAL DRAIN THEORY

The basic theory of radial consolidation around a vertical sand drain system is an extension of Terzaghi's (1925) one-dimensional consolidation theory. Since it is obvious that the coefficient of consolidation in the horizontal direction is much higher than that in the vertical direction, and that the vertical drains reduce the drainage path considerably in the radial direction, the effectiveness of vertical drains in accelerating the rate of consolidation and improving the strength of soft soil is remarkably improved. The theory of vertical drain was probably first solved by Kjellman (1948). His solution based on *equal vertical strain hypothesis*, was developed on the assumption that horizontal sections remain horizontal throughout the consolidation process. However, Barron (1948) presented the most comprehensive solution to the problem of radial consolidation by drain wells. He studied the two extreme cases of free strain and equal strain and showed that the average consolidation obtained in these cases is nearly the same. The 'free strain hypothesis' assumes that the load is uniform over a circular zone of influence for each vertical drain, and that the differential settlements occurring over this zone have no effect on the redistribution of stresses by arching of the fill load. The 'equal vertical strain hypothesis' on the other hand, assumes that arching occurs in the upper layer during the consolidation process without any differential settlement in the clay layer. The arching effect implies a more or less rigid boundary at the surface of the soil layer being consolidated with vertical drains. It means that the vertical strain is uniform in the horizontal section of the soil.

Barron (1948) also considered the influence of well resistance and smear on the consolidation process due to vertical well drains. Takagi (1957) extended Barron's solution to incorporate a variable rate of loading, whereas Richart (1959) presented a convenient design chart for the effect of smear, where the influence of variable void ratio was also considered. Hansbo (1960) presented a solution by pointing out that the Darcy's law might not be valid when the hydraulic gradient is in the range of magnitudes prevailing during most consolidation processes in practice. However, in this equal strain solution, the effect of smear and well resistance were not considered. Later, a simplified solution to the problem of smear and well resistance was proposed by Hansbo (1979, 1981), giving results almost identical with those given by Barron (1948) and Yoshikuni and Nakanodo (1974). Onoue (1988b) presented a rigorous solution based on the free strain hypothesis.

Equal Vertical Strain Hypothesis (Barron, 1948)

The first conventional procedure for predicting radial consolidation was introduced by Barron (1948). This procedure was based on the extension of the theory of consolidation initially presented by Terzaghi (1925). Barron's solution is based on the following assumptions: (a) all vertical loads are initially carried by excess pore water pressure, u, which means that the soil is saturated, (b) the applied load is assumed to be uniformly distributed and all compressive strain within the soil occurs in the vertical direction, (c) the zone of influence of the drain is assumed to be circular and axisymmetric, (d) permeability of the drain is infinite in comparison with that of the soil, and (e) Darcy's law is valid.

Barron (1948) developed the exact (rigorous) solution of vertical drain based on 'free strain hypothesis' and an approximate solution based on 'equal strain hypothesis'. The difference in the predicted pore water pressures calculated using the free strain and equal strain assumptions are shown to be very small. Therefore, the approximate solution based on the 'equal strain hypothesis' gives satisfactory results compared to the rigorous free strain hypothesis.

Figure 7 shows the schematic illustration of a soil cylinder with a central vertical drain, where, r_w = the radius of the drain, r_s = the radius of smear zone, R = the radius of soil cylinder and l = the length of the drain installed into the soft ground. The coefficient of permeability in the vertical and horizontal directions are k_v and k_h, respectively and k'_h is the coefficient permeability in the smear zone. The three-dimensional consolidation of

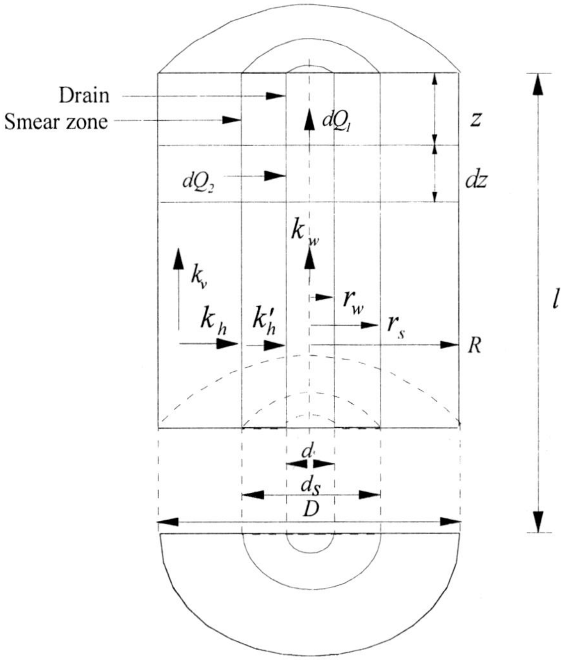

Fig. 7 *Schematic of soil cylinder with vertical drain (after Hansbo, 1979).*

radial drainage (Barron, 1948) is given by:

$$\frac{\partial \bar{u}}{\partial t} = c_v \left(\frac{\partial^2 u}{\partial z^2} \right) + c_h \left(\frac{\partial^2 u}{\partial r^2} + \frac{1}{r} \frac{\partial u}{\partial r} \right) \tag{3}$$

where t is the time elapsed after the load is applied, u is the excess pore water pressure at radius r and at depth z.

For radial flow only, the above equation becomes:

$$\frac{\partial \bar{u}}{\partial t} = c_h \left(\frac{\partial^2 u}{\partial r^2} + \frac{1}{r} \frac{\partial u}{\partial r} \right) \tag{4}$$

The solution of the excess pore pressure for radial flow only, u_r of the above equation based on "equal strain" assumption is given by:

$$u_r = \frac{4\bar{u}}{D^2 F(n)} \left[R^2 \, ln\left(\frac{r}{r_w} \right) - \frac{r^2 - r_w^2}{2} \right] \tag{5}$$

where, D is the diameter of soil cylinder, the drain spacing factor, $F(n)$, is given by:

$$F(n) = \frac{n^2}{n^2 - 1} \ln(n) - \frac{3n^2 - 1}{4n^2} \tag{6}$$

where, $n = \dfrac{R}{r_w}$ is drain spacing ratio, and the average excess pore water pressure is given by:

$$\bar{u}\left(=u_{av}\right) = u_o \exp\left(\frac{-8T_h}{F(n)}\right) \tag{7}$$

The average degree of consolidation, $\overline{U}_h$, in the soil body is given by:

$$\overline{U}_h = 1 - \exp\left(-\frac{8T_h}{F(n)}\right) \tag{8}$$

where the time factor T_h is defined as:

$$T_h = \frac{c_h t}{D^2} \tag{9}$$

The coefficient of radial drainage consolidation, c_h, is represented by:

$$c_h = \frac{k_h(1 + e)}{a_v \gamma_w} \tag{10}$$

where γ_w is unit weight of water and a_v is the coefficient of compressibility of the soil, e is the void ratio, and k_h is the horizontal permeability of the soil.

The solution of Eqn. (4) taking account of smear effect is given by:

$$u_r = \bar{u}_r \frac{1}{\nu}\left[\ln\frac{r}{r_s} - \frac{r^2 - r_s^2}{2R^2} + \frac{k_h}{k_h'}\left(\frac{n^2 - s^2}{n^2}\right)\ln s\right] \tag{11}$$

where the smear factor ν is given by:

$$\nu = F(n, s, k_h, k_h') = \left[\frac{n^2 - s^2}{n^2}\ln\frac{n}{s} - \frac{3}{4} + \frac{s^2}{4n^2} + \frac{k_h}{k_v}\left(\frac{n^2 - s^2}{n^2}\right)\ln s\right] \tag{12}$$

and

$$\bar{u}\left(=u_{av}\right)=u_o\ \exp\left(\frac{-8T_h}{\nu}\right) \tag{13}$$

In the above expression, s is the extent factor of the smear zone with respect to the size of the drain and is given by: $s = r_s/r_w$.

The average degree of consolidation including smear is now given by:

$$\overline{U}_h = 1-\exp\left(-\frac{8T_h}{\nu}\right) \tag{14}$$

Curves of average radial excess pore water pressure, $\overline{u}_r$, and average degree of consolidation, $\overline{U}_h$ (purely radial flow) vs. time factor T_h for various values of n are shown in Fig. 8. The average degree of consolidation, $\overline{U}_h$ due to vertical flow versus time factor T_v is also indicated.

Worked-out example

A road embankment is constructed on top of a 9.2 m thick layer of clay, sandwiched between silty sand at the top and dense sand at the bottom. The required degree of consolidation before the embankment construction is 90%, within 9 months. For this purpose, sand drains of 450 mm diameter, need to be installed in a square arrangement. Estimate the spacing of the drain. From laboratory tests, assume that c_h = 0.288 m²/month and c_v = 0.187 m²/month

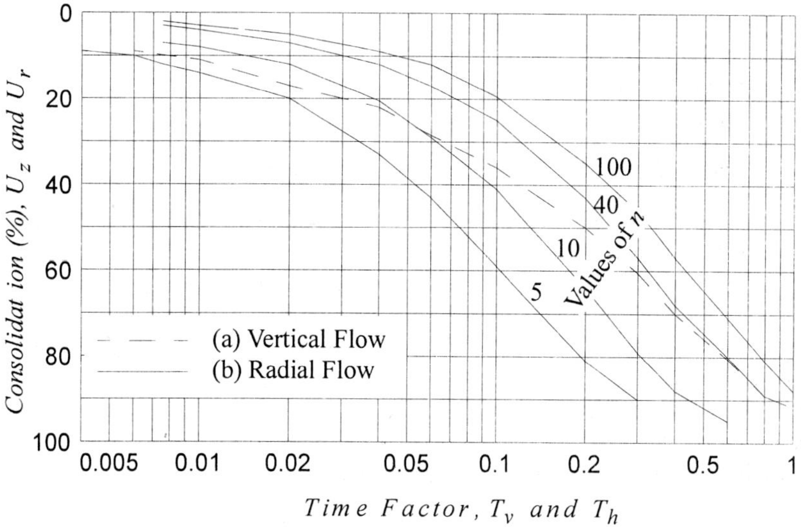

Fig. 8 *Average consolidation rates a) for vertical flow b) for radial flow (after Barron, 1948).*

Solution

$$T_v = \frac{c_v t}{H^2}$$

where, H = drainage path = 9.6/2 = 4.6 m

$$\text{Hence, } T_v = \frac{0.187}{4.6^2} \times 9$$

$$= 0.08$$

From Fig. 8, $U_v = 0.32$

$1 - U = (1 - U_r)(1 - U_v)$ - Refer Section 8

$1.0 - 0.9 = (1\text{-}U_r)(1.0\text{-}0.32)$

$U_r = 0.85$

From Figure 8, assuming $n_1 = 5$ $T_h = 0.25$, and from Eqn. (9),

$$D = \left(\frac{c_h t}{T_h}\right)^{1/2}$$

$$= \left(\frac{0.288 \times 9}{0.25}\right)^{1/2}$$

$$= 3.219 \text{ m}$$

$$n_2 = R/r_w = 3.219/2 \text{ x } .225 = 7$$

If n_1 and n_2 are not very close, assume another value for n and repeat the calculation.

If $n = 7$, then $D = 3.219$ m

Since the vertical drains are installed in a square pattern,

From Eqn. 40 (Refer Section 7),

Drain spacing, S =3.219/1.13 = 2.85 m

Approximate Equal Strain Solution (Hansbo, 1981)

Hansbo (1981) derived an approximate solution for vertical drain based on the 'equal strain hypothesis' by taking both smear and well resistance into consideration. The general concept of this solution is the same as illustrated previously in Fig. 7. By applying Darcy's law, the rate of flow of internal pore water in the radial direction can be estimated. The total flow of water from the slice, dz, to the drain, dQ_1, is equal to the change of flow of water from the surrounding soil, dQ_2, which is proportional to the change of volume of the soil mass. The average degree of consolidation, $\overline{U}$, of the soil cylinder with vertical drain is given by:

$$\overline{U}_h = 1 - \exp\left(-\frac{8T_h}{\mu}\right) \tag{15}$$

$$\mu = ln\left(\frac{n}{s}\right) + \left(\frac{k_h}{k'_h}\right) ln\,(s) - 0.75 + \pi z(2l - z)\frac{k_h}{q_w}\left\{1 - \frac{k_h / k'_h - 1}{(k_h / k'_h)(n/s)^2}\right\} \tag{16}$$

Or, in a simplified form:

$$\mu = ln\left(\frac{n}{s}\right) + \left(\frac{k_h}{k'_h}\right) ln\,(s) - 0.75 + \pi z(2l - z)\frac{k_h}{q_w} \tag{17}$$

The effect of smear only is given by:

$$\mu = ln\left(\frac{n}{s}\right) + \left(\frac{k_h}{k'_h}\right) ln\,(s) - 0.75 \tag{18}$$

The effect of well resistance only is given by:

$$\mu \approx ln(n) - 0.75 + \pi z(2l - z)\frac{k_h}{q_w} \tag{19}$$

If both smear and well resistance are ignored, this parameter becomes,

$$\mu = ln\,(n) - 0.75 \tag{20}$$

Plane Strain Consolidation Model (Indraratna & Redana, 1997)

Most finite element analyses on embankments are conducted based on the plane strain assumption. However, this kind of analysis poses a problem, because the consolidation around vertical drains is axisymmetric. Therefore, to employ a realistic 2-D finite element analysis for vertical drains, the equivalence between the plane strain and axisymmetric analysis needs to be established.

The matching of axisymmetric and plane strain conditions can be done in three ways:

1. Geometric matching approach—the spacing of the drains is matched while keeping the permeability the same.
2. Permeability matching approach—permeability coefficient is matched while keeping the spacing of drains to be the same.
3. Combination of permeability and geometric matching approach—plane strain permeability is calculated for a convenient drain spacing.

Indraratna & Redana (1997) converted the vertical drain system shown in Fig. 7 into equivalent parallel drain well by adjusting the coefficient of permeability of the soil and by assuming the plane strain cell to have a width of 2B as shown in Fig. 9. The half width of drains b_w and half width of smear zone b_s may be taken to be the same as their axisymmetric radii r_w and r_s respectively, which gives:

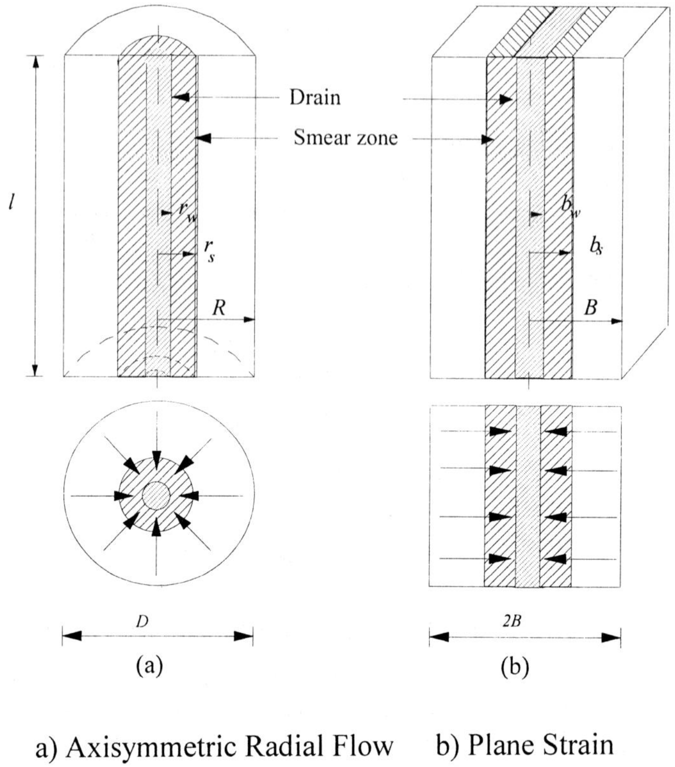

Fig. 9 *Conversion of an axisymmetric unit cell into plane strain condition (Indraratna & Redana, 1997)*

$$b_w = r_w \quad \text{and} \quad b_s = r_s \tag{21}$$

The equivalent drain diameter (d_w) or radius r_w for band drains can be determined by 'perimeter equivalence' (Hansbo,1979) to give:

$$d_w = 2\frac{(a+b)}{\pi} \quad \text{or} \quad r_w = \frac{(a+b)}{\pi} \tag{22}$$

Rixner et al. (1986) presented the equivalent drain diameter d as the average of drain thickness and width by considering the shape of the drain and effective drainage area.

$$d = \frac{a+b}{2} \tag{23}$$

where, a = the width of the PVD and b = the thickness of the PVD.

Indraratna & Redana (1997) represented the average degree of consolidation in plane strain condition as:

$$\overline{U}_{hp} = 1 - \frac{\overline{u}}{\overline{u}_o} = 1 - exp\left(\frac{-8T_{hp}}{\mu_p}\right) \tag{24}$$

where $\overline{u}_0$ = initial pore pressure, $\overline{u}$ = pore pressure at time t (average values), T_{hp}= time factor in plane strain, and

$$\mu_p = \left[\alpha + (\beta)\frac{k_{hp}}{k'_{hp}} + (\theta)\left(2lz - z^2\right)\right] \tag{25}$$

where, k_{hp} and k'_{hp} are the undisturbed horizontal and corresponding smear zone permeabilities, respectively. The geometric parameters α, β and the flow term θ are given by:

$$\alpha = \frac{2}{3} - \frac{2b_s}{B}\left(1 - \frac{b_s}{B} + \frac{b_s^2}{3B^2}\right) \tag{25a}$$

$$\beta = \frac{1}{B^2}\left(b_s - b_w\right)^2 + \frac{b_s}{3B^3}\left(3b_w^2 - b_s^2\right) \tag{25b}$$

$$\theta = \frac{2k_{hp}^2}{k'_{hp}\, q_z\, B}\left(1 - \frac{b_w}{B}\right) \tag{25c}$$

where q_z = the equivalent plane strain discharge capacity.

At a given stress level and at each time step, the average degree of consolidation for both axisymmetric $\left(\overline{U}_h\right)$ and equivalent plane strain $\left(\overline{U}_{hp}\right)$ conditions are made equal, hence:

$$\overline{U}_h = \overline{U}_{hp} \tag{26}$$

Combining Eqs. (24), (26) with the original Hansbo (1981) theory, Eqn. (15), the time factor ratio can be given by following equation:

$$\frac{T_{hp}}{T_h} = \frac{k_{hp}}{k_h}\frac{R^2}{B^2} = \frac{\mu_p}{\mu} \tag{27}$$

By assuming the magnitudes of R and B to be the same, Indraratna & Redana (1997) presented the relationship between k_{hp} and k'_{hp} as follows:

$$k_{hp} = \frac{k_h\left[\alpha + (\beta)\frac{k_{hp}}{k'_{hp}} + (\theta)\left(2lz - z^2\right)\right]}{\left[ln\left(\frac{n}{s}\right) + \left(\frac{k_h}{k'_h}\right)ln(s) - 0.75 + \pi\left(2lz - z^2\right)\frac{k_h}{q_w}\right]} \tag{28}$$

If well resistance is ignored in Eqn. (28) by omitting all terms containing l and z, the influence of smear effect can be represented by the ratio of the smear zone permeability to the undisturbed permeability as follows:

$$\frac{k'_{hp}}{k_{hp}} = \frac{\beta}{\frac{k_{hp}}{k_h}\left[ln\left(\frac{n}{s}\right) + \left(\frac{k_h}{k'_h}\right) ln(s) - 0.75\right] - \alpha} \tag{29}$$

If both smear and well resistance effects are ignored in Eqn. (28), then the simplified ratio of plane strain to axisymmetric permeability is readily obtained, as proposed earlier by Hird et al. (1992):

$$\frac{k_{hp}}{k_h} = \frac{0.67}{\left[ln(n) - 0.75\right]} \tag{30}$$

The well resistance is derived independently and yields an equivalent plane strain discharge capacity of drains as also proposed earlier by Hird et al. (1992):

$$q_z = \frac{2}{\pi B} q_w \tag{31}$$

DRAIN PROPERTIES

Equivalent drain diameter for band shaped drain

The conventional theory of consolidation with vertical drains assumes that the vertical drains are circular in cross-section. Therefore, a band-shaped drain needs to be converted to an equivalent circular diameter, which implies that the equivalent diameter of a circular drain has the same theoretical radial drainage capacity as the band-shaped drain. Kjellman (1948) stated, "the draining effect of a drain depends to a great extent upon the circumference of its cross-section, but very little upon its cross-sectional area". Based on Kjellman's consideration, Hansbo (1981) introduced the equivalent diameter for a pre-fabricated band-shaped drain, as given earlier in Eqn. (22).

Another study (Rixner et al., 1986) suggested that the more appropriate d_w is given by the less complex relationship as in Eqn. (23).

Pradhan et al. (1993) suggested that the equivalent diameter of band-shaped drains should be estimated by considering the flow net around the soil cylinder of diameter d_e (Fig. 10). The mean square distance of their flow net is calculated as:

$$\bar{s}^2 = \frac{1}{4} d_e^2 + \frac{1}{12} a^2 - \frac{2a}{\pi^2} d_e \tag{32}$$

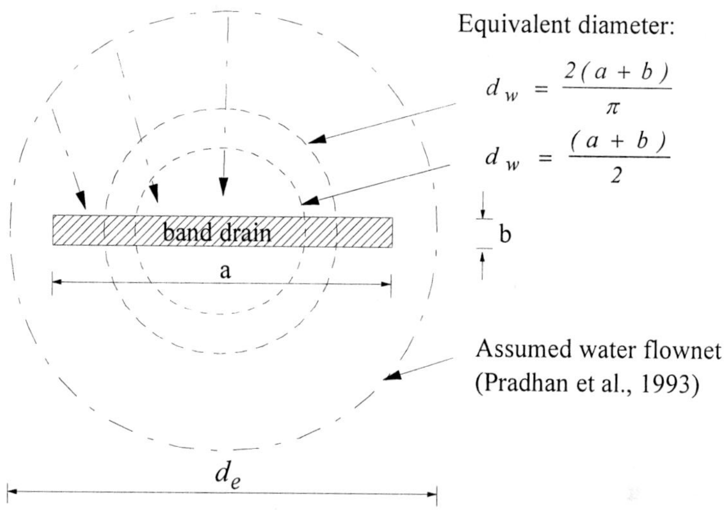

Fig. 10 *Equivalent diameter of band-shaped vertical drain.*

$$\text{Then, } d_w = d_e - 2\sqrt{\left(\overline{s}^2\right)} + b \tag{33}$$

Based on finite element studies and by comparing the equivalent band shaped drains diameter represented by Eqns. 22 and 23, it was found that the equivalent diameter given by Eqn. (22) should be reduced by 20%. It was suggested by Rixner et al. (1986) that the use of Eqn. (23) is more appropriate than Eqn. (22), based on numerical modelling.

Filter and apparent opening size (AOS)

The drain material (sand drain) and the filter jacket of PVD have to perform two basic but contrasting requirements, which are retaining the soil particles and at the same time allowing the pore water to pass through. The general guideline of the drain permeability is given by:

$$k_{filter} > 2\, k_{soil} \tag{34}$$

An effective filtration can minimize soil particles from moving through the filter. A commonly employed filtration requirement is given by:

$$\frac{O_{95}}{D_{85}} \leq 3 \tag{35}$$

where, AOS, O_{95} indicates the approximate largest particle that would effectively pass through the filter. Sieving is done using glass beads of successively larger diameter until 5% passes through the filter, and this size in millimetres defines the AOS, O_{95} based on ASTM D 4751 (ASTM,

1993). This apparent opening size (AOS) is usually taken to be less than 90 microns based on Eqn. (35). D_{85} indicates the diameter of clay particles corresponding to 85% passing.

The retention ability of the filter is given by:

$$\frac{O_{50}}{D_{50}} \leq 24 \tag{36}$$

Filter material can also become clogged if the soil particles become trapped within the filter fabric structure. Clogging is prevented by ensuring that (Christopher and Holtz, 1985):

$$\frac{O_{95}}{D_{15}} \geq 3 \tag{37}$$

Discharge capacity

The discharge capacity of the pre-fabricated vertical drain is required to analyse the drain (well) resistance factor. However, well resistance factor is always less significant than the drain spacing and the disturbance (smear effect). In order to measure the discharge capacity of drains, it is required to simulate as closely as possible the conditions in the field. In this case, the discharge capacity will be a function of the volume of the core or the drain channel, the lateral earth pressure acting on the drains, possible folding, bending and twisting of the drain due to large settlement, infiltration of fine soil particles through the filter, and the biological and chemical degradation. Incorporating the above factors, the actual discharge capacity, q_w, is then given by:

$$q_w = (F_t)\,(F_c)\,(F_{fc})\,q_{req} \tag{38}$$

where F_t, F_c and F_{fc} are the influence factors due to time, drain deformation and clogging, respectively. The term q_{req} is the theoretical discharge capacity calculated from Barron's theory of consolidation, which is given by:

$$q_{req} = \frac{\varepsilon_f U_{10} l \pi c_h}{4T_h} \tag{39}$$

where, ε_f is the final settlement of the soft soil equivalent to 25% of the length of the drain installed to the soft ground, U_{10} is the 10% degree of consolidation, l is the depth of the vertical drain, c_h is horizontal coefficient of consolidation and T_h is the time factor for horizontal (lateral) consolidation.

From laboratory tests, the influence factor of time, F_t, has been estimated to be less than 1.2, and it is usually conservatively taken to be 1.25. The reduction of the discharge capacity under the worst condition of bending, folding and twisting has been suggested to be about 48%, which gives an influence factor of deformation, F_c of about 2. The filtration tests show that the trapped fine soil particles decrease the permeability of the PVD, and in turn decrease the PVD discharge capacity. This deterioration is complicated by the biological and chemical growth in the geotextile filter. From filtration tests, the value of F_{fc} is suggested to vary between 2.8 and 4.2 with an average of about 3.5. After considering all the worst conditions that may occur in the field, the discharge capacity, q_w of the PVD could be as high as 500-800 m^3/year, but reduced to 100-300 m^3/year where the hydraulic gradient is unity under elevated lateral pressure (Rixner et al., 1986).

The discharge capacity of various types of drains is shown in Fig. 11, where the discharge capacity is influenced by lateral confining pressure. It is also suggested, in lieu of laboratory test data, that the discharge capacity can be conservatively assumed to be 100 m^3/year. Hansbo (1981) based on laboratory test results suggested a much smaller discharge capacity of drains, as summarised in Table 1.

Table 1

Short-term discharge capacity, in m^3/year, of eight band drains measured in laboratory (Hansbo, 1981)

Drain type	*Lateral pressure in kPa*			
	40	80	250	500
Geodrain	26	20	20	16
Other drain types	21	20	18	10
	24	22	14	12
	15	14	14	13
	10	5	1	Clogged
	21	19	17	15
	—	17	13	12
	19	17	9	4

Holtz et al. (1991) reported that the discharge capacity of prefabricated vertical drains could vary from 100-800 m^3/year. The discharge capacity of PVD is a function of its filter permeability, core volume or cross-section area, lateral confining pressure, drain stiffness controlling its deformation characteristics, among other factors (Hansbo, 1979; Holtz et al., 1991). For long vertical drains that are vulnerable to well resistance, Hansbo (1981) and Holtz et al. (1988, 1991) pointed out that in the field, the actual reduction of the discharge capacity can be attributed to: (a) reduced flow in drain core due to increased lateral earth pressure, (b) folding and crimping of drain due to excessive settlements, and (c) infiltration of fine silt or clay particles through the filter (siltation).

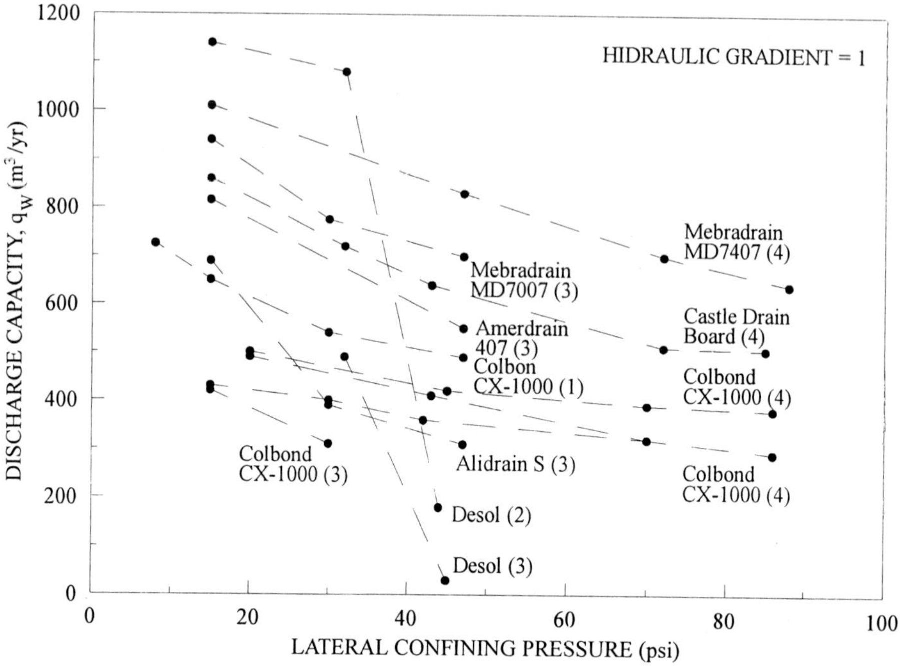

Fig. 11 *Typical values of vertical discharge capacity (after Rixner et al., 1986).*

Based on a number of experimental studies, Holtz et al. (1988) concluded that as long as the initial discharge capacity of PVD exceeds 100-150 m^3/year, some reduction in discharge capacity due to installation should not seriously influence the consolidation rates. However, discharge capacity q_w can fall below this desired minimum value due to the three reasons mentioned earlier. For certain types of PVD, affected by significant vertical compression and high lateral pressure, q_w values may be reduced to 25-100 m^3/year (Holtz et al., 1991). Clearly, the 'clogged' drains are associated with q_w values approaching zero.

INFLUENCE ZONE OF DRAINS

Vertical drains are commonly installed in square or triangular patterns (Fig. 12). As illustrated in Fig. 12, the influence zone (D) is a controlled variable, since it is a function of the drain spacing (S) as given by:

$$D = 1.13\ S \text{ for drains installed in a square pattern, and} \quad (40)$$

$$D = 1.05\ S \text{ for drains installed in a triangular pattern} \quad (41)$$

A square pattern of drains may be easier to lay out and control during installation in the field, however, a triangular pattern is usually preferred since it provides a more uniform consolidation between drains.

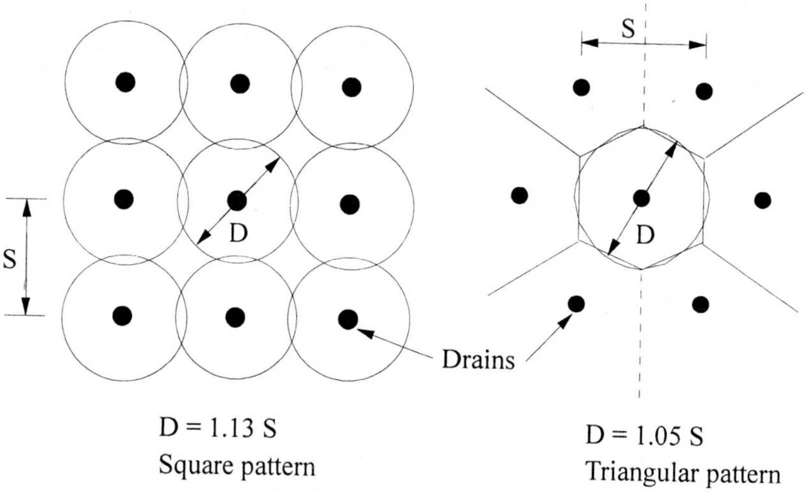

Fig. 12 *Plan of drain well pattern and zone of influence of each well.*

CONSOLIDATION AROUND VERTICAL DRAINS

Rate of consolidation

The main reason for using pre-fabricated vertical drain is to reach the desired degree of consolidation within a specified time period. But in a vertical drain system, both radial and vertical consolidation should be considered in calculating the specified time period. Carillo (1942) gave the combined effect as:

$$1 - U = (1 - U_r)(1 - U_v) \tag{42}$$

where U – Overall degree of consolidation
U_r – Average degree of consolidation due to radial drainage
U_v – Average degree of consolidation due to vertical drainage

Coefficient of Consolidation with Radial Drainage

Log U vs t approach

Aboshi and Monden (1963) presented a curve fitting method using log U and linear t. This method is developed by taking 'log' on both sides of Barron's solution (Eqn. 8), which results in the following expression:

$$T_h = -\frac{F(n)}{8} \ln\left(1 - \overline{U}\right) \tag{43}$$

Equation (43) represents the theoretical time factor for radial consolidation of perfect drains without considering the effect of smear. The coefficient of radial consolidation c_h is determined by plotting the logarithm of the average degree of consolidation against the linear consolidation time (log U vs. T_h), where a linear slope provides the c_h value (Fig. 13).

In order to include the effect of smear, 'log' is taken on both sides of Eqn. (14), which yields a new time factor T_h as given by:

$$T_h = -\frac{v}{8} \ln\left(1 - \overline{U}\right) \tag{44}$$

By substituting $c_h = \frac{T_h H_d^2}{t}$ yields,

$$c_h = -\frac{D^2 v}{8} \frac{H_d \ln\left(1 - \overline{U}\right)}{dt} \tag{45}$$

By using Eqn. (14), the coefficient of radial drainage consolidation c_h can be defined from the slope of log U-t curve with settlement data. The pore water pressure data can also be plotted other than settlement, in this method.

Plotting settlement data (Asaoka, 1978)

Asaoka (1978) developed a method where a series of settlements (ρ_1, ... $\rho_{I-1}, \rho_i, \rho_{i+1}$ etc.) which are observed at constant time intervals are plotted as

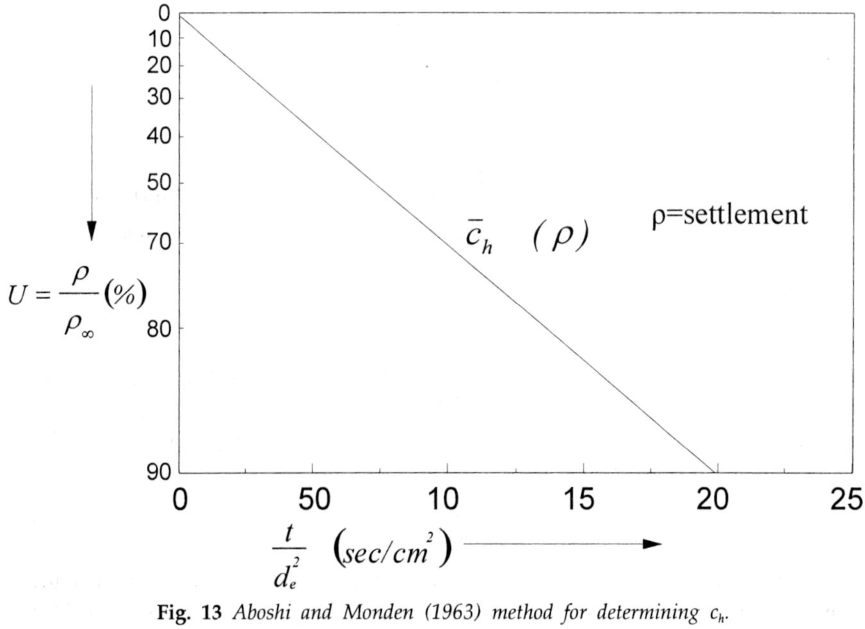

Fig. 13 *Aboshi and Monden (1963) method for determining c_h.*

shown in Fig. 14. The coefficient of radial drainage consolidation in this method is derived using Barron's (1948) solution, which is given by:

$$c_h = -\frac{D^2 v}{8}\frac{\ln \beta}{\Delta t} \tag{46}$$

where, v is expressed in Eqn. (12), β is the slope of the line formed by the observed displacement data, and Δt is the time interval between observations. To obtain the average coefficient of consolidation, v (smear factor) is replaced by drain spacing factor, $F(n)$ as expressed in Eqn. (5) which gives:

$$c_h = -\frac{D^2 F(n)}{8}\frac{\ln \beta}{\Delta t} \tag{47}$$

EFFECT OF HORIZONTAL TO VERTICAL PERMEABILITY RATIO

Generally, the coefficient of permeability can be determined indirectly from conventional oedometer test using Terzaghi's one-dimensional consolidation theory. This indirect method of determining coefficient of permeability leads to some error due to the assumption of constant coefficient of permeability, constant coefficient of volume change and constant coefficient of consolidation (k, m_v and c_v), respectively, during the

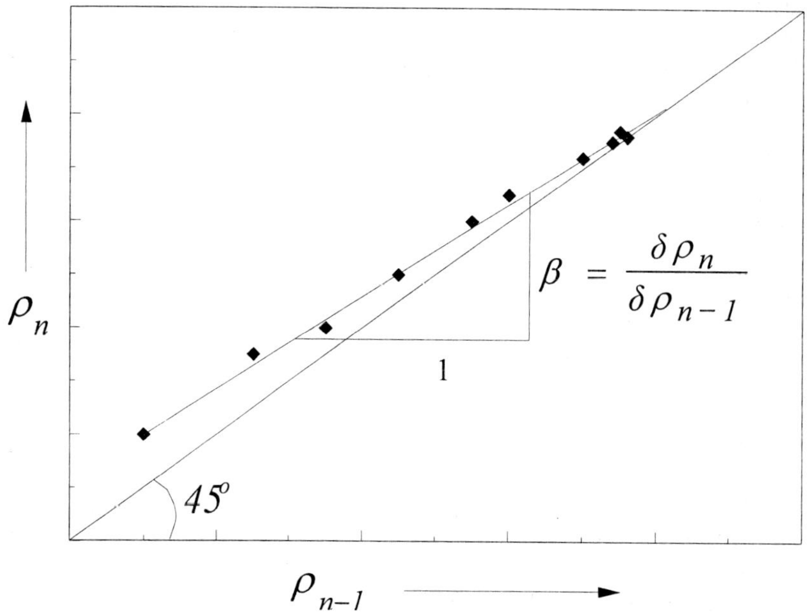

Fig. 14 *Asaoka (1978) method to determine* c_h.

consolidation process. The modified triaxial and oedometer tests equipped with the permeameter produce more reliable values of permeability. The falling head permeability test conducted on the modified oedometer appears to be the best (simpler and faster) to determine permeabilities of natural clays.

The permeability characteristics of a number of intact clays determined using the modified apparatus explained above is reported later by Tavenas et al. (1983b). In these tests, the horizontal permeability was also determined using samples rotated horizontally (90°) and of intermediate inclination, 45°. For marine clays recovered from Champlain sea formation (Canada), the anisotropy ratio ($r_k = k_h/k_v$) estimated using the modified oedometer test was found to vary between 0.91 and 1.42 with an average of 1.1. In triaxial testing, this ratio was found to be in the range of 0.81-1.16 with an average of 1.03, which implies that anisotropy was not an influence factor in this soil.

According to the experimental results plotted in Fig. 6 (Indraratna & Redana, 1995), the value of k_h' / k_v' in the smear zone varies between 0.9 and 1.3 with an average of 1.15. Hansbo (1987) argued that for extensive smearing, the horizontal permeability coefficient in the smear zone (k_h') should approach that of the vertical permeability coefficient (k_v'), thus suggesting that the ratio k_h' / k_v' could approach 1. The experimental results shown in Fig. 6 (Indraratna & Redana,1995) seem to be in agreement with Hansbo (1987). For the applied consolidation pressures, it is observed that the value of k_h/k_v varies between 1.4 and 1.9 with an average of 1.63 in the undisturbed zone. Shogaki et al. (1995) reported that the average values of k_h/k_v were in the range of 1.36-1.57 for undisturbed isotropic soil samples taken from Hokkaido to Chugoku region in Japan. Tavenas et al. (1983b) reported that for the soil tested in conventional oedometer, the k_h/k_v ratio was found to vary between 0.91 and 1.42 for intact natural clays, and from 1.2 to 1.3 for Matagami varved clay. Bergado et al. (1991) conducted a thorough laboratory study on the development of smear zone in soft Bangkok clay, and they reported that the horizontal permeability coefficient of undisturbed zone to the smear zone varied between 1.5 and 2 with an average of 1.75. More significantly, the ratio of k_h' / k_v' was found to be almost unity within the smear zone.

INSTALLATION AND MONITORING OF VERTICAL DRAINS

Figure 15 illustrates a typical scheme of vertical drains installation and monitoring instruments required to monitor the performance of the soil foundation beneath an embankment. Prior to vertical drain installation, it is necessary to conduct general site preparation. This work may include the removal of vegetation and surficial debris, establishing site grading

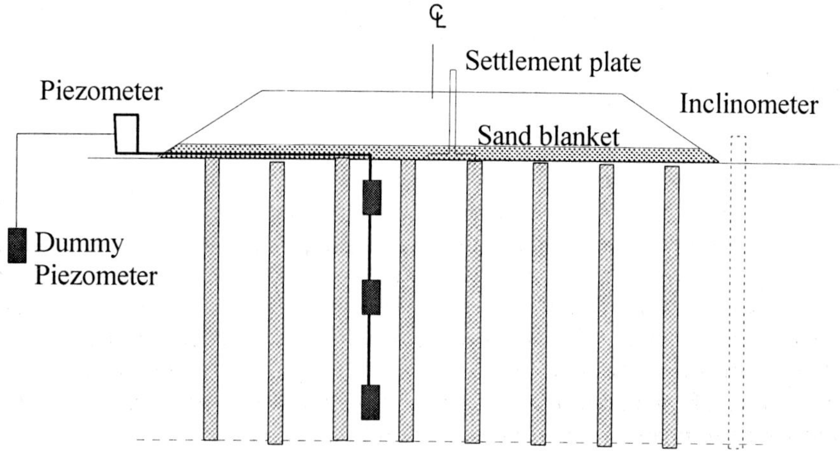

Fig. 15 *Basic instrumentation of embankments.*

and constructing a sand blanket. The purpose of the sand blanket is to conduct the expelled water away from the drains and to provide a sound-working mat.

Several types of geotechnical instrumentation are required to install before and after construction, in order to monitor the performance of the embankment. Monitoring is essential to prevent sudden failures, to record changes in the rate of settlement and to verify the design parameters. Performance evaluation is also important to improve settlement predictions and to provide sound guidelines for future projects.

Settlement gauges (e.g. of hydraulic type) are used to measure the long-term settlements at the original ground surface and should be placed immediately after the installation of vertical drains. Settlement plates are suitable for initial reading and should be installed at the bottom of the sand blanket, at intermediate depths, and at the bottom of the compressible layer to monitor its settlement. A benchmark is usually set up on a stable ground at a reasonable distance from the fill. Piezometers are used to monitor the complete pore pressure profiles and should be installed at the bottom of sand blanket, at intermediate depths and at the bottom of the compressible layer. A dummy or remote piezometer should also be installed at an adequate distance from the embankment to record the original or natural ground water level, and in general, the pore water pressure at a particular depth. Alignment stakes can be installed parallel to the embankment slope prior to the placement of the fill. The alignment stake is a simple means of measuring the lateral displacement of the foundation during construction of the embankment. In fact, they could provide an early warning of bearing capacity failure. A more sophisticated equipment to measure lateral displacement is an inclinometer, which is usually placed set around the toe of the embankment.

SOFT CLAY MODELLING

In order to predict the behaviour of an engineering structure, it is necessary to use an appropriate constitutive model, which represents the stress-deformation response of the material. Deformation analysis in geotechnical engineering often assumes a linear elastic material at small stresses. This assumption probably holds true for over-consolidated clay, however, most soils exhibit plastic behaviour at increased stresses. As illustrated in Fig. 16, the initial part of the stress-strain curve (OA) is linear, which means that if the material is unloaded at any point on OA, the stress-strain behaviour of the material follows the same path to return to the origin O. However, if the material is loaded to point B and then unloaded, the stress-strain curve may still return to origin O but following a different path, OB, where the material exhibits non-linear elasticity. If the time is a variable in both these processes, the material is called *visco-elastic*. Most soils are visco-elastic and they follow a non-linear elastic behaviour, where their true behaviour is difficult to model exactly. Beyond point B, the stress-strain behaviour is not reversible, and permanent strains occur on the material as represented by OD.

The theories of critical state soil mechanics have been developed based on the application of the theory of plasticity. Utilizing the critical state concept based on the theory of plasticity in soil mechanics, a more sophisticated Cam-clay model has been introduced to represent the behaviour of clay (Schofield & Wroth, 1967). The Cam-clay model has received wide acceptance due to its simplicity and accuracy to model clay behaviour, especially for normally and lightly over-consolidated clay. In

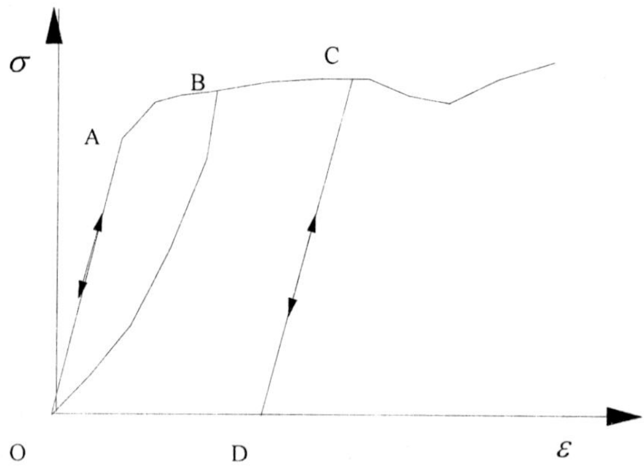

Fig. 16 *Stress-strain relationship.*

this model, the shear strength of the soil is related to the void ratio. To describe the state of soil during triaxial testing, the following critical state parameters are defined by:

$$p' = \frac{\sigma_1' + 2\sigma_3'}{3} = \frac{\sigma_1 + 2\sigma_3}{3} - u \tag{48}$$

$$q = \sigma_1' - \sigma_3' = \sigma_1 - \sigma_3 \tag{49}$$

where, σ_1' represents the effective axial stress, σ_3' represents the effective confining stress and u is the pore water pressure.

In critical state theory, the virgin compression, swelling and recompression lines are assumed to be straight lines in ($ln\ p'$-V) plots with slope $-\lambda$ and $-\kappa$ respectively, as shown in Fig. 17. The isotropic virgin compression line or isotropic normal consolidation line (NCL) is expressed as:

$$V = N - \lambda \ln (p') \tag{50}$$

where, N (V_λ) is the value of V when $lnp' = 0$ or $p' = 1$.

The swelling or recompression line is expressed as:

$$V = V_\kappa - \kappa \ln (p') \tag{51}$$

The relation between C_c (compression index) and λ may be expressed as:

$$\lambda = \frac{C_c}{2.307} \tag{52}$$

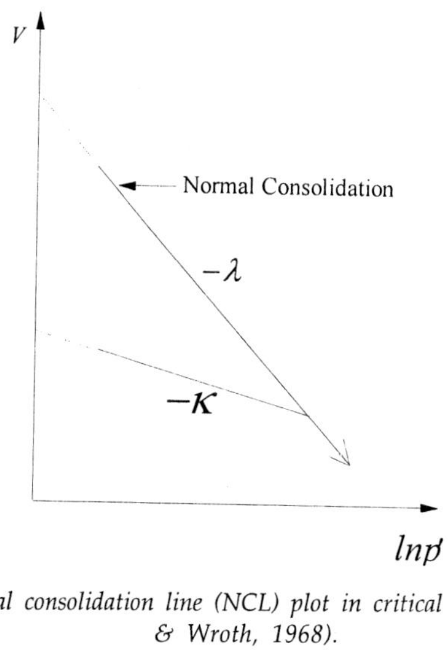

Fig. 17 *Isotropic normal consolidation line (NCL) plot in critical state theory (after Schofield & Wroth, 1968).*

The slope of the straight line in q-p' plot is called Critical State Line (CSL) as shown in Fig. 18. The slope of the critical state line, M is expressed as:

$$q = Mp' \tag{53}$$

In the V-$\ln p'$ plot, if Γ is used to represent the value of V_λ which corresponds to a $\ln p' = 0$ (i.e. $\Gamma = e_{cs}+1$), then the equation of the straight line is given by:

$$V = \Gamma - \lambda \ln p', \text{ or } p' = \exp\frac{\Gamma - V}{\lambda} \tag{54}$$

Hence, the critical state line must satisfy both Eqs. (48) and (49).

Combining the CSL equation into the Mohr circle plot, the relationship between drained angle of friction (ϕ') and M may be given by:

$$M = \frac{6 \sin \phi'}{3 - \sin \phi'} \tag{55}$$

The initial void ratio can be estimated at any given depth below the ground level once p', q and P'_c are known, and the e-$\ln p'$ plot is shown in Fig. 19. The parameter e_{cs} is defined as the voids ratio on the critical state line for a value of $p' = 1$.

The intersection between the swelling line and the in-situ-stress line is assumed to be at point A and given by coordinates e_A and P'_A. Point P

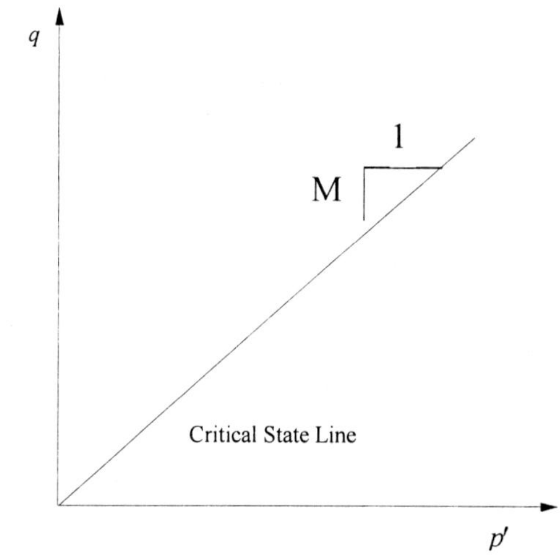

Fig. 18 *Position of the critical state line.*

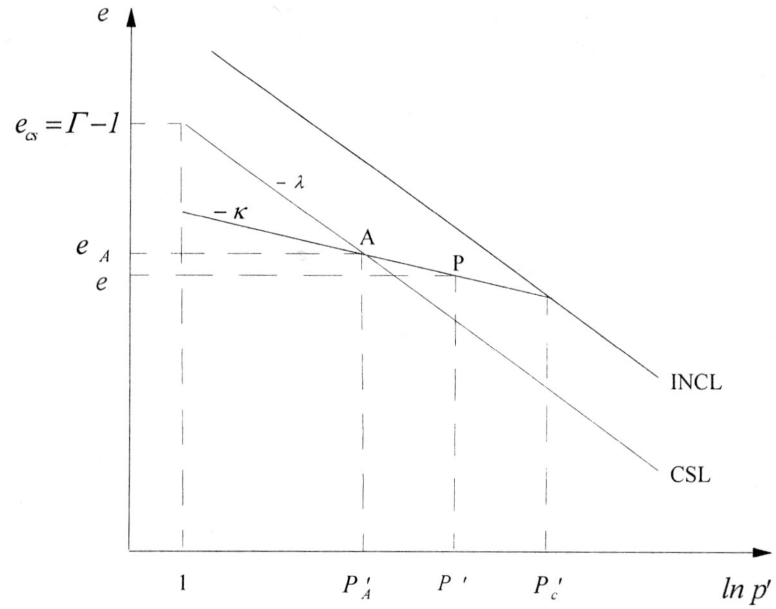

Fig. 19 *Position of the initial void ratio on critical state line.*

represents the intersection between the initial void ratio, e and the effective mean normal stress p'. Then, the following relation may be established:

$$e_A = e_{cs} - \lambda \ln P'_A \tag{56}$$

where, $P'_A = \dfrac{P'_c}{2}$ for Modified Cam-clay and $P'_A = \dfrac{P'_c}{2.718}$ for Cam-clay.

Along the swelling line (κ line) passing through the initial stress state at point P, the following relation can be applied:

$$e - e_A = \kappa(\ln p' - \ln P'_A) \tag{57}$$

Substituting e_A from Eq. (56) gives:

$$e_{cs} = e + (\lambda + \kappa) \ln P'_A - \kappa \ln p' \tag{58}$$

Cam-clay models

Modified Cam-clay

It was found that the original Cam-clay model was deficient in some aspects of modelling the stress-strain behaviour of soil. Two aspects of dissatisfaction were: the shape of the yield locus at increased p'_c and the predicted value of K_0 (the coefficient of earth pressure at rest). Therefore, the modified Cam-clay was introduced to address those problems (Roscoe and Burland, 1968). The obvious difference between modified Cam-clay and Cam-clay model is the shape of the yield locus, where the yield locus of modified Cam-clay is elliptical as shown in Fig. 20(b). The flow rule for modified Cam-clay is given by:

$$\frac{\delta v^p}{\delta \varepsilon^p} = \frac{M^2 - \eta^2}{2\eta} \tag{59}$$

where, δv^p and $\delta \varepsilon^p$ are volumetric and deviatoric plastic strain increments, respectively, and $\eta = \frac{q}{p'}$ represents the stress ratio.

The modified Cam-clay yield locus is given by:

$$q + M^2 p'^2 = M^2 p' p'_c \tag{60}$$

The equation of the Stable State Boundary Surface (SSBS) is given by:

$$V_\kappa = \Gamma + (\lambda - \kappa)\{ln\,(2) - ln\,(1 + (\eta / M)^2)\} \tag{61}$$

TYPES OF FINITE ELEMENTS

The finite element software code CRISP is based on the modified Cam-clay model (Britto and Gunn, 1987). This programme has been used in the past by many researchers for soft clay embankment modelling.

The basic element type in CRISP is the linear strain triangle (LST-element type 2) consisting of six displacement nodes. The higher order of this element is the cubic strain triangle (CuST-element type 6) which contains 15 unknown displacement nodes. For consolidation analysis,

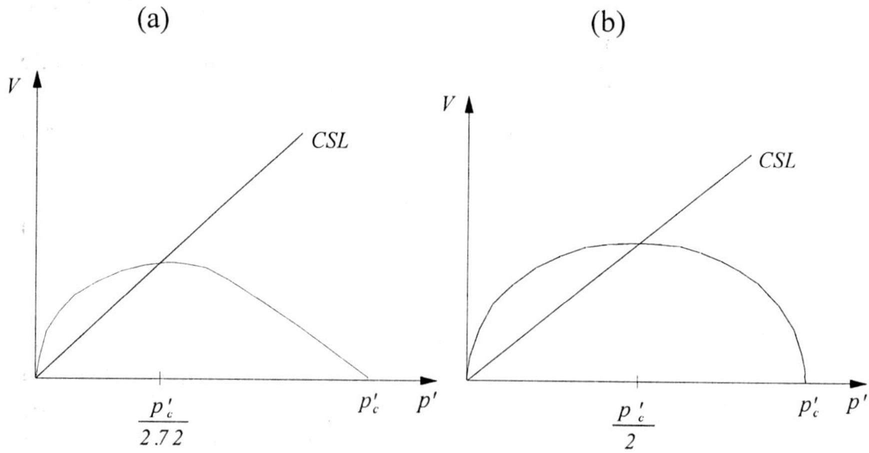

Fig. 20 *The yield locus of Cam-clay and modified Cam-clay.*

additional porewater pressures nodes are deployed in element types 3, 5 and 7 (Fig. 21).

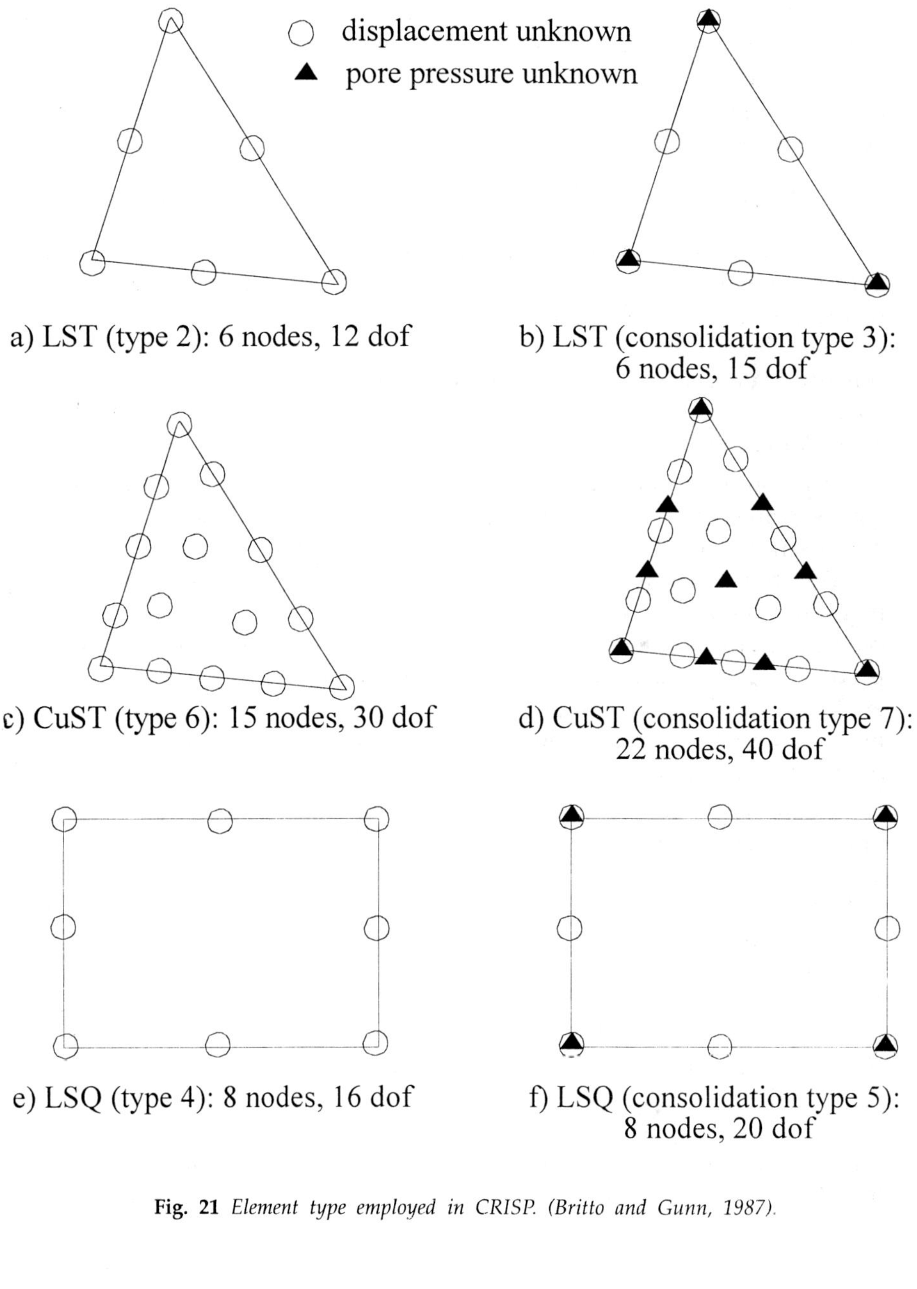

Fig. 21 *Element type employed in CRISP. (Britto and Gunn, 1987).*

SALIENT ASPECTS OF NUMERICAL MODELLING

Currently, pore pressures, settlements, lateral displacements and stresses of the vertical drain installed field site can be analysed as accurately as possible using sophisticated finite element software. Commercial packages such as ABAQUS, PLAXIS and SAGE-CRISP, are capable of performing fully coupled consolidation analysis. According to past experience, finite element analysis of lateral deformation has been relatively poor in contrast to settlements (Indraratna et al., 1994). The recent finite element models applied to vertical drains are described below.

Drain efficiency by pore pressure dissipation (Indraratna et al., 1994)

In this study, the performance of embankment stabilized with vertical drains at Muar clay, Malaysia was analyzed using the finite element code, CRISP (Britto and Gunn, 1987). The effectiveness of the pre-fabricated drains was evaluated according to the rate of excess-pore pressure dissipation at the soil drain interface. Both single and multi-drain (whole embankment) analyses were carried out to predict the settlement and lateral deformation beneath the embankment.

A plane strain analysis was applied to a single drain and to the whole PVD scheme. An axisymmetric horizontal permeability was used in the plane strain model. As explained in detail by Ratnayake (1991), the prediction of settlement using the single drain analysis over-predicts the measured settlement, even though the effect of smear is included. In the case of multi-drain analysis underneath the embankment, the over-prediction of settlement is more significant compared to the single drain analysis. Therefore, to enable better predictions, it was necessary to consider more accurately, the dissipation of the excess pore pressures at the drain boundaries at a given time.

In order to elaborate this technique, the average undissipated excess pore pressures could be estimated by finite element back-analysis of the settlement data at the centreline of the embankment as shown in Fig. 16. In this figure, 100% represents zero dissipation when the drains are fully loaded. Accordingly, at the end of the first stage of consolidation (i.e., 2.5 m of fill after 105 days), the undissipated pore pressures decrease from 100% to 16%. For the second stage of loading, the corresponding magnitude decreases from 100% to 18% after a period of 284 days, during which the height of the embankment has already attained the maximum of 4.74 m. It can be deduced from Fig. 22 that perfect drain conditions are approached only after a period of 400 days. Although the general trends between the finite element results and field data are in agreement especially during the initial stages, the marked discrepancy beyond 100 days is too large to be

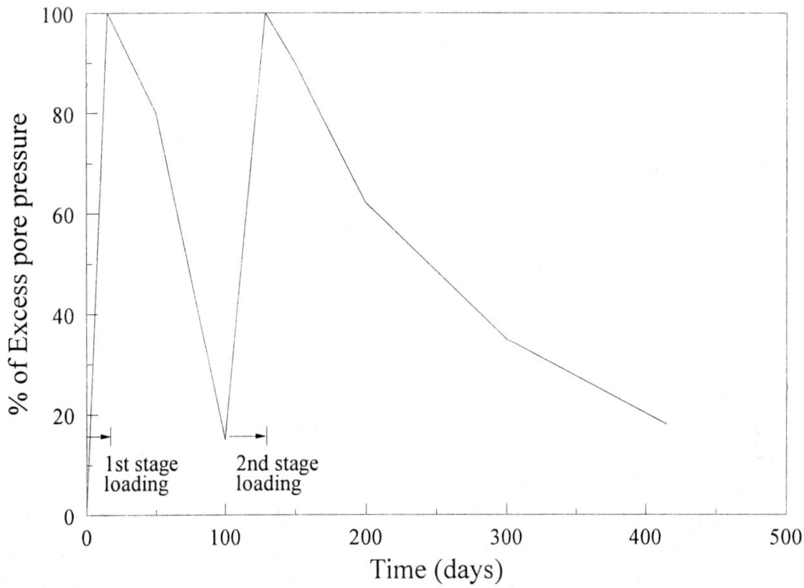

Fig. 22 *Percentage of undissipated excess pore pressures at drain-soil interfaces (Indraratna et al., 1994).*

attributed solely to the plane strain assumption. These excess pore-pressures reflect the retarded efficiency of the vertical drains (partial clogging). A better prediction was obtained for settlement, pore pressure and lateral deformation when 'non-zero' excess pore pressures at drain interface were input into the finite element model, simulating 'partially clogged' conditions. The 'non-zero' excess pore pressures can also represent the smear effect that contributes to decreased efficiency, as discussed later.

Matching Permeability & Geometry (Hird et al., 1995)

Hird et al. (1992, 1995) presented a modelling technique of vertical drains in two-dimensional finite element analysis using CRISP (Britto and Gunn, 1987). In this analysis, the permeability and geometry matching were applied to several embankments stabilized with vertical drains in Porto Tolle (Italy), Harlow (UK) and Lok Ma Chau (Hongkong).

An acceptable prediction of settlements was obtained, although the pore water pressure dissipation was more difficult to predict. However, at Lok Ma Chau (Hongkong), the settlements were significantly over-predicted. This was because, in this case history analysis, the effect of smear was not considered although the plane strain model proposed by Hird et al. (1992) allows the smear effect to be modelled.

At Porto Tolle embankment, pre-fabricated vertical geodrains were installed on a 3.8 m triangular grid to a depth of 21.5 m below ground level. The equivalent radius of geodrain was 31 mm, and its discharge capacity was conservatively estimated at about 140 m^3/year. The embankment, which was constructed over a period of 4 months, had a height of 5.5 m, a crest width of 30 m, a length of over 300 m, and a side slope of about 1 in 3. For the purpose of finite element analysis, the clay was modelled using the modified Cam-clay model of Roscoe and Burland (1968) and their parameters are as follows: $\lambda = 0.16$, $\kappa = 0.032$, $\Gamma = 2.58$, $M_{ax} = 1.16$, $M_{pl} = 0.92$, $v = 0.3$, and permeability was $k_h = 4.1\text{x}10^{-9}$ m/s and $k_v = 3.5\text{x}10^{-10}$ m/s. The equivalent plane strain permeability is estimated according to Hird et al., (1992).

In this study, a single-drain analysis at embankment centreline was considered. The typical results of finite element analysis are compared in Figs. 23(a) and 23(b). Field settlement data are also plotted in Fig. 23(a) and these show that both during and after construction, the magnitude of settlement was reasonably well modelled. In Fig. 23(b) the observed and computed excess pore pressures mid-way between the drains are compared; only the axisymmetric computed results are shown, since in plane strain analysis, pore pressures are not matched at corresponding points, but merely on average. During construction, the observed and computed pore pressures compared tolerably well. Afterwards, they differed greatly, although the field data may have been unreliable because of the presence of organic gas in the soil (Jamiolkowski and Lancellotta, 1984).

Modelling of Discharge Capacity (Chai et al. 1995)

Chai et al. (1995) extended the method proposed by Hird et al. (1992) to include the effect of well resistance and clogging. Four types of analyses were presented considering: (1) no vertical drains, (2) embankment with vertical drains and discharge capacity of the drains increasing with depth, (3) drains with constant discharge capacity and (4) the same as (3) but assuming that the drains were clogged below 9 m depth.

The discharge capacity of the drain in plane strain for matching the average degree of horizontal consolidation is given by:

$$q_{wp} = \frac{4k_h l^2}{3B\left[ln\left(\frac{n}{s}\right) + \frac{k_h}{k_s} ln(s) - \frac{17}{12} + \frac{2l^2\,\pi k_h}{3q_{wa}}\right]} \qquad (62)$$

The model developed in this study was refined using a single drain model of 5 m long, and both elastic and elasto-plastic analyses were applied to predict its performance. Excellent agreement was obtained between axisymmetric and plane strain models especially with varied discharge capacity q_{wp} as shown in Fig. 24, for elasto-plastic analysis. The well

(a)

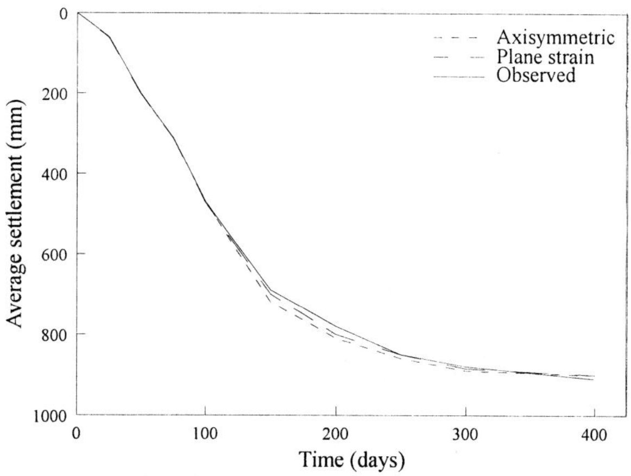

(b)

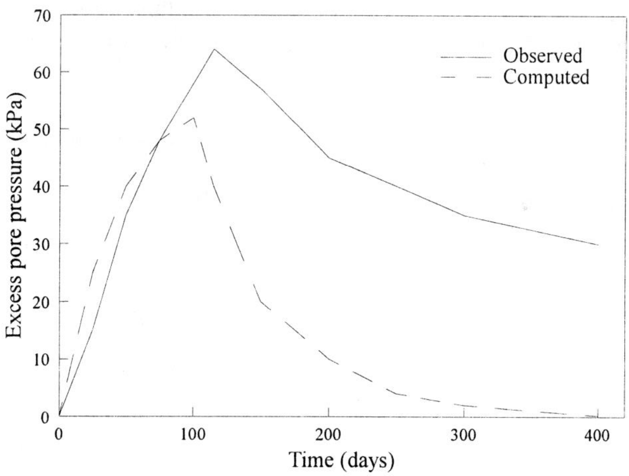

Fig. 23 *Result of axisymmetric and matched plane strain for Porto Tolle embankment: a) average surface settlement and b) excess pore pressure (Hird et al., 1995).*

resistance matching also results in a more realistic excess pore water pressure variation with depth as shown in Fig. 25, for elastic analysis. It can be seen that the varied discharge capacity yielded a more uniform and closer match between axisymmetric and plane strain methods compared to constant discharge capacity assumption.

The model described above was calibrated verified with the performance of an embankment stabilised with vertical drains founded in

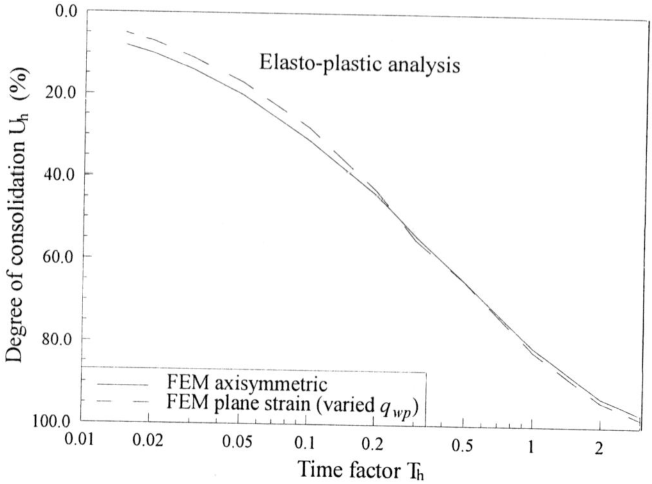

Fig. 24 *Comparison of average degree of horizontal consolidation (Chai et al., 1995).*

Muar clay, Malaysia. This study shows that the vertical drains not only increase the settlement rate, but also reduce the lateral deformation. A more realistic excess pore pressure distribution was also obtained when the well resistance and clogging were introduced in the analysis.

Deformation as a Stability Indicator

Indraratna et al. (1997) investigated the effect of ground improvement by

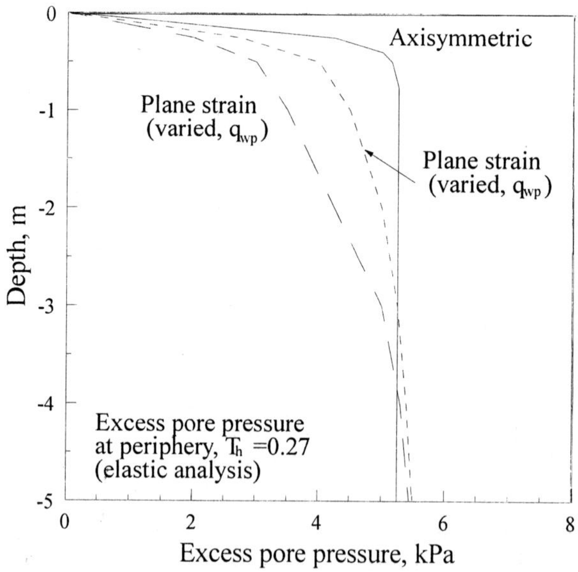

Fig. 25 *Comparison of excess pore pressure variation with depth (Chai et al., 1995).*

preloading together with geogrid and vertical band drains, as well as sand compaction piles constructed on Muar clay, Malaysia. The settlement and lateral displacement of the soft clay foundation were analysed using the plane strain finite element formulation, and the findings were compared to the field measurements. In order to conduct a two-dimensional plane strain analysis, the vertical drain system was converted into an equivalent drainage wall as explained earlier (Fig. 9).

The analysis employed critical state soil mechanics, and the deformations were predicted on the basis of the fully coupled (Biot) consolidation model incorporated in the finite element code CRISP (Britto and Gunn, 1987). In the analysis, the soil underneath the embankment was discretised using linear strain quadrilateral (LSQ) elements. The vertical drains were modelled as ideal and non-ideal drains, where in the former, the well resistance factor was ignored. This study shows that the accurate prediction of lateral displacement depends on the correct assessment of the value of the Cam-clay parameter λ, the shear resistance at the embankment-foundation interface and the nature of assumptions made in the modelling of drains and sand piles. The actual soil properties are influenced by the working stress range and the assumed stress path of the sub-soil at a given depth. The normally consolidated parameters associated with the Cam-clay theories over-estimate lateral displacement and settlements, if the applied stresses are smaller than the pre-consolidation pressure.

The performance of vertical band drains and sand compaction piles was compared based on normalized deformation as shown in Table 2. The ratio of maximum lateral displacement to fill height (β_1) and the ratio of maximum settlement to fill height (β_2) were considered as stability indicators.

Table 2

Effect of ground improvement on normalised deformation factors (Indraratna et al., 1997).

Ground improvement scheme	α	β_1	β_2
Sand compaction piles for pile/soil stiffness ratio 5 (h = 9.8 m, including 1 m sand layer	0.185	0.018	0.097
Geogrids + vertical band drains in square pattern at 2.0 m spacing (h = 8.7 m)	0.141	0.021	0.149
Vertical band drains in triangular pattern at 1.3 m spacing (h = 4.75 m)	0.123	0.034	0.274
Embankment rapidly constructed to failure on untreated foundation (h = 5.5 m)	0.634	0.104	0.164

Where

α = ratio of maximum lateral displacement at the toe to the maximum settlement at centre line of embankment.

β_1 = ratio of maximum lateral displacement to the corresponding fill height.

β_2 = ratio of maximum settlement to the corresponding fill height.

In comparison with an unstabilized embankment constructed to failure (Indraratna et al., 1992), the stabilized foundations are characterized by considerably smaller values for α and β_1, highlighting their obvious implication on stability. The normalized settlement (β_2) on its own does not seem to be a proper indicator of instability. The foundation having SCP gives the lowest values of β_1 and β_2, clearly suggesting the benefits of sand compaction piles over the band drains.

Application of Vacuum Pressure (Bergado et al., 1998)

Finite element analysis was applied by Bergado et al. (1998) to analyze the performance of embankment stabilized with vertical drains, where a combined preload and vacuum pressure were utilized at the Second Bangkok International Airport site. A simple approximate method for modelling the effect of PVD as proposed by Chai and Miura (1997) was incorporated in this study. PVD increases the mass permeability in the vertical direction. Consequently, it is possible to establish a value of permeability of the natural subsoil and the radial permeability towards the PVD. This equivalent vertical permeability (K_{ve}) is derived, based on equal average degree of consolidation.

The approximate average degree of vertical consolidation U_v is given by:

$$U_v = 1 - \exp\,(-3.54)\,T_v \tag{63}$$

where, T_v is the dimensionless time factor.

The equivalent vertical permeability, K_{ve} can be expressed as:

$$K_{ve} = \left(1 + \frac{2.26 L^2 K_h}{F D_e^2 K_v}\right) K_v \tag{64}$$

where

$$F = \ln\left(\frac{D_e}{d_w}\right) + \left(\frac{K_h}{K_s} - 1\right) \ln\left(\frac{d_s}{d_w}\right) - \frac{3}{4} + \frac{\pi 2 L^2 K_h}{3 q_w} \tag{65}$$

In Eqn. (65), D_e is the equivalent diameter of a unit PVD influence zone; d_s is the equivalent diameter of the disturbed zone; d_w is the equivalent diameter of PVD; K_h and K_s are the undisturbed and disturbed horizontal permeability of the surrounding soil, respectively; L is the length for one-way drainage, and q_w is discharge capacity of PVD. The effects of smear and well resistance have been incorporated in the derivation of the equivalent vertical permeability.

Two full-scale test embankments, TV1 and TV2 each with base area of 40 x 40 m were analyzed by Bergado et al. (1998). The PVDs were installed to a depth of 15 m and 12 m for embankment TV1 and TV2, respectively.

The design parameters are shown in Table 3, and the typical cross-section of embankment TV1 is shown in Fig. 26. It can be concluded from this study that the vacuum assisted consolidation has been effectively utilized for both embankments TV1 and TV2. The performance of embankment TV2 with vacuum preloading (compared to the embankment at the same site without vacuum preloading), showed an acceleration in the rate of settlement of about 60 % and a reduction in the period of preloading by about 4 months.

Table 3
Parameters of vertical drains

Spacing, S	1.0 m (triangular pattern)
Diameter of drain, d_w	50 mm
Diameter of smear zone, d_s	300 mm
Ratio of K_h/K_s	10
Drainage length, l	15 m for TV1 and 12 m for TV2
Discharge capacity, q_w (per-drain)	50 m^3/year

Equivalent Plane Strain Modelling

An attempt was made by Indraratna & Redana (1997) to analyze the effect

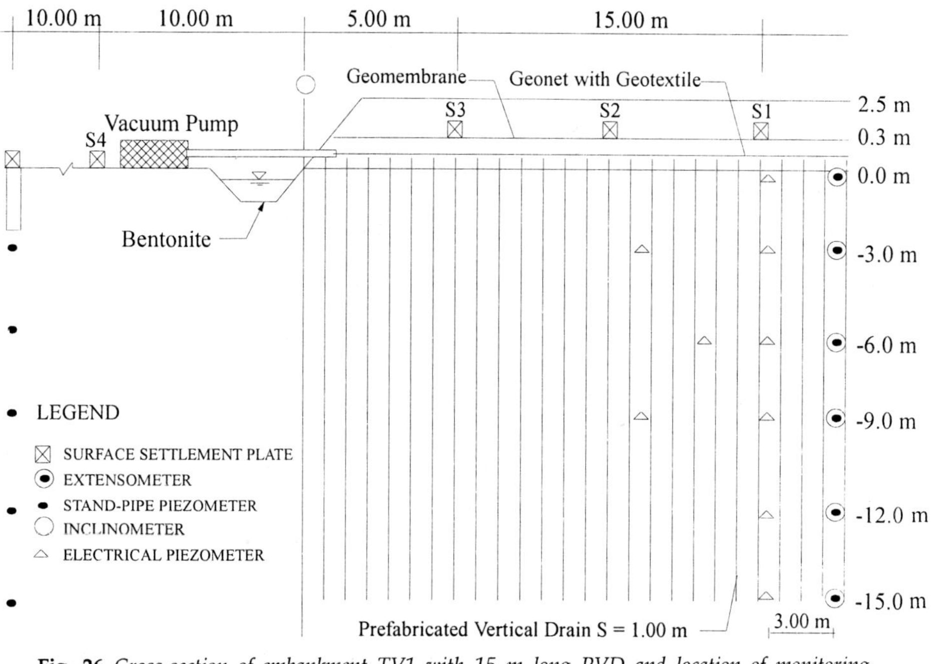

Fig. 26 *Cross-section of embankment TV1 with 15 m long PVD and location of monitoring instruments (Bergado et al., 1998).*

of smear zone and well resistance in a vertical drain using a 2D plane strain finite element model employing a modified Cam-clay theory. This was executed by converting the vertical drain system into equivalent parallel drain walls by adjusting the spacing of the drains and the coefficient of permeability of the soil as discussed earlier (Equations 21-31). The transformed permeability coefficient was then incorporated in the finite element code, CRISP through appropriate subroutines.

In order to verify the proposed model, a finite element analysis was executed for both axisymmetric and equivalent plane strain models. As an example, a unit drain was considered with a drain installed to a depth of 5 m under the ground surface at 1.2 m spacing. The model parameters and the soil properties are given in Table 4.

Table 4. *Model parameters and soil properties*

Spacing, S	1.2 m
Radius of drain, r_w	0.03 m
Radius of mandrel	0.05 m
Ratio of K_h/K_v in undisturbed zone	2
Ratio of K_h/K_v in smear zone	1
Plane strain permeability in undisturbed zone, k_{hp}	2.97 x 10^{-9} m/s
Plane strain permeability in smear zone, k'_{hp}	5.02 x 10^{-10} m/s
Radius of unit cell	0.6 m
Gradient of volume vs log pressure relation for consolidation, λ	0.2
Gradient of volume vs log pressure relation for swelling, κ	0.04
Slope of critical state line, M	1.0
Void ratio at unit consolidation pressure, e_{cs}	2
Poisson's ratio, ν	0.25
Saturated unit weight of the soil, γ_s	18 kN/m^3

In the analysis with smear, the size of the smear zone was taken to be 5 times the size of the mandrel based on the experimental results. For the verification of the model a simplified permeability variation was assumed as shown in Fig. 27, where the coefficient of permeability within the smear zone was taken to be constant (i.e. k'_{hp} = 5.02 × 10^{-10} m/s). A higher permeability coefficient was used for the undisturbed zone (k_{hp} = 2.97 × 10^{-9} m/s). Figure 27 shows both the assumed axisymmetric and the converted plane strain permeability values.

The results of both axisymmetric and plane strain analysis are plotted in Fig. 28, where the average degree of radial consolidation U_h (%) is plotted against the time factor T_h for perfect drain conditions. As illustrated in Fig. 28, the proposed plane strain analysis gives a good agreement with the axisymmetric analysis. The maximum difference between the two methods is less than 5%. Figure 29 indicates the settlement-time plots for both axisymmetric and equivalent plane strain conditions, which again reveal

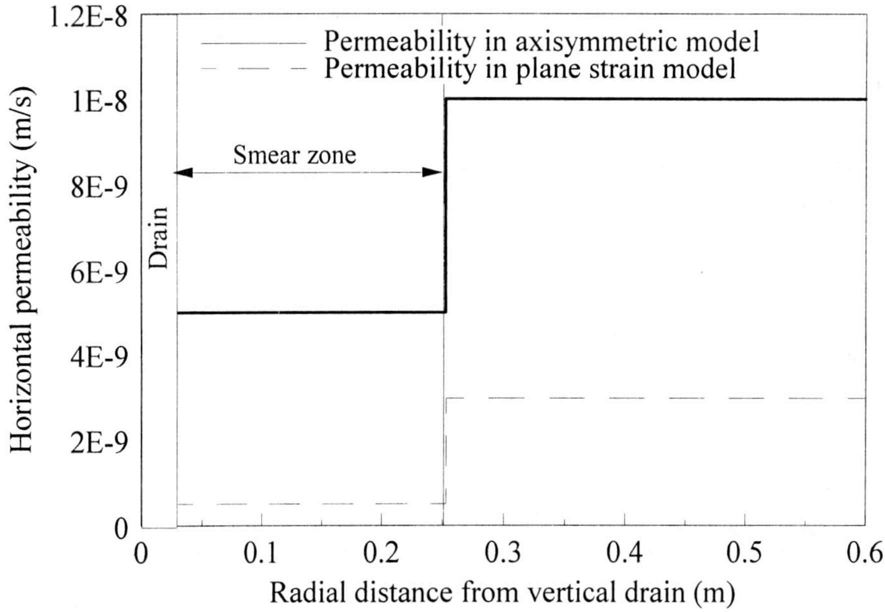

Fig. 27 *Simplified variation of permeability within and outside smear zone.*

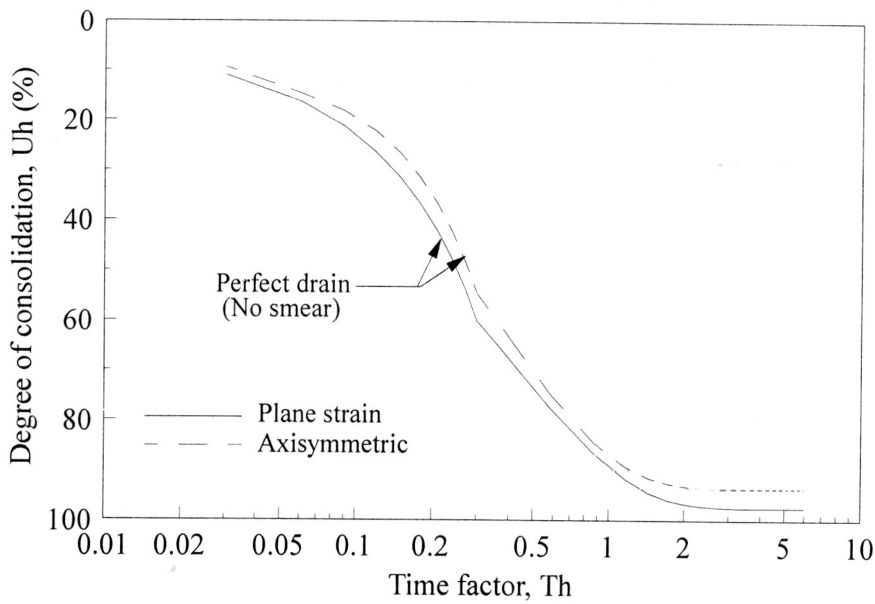

Fig. 28 *Average degree of consolidation vs time factor.*

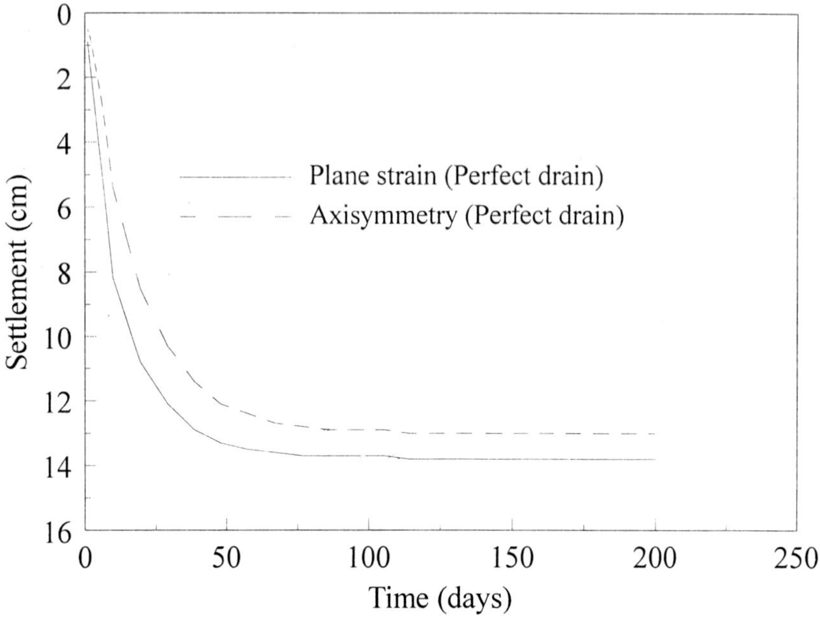

Fig. 29 *Comparison of surface settlement vs time for axisymmetric and equivalent plane strain analyses for perfect drain condition.*

good agreement. Excess pore pressures are zero in both cases, hence not shown on a separate plot.

Figures 30 and 31 illustrate the settlement vs. time and excess pore

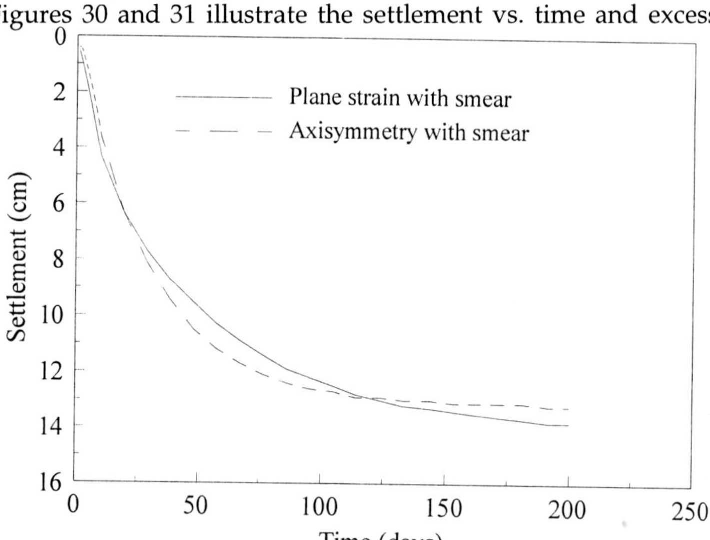

Fig. 30 *Comparison of the average surface settlement for axisymmetric and equivalent plane strain analyses with smear (Indraratna et al., 2000).*

pressure against time plots for drains with smear only (i.e. no well resistance). The comparison between axisymmetric and equivalent plane strain models is in good agreement. It is noted that the pore pressures are more difficult to match than the settlements. The maximum error in settlement plot is 7 %, whereas the maximum deviation in excess pore water pressure is about 15 %.

The well resistance can also be included using Eqn. (24). Figures 30 and 31 illustrate the settlements and excess pore pressure variations with time for single drains including smear plus well resistance. Once again, very good agreement between the axisymmetric model and the equivalent plane strain model is found. It is important to note that the inclusion of well resistance reduces the errors, notably in excess pore pressures.

Based on the above single drain analysis, Figs. 28–33 all provide concrete evidence that the equivalent (converted) plane strain model is an excellent substitute for the axisymmetric model. In finite element modelling, the 2 D, plane strain analysis is expected to cut down the computational time considerably, in comparison with the time taken by a 3 D, axisymmetric model, especially in the case of multi-drain analysis.

APPLICATION OF NUMERICAL MODELLING IN PRACTICE AND FIELD OBSERVATION

Indraratna & Redana analysed (1995, 1999, 2000) the performance of several test embankments using their plane strain model. Figure 34 summarizes

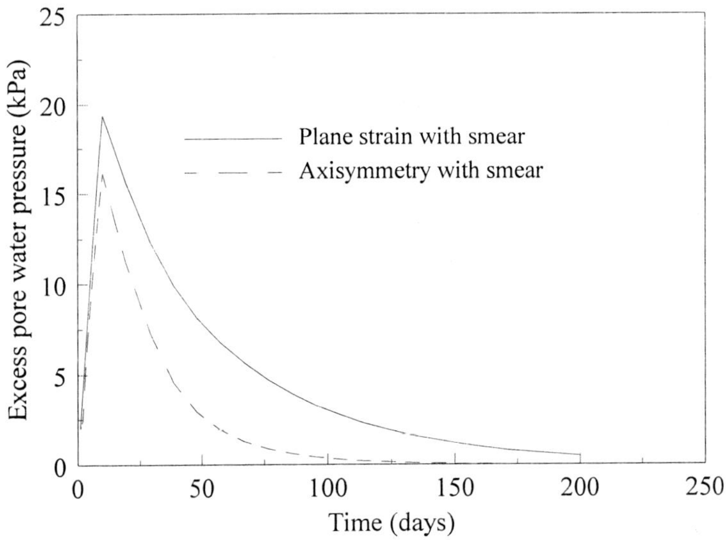

Fig. 31 *Comparison of the average pore water pressure at mid depth for axisymmetric and equivalent plane strain analyses with smear (Indraratna et al., 2000).*

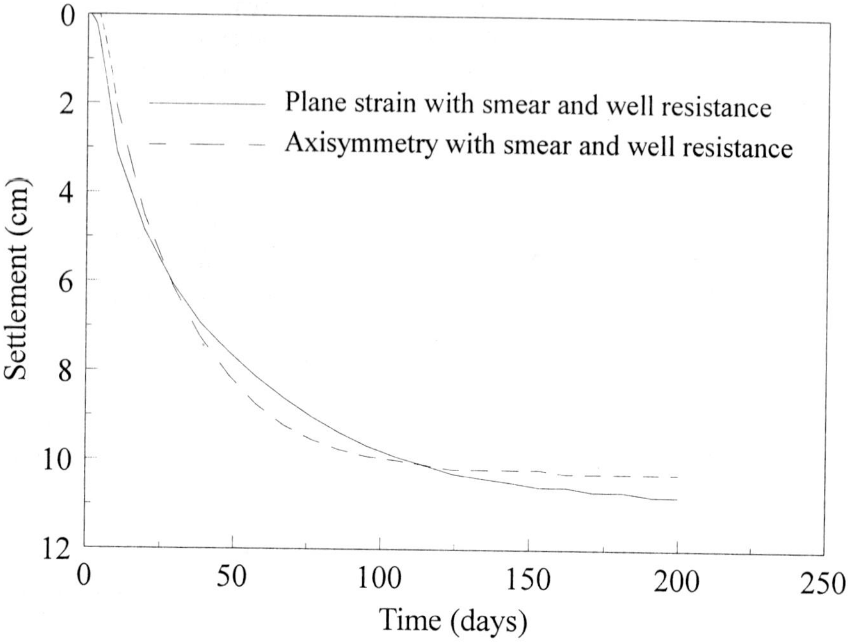

Fig. 32 *Comparison of the average surface settlement for axisymmetric and equivalent plane strain analyses with smear and well resistance (Indraratna et al., 2000).*

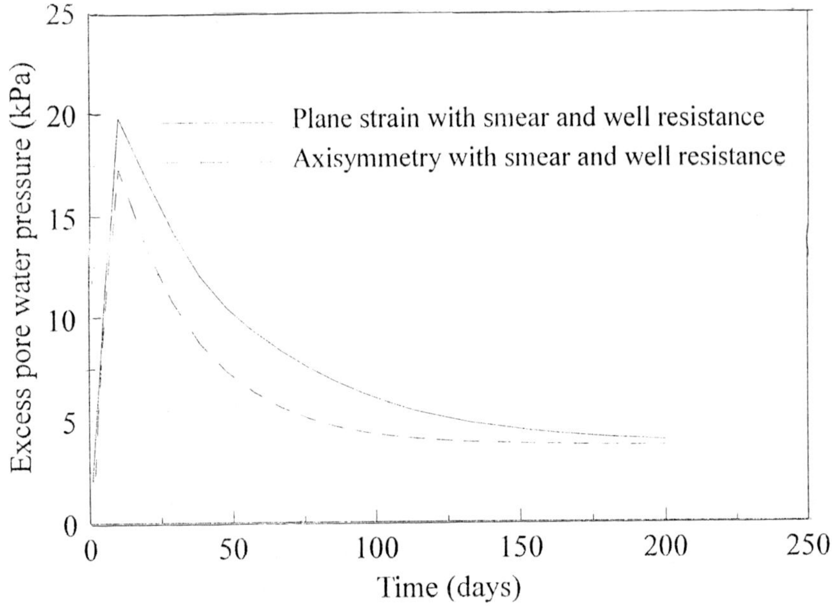

Fig. 33 *Comparison of the excess pore pressure at mid depth for axisymmetric and equivalent plane strain analyses with smear and well resistance (Indraratna et al., 2000).*

the typical subsoil profile, modified Cam-clay parameters and the effective stress conditions in the site. The unit weight of the weathered crust is about 18 kN/m^3 and the lowest unit weight of the soil is about 14.3 kN/m^3 at a depth of 7 m. The typical finite element meshes for embankments employing the multi-drain analysis are given in Fig. 35 and 36. The foundation is discretized into linear strain quadrilateral (LSQ) elements. For the zone stabilized with Prefabricated Vertical Drains (PVD), a finer mesh was used so that each drain element includes the smear zone on either side of the PVD. The location of inclinometers and piezometers is accurately defined in the mesh, with the measurement points placed on the mesh nodes. The piezometer locations are shown in the insert of each mesh. The embankment load is applied in stages (i.e. sequential construction). Figure 37 indicates the rate of loading and the construction history of a typical embankment.

The numerical analysis was based on the modified Cam-clay model (Roscoe and Burland, 1968) incorporated in the finite element code, CRISP92 (Britto and Gunn, 1987). The equivalent plane strain values were calculated based on Eqns. (29) and (30) (Indraratna & Redana, 1997). After a few trials to include the effect of well resistance, the minimum discharge capacity (q_w) was estimated to model the settlements and pore water pressure dissipation.

The results of the plane strain analysis of a typical embankment together with the measured settlements are plotted in Fig. 38. The analysis based on

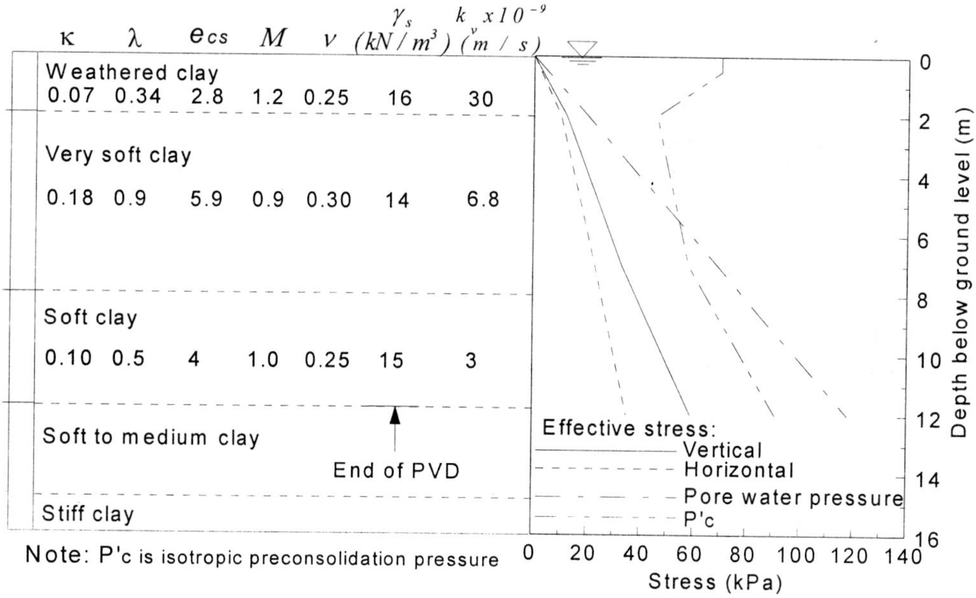

Fig. 34 *Sub-soil profile, Cam-clay parameters and stress condition used in numerical analysis, second Bangkok international airport, (after AIT, 1995).*

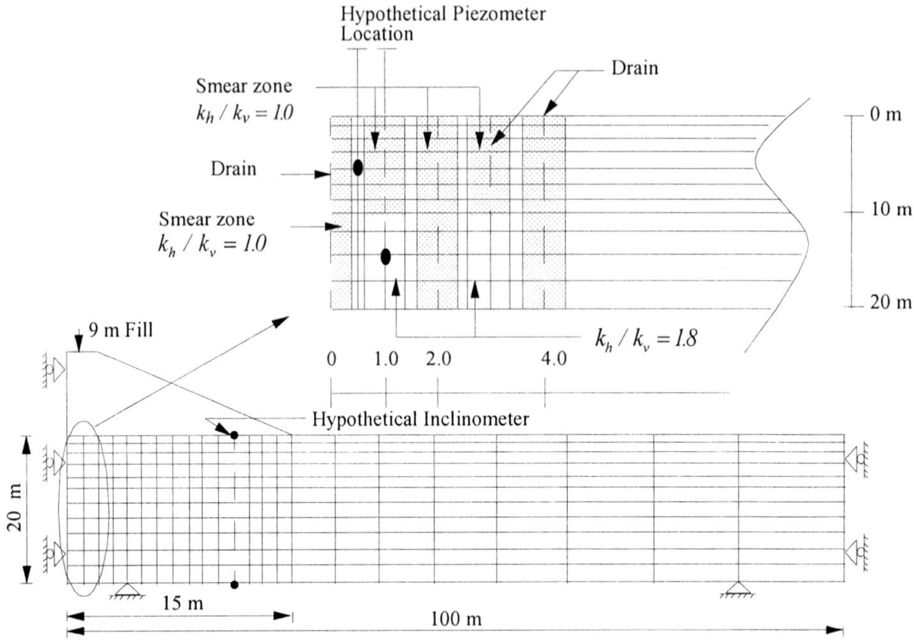

Fig. 35 *Finite element mesh of embankment for plane strain analysis with variable drain lengths (Redana, 1999).*

'perfect drain' conditions (i.e. no smear, complete pore pressure dissipation) overpredicts the measured settlement, but the inclusion of smear effect improves the accuracy of the predictions. The inclusion of the effects of smear and well resistance underestimate the measured settlements. In terms of settlements, the role of well resistance could be regarded as insignificant, in comparison with the smear effect.

The measured and predicted excess pore pressures along the centerline of the embankment at a depth of 2 m below the ground surface are compared in Fig. 39. In the 'smear only' analysis, the pore water pressure increase is well predicted during Stage 1 and Stage 2 loading. However after Stage 3 loading, the predicted pore pressure is significantly smaller than the field measurements. As expected, the perfect drain predictions underestimate the actual pore pressures. The inclusion of the effects of smear and well resistance gives a better prediction of the pore water pressure dissipation for all stages of loading.

The prediction of settlement along the ground surface from the centerline of a typical embankment after 400 days is shown in Fig. 40. At the embankment centerline, the limited available data agree well with the settlement profile. Also heave could be predicted beyond the toe of the embankment, i.e. at about 42 m away from the centerline.

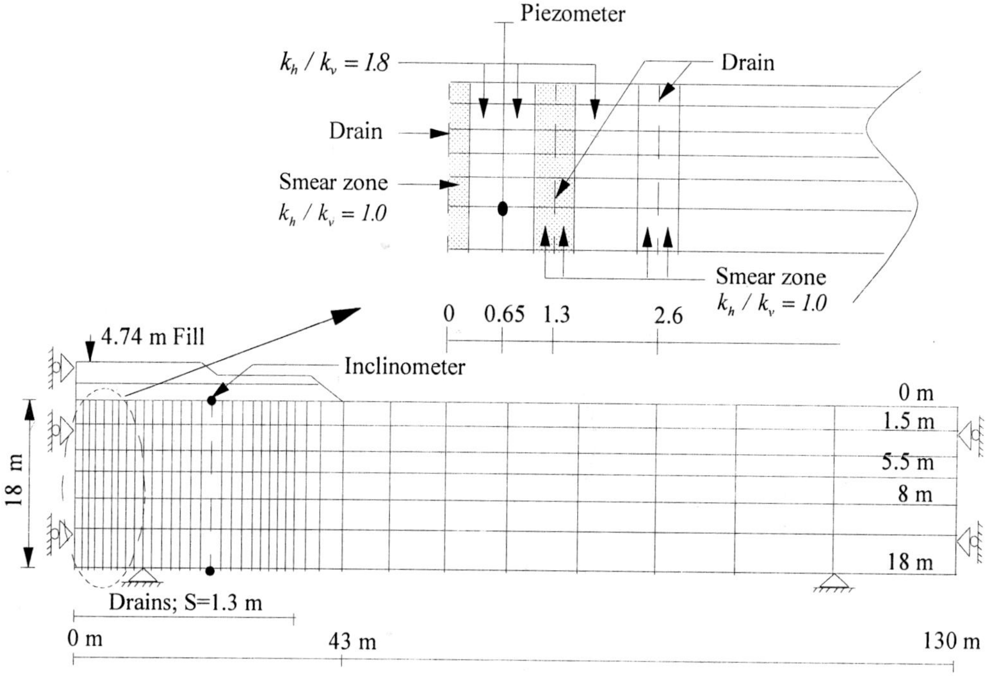

Fig. 36 *Finite element mesh of the embankment for plane strain analysis with constant drain length (Indraratna & Redana, 1999).*

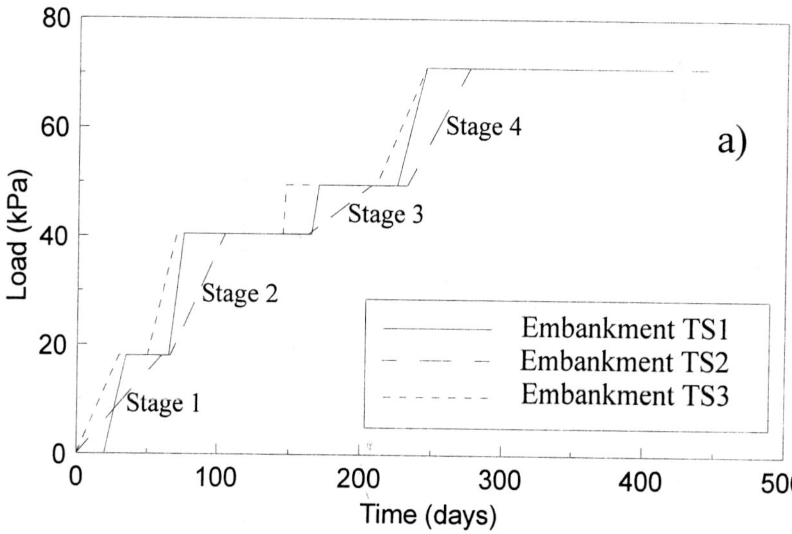

Fig. 37 *Construction loading history for embankments TS1, TS2 and TS3 at second Bangkok international airport (AIT, 1995).*

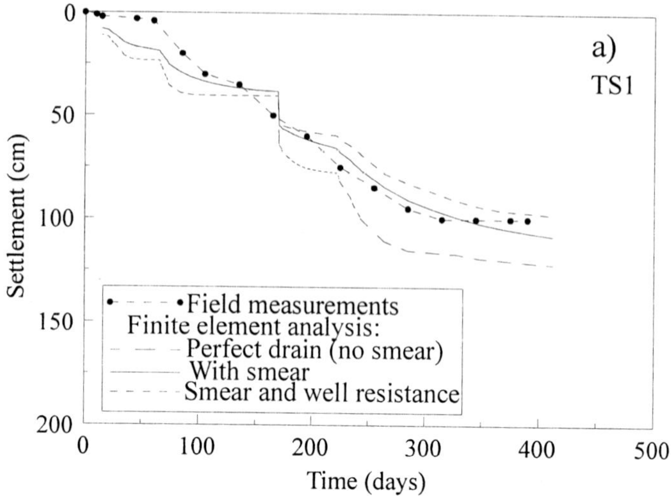

Fig. 38 *Surface settlement at the centerline for embankment TS1, second Bangkok international airport (Indraratna & Redana, 2000).*

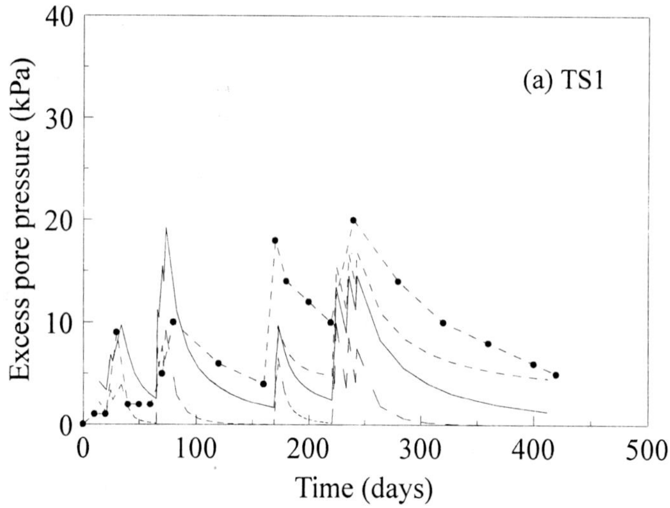

Fig. 39 *Variation of excess pore water pressures at 2 m depth below ground level at the centerline for embankment TS1 (Indraratna & Redana, 2000).*

Measured and predicted lateral deformation for the inclinometer installed away from the centerline of the embankment is shown in Fig. 41. The lateral displacements at 44 days and 294 days after loading are well predicted when both the effects of smear and well resistance are included. As shown in Fig. 41(b) the inclusion of smear effect by itself underestimates

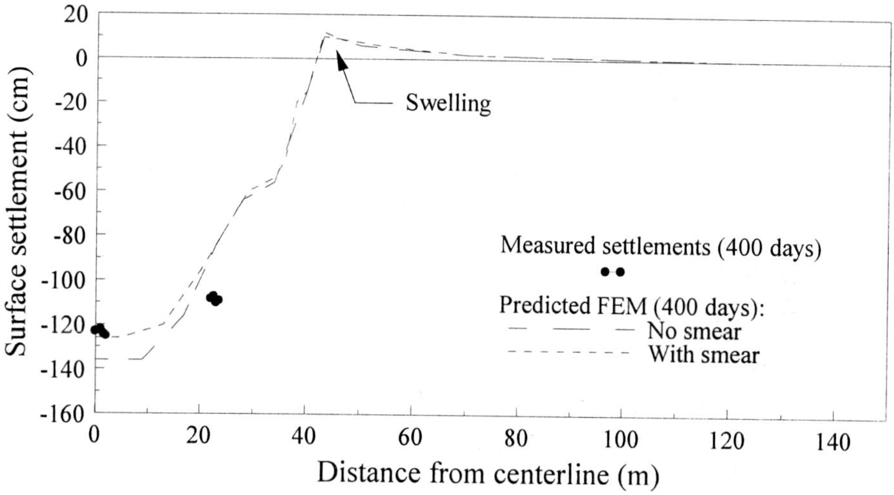

Fig. 40 *Surface settlement profiles after 400 days, Muar clay, Malaysia (Indraratna & Redana, 1999).*

the magnitude of lateral displacement. The 'perfect drain' condition yields the smallest lateral deformation. The predicted lateral deformation for 'no drains' is plotted for comparison. It is verified that the presence of PVD is expected to reduce the lateral movement of soft clay under embankment loading.

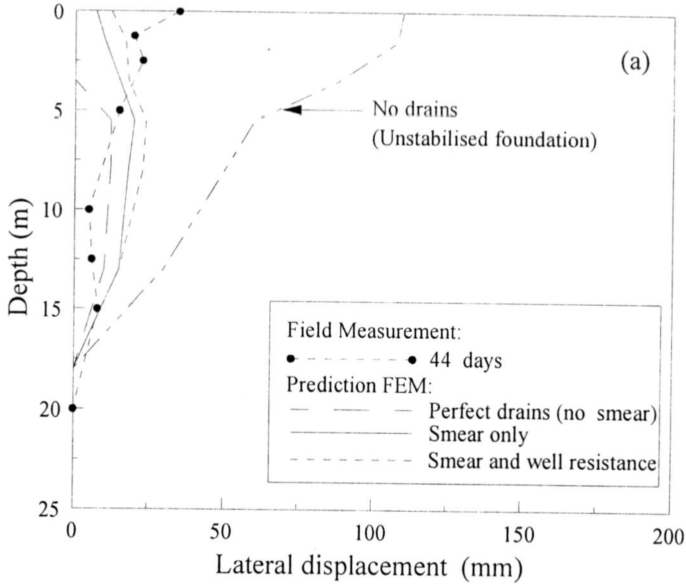

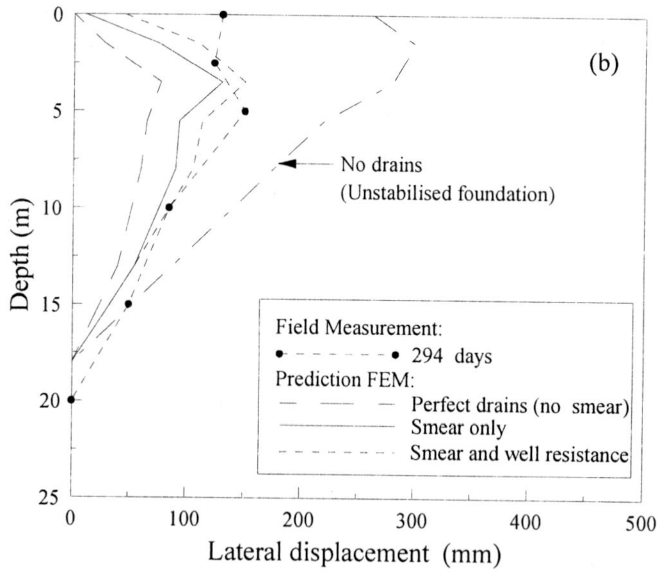

Fig. 41 *Lateral displacement profiles at 23 m away from centreline of embankments a) after 44 days, and b) after 294 days (Indraratna & Redana, 1999).*

REFERENCES

Aboshi, H., 1992. *Applied Ground Improvement Techniques, Workshop*. Southeast Asian Geotechnical Society (SEAGS), Asian Institute of Technology, Bangkok, Thailand, 40 p.

Aboshi, H. and Monden, H., 1963, *Determination of the horizontal coefficient of consolidation of an alluvial clay*. Proc. 4th Australian-New Zealand Conf. SMFE, pp. 159-164.

Akagi, T., 1977, *Effect of mandrel-driven sand drains on strength*. Proc. 9th Int. Conf. Soil Mech. and Found. Eng., Tokyo, Vol. 1, pp.3-6.

Akagi, T., 1981, *Effect of Mandrel-driven sand drains on soft clay*. Proc. 10th Int. Conf. Soil Mech. and Found. Eng., Stockholm, pp. 581-584.

American Society of Testing and Materials, 1993, *Annual Book of ASTM Standards: Soil and Rock; dimension Stone; Geosynthetics*. Vol. 04.08, Philadelphia, USA, 1469 p.

Asaoka, A., 1978, *Observation procedure of settlement prediction*. Soils and Foundations, Vol. 18, No. 4, pp. 87-101.

Asian Institute of Technology, 1995, *The Full Scale Field Test of Prefabricated Vertical Drains for The Second Bangkok International Airport (SBIA), Final Report*. Div. of Geotech. & Trans. Eng. AIT, Bangkok, Thailand, 259 p.

Atkinson, J., 1993, *An Introduction to The Mechanics of Soils and Foundations: Through Critical Soil Mechanics*. McGraw-Hill International Series in Civil Engineering, Berkshire, UK, 337 p.

Atkinson, J.H. and Bransby, P.L., 1978, *The Mechanics of Soils*. McGraw-Hill, London, 375 p.

Barron, R.A., 1948, *Consolidation of fine-grained soils by drain wells*. Transactions ASCE, Vol. 113, paper 2346, pp. 718-724.

Bergado, D.T., Asakami, H., Alfaro, M.C., and Balasubramaniam, A.S., 1991, Smear effects of vertical drains on soft Bangkok clay. *J. Geotech. Eng.*, ASCE, Vol. 117, No. 10, pp. 1509-1530.

Bergado, D.T., Chai, J.C., Miura, N., and Balasubramaniam, A.S., 1998, *PVD improvement of soft Bangkok clay with combined vacuum and reduced sand embankment preloading*. J. Geotech. Eng., Southeast Asian Geotech. *Soc.*, Vol. 29, No. 1, pp. 95-122.

Biot, M., 1941, General theory of three-dimensional consolidation. *J. Applied Physics*, Vol. 12, pp. 155-164.

Britto, A.M., and Gunn, M.J., 1987, *Critical State Soil Mechanics via Finite Elements*. Elis Horwood Limited, England, 488 p.

Burland, J.B., 1965, The yielding and dilation of clay. Correspondence, *Geotechnique*, No. 15, pp. 211-214.

Carillo, N., 1942, Simple Two and Three Dimensional Cases in The Theory of Consolidation of Soils. *J. Math. and Phys.*, Vol. 21, No. 1, pp. 1-5.

Chai, J.C., and Miura, N., 1997, *Method of modelling vertical drain-improved sub-soil. Proc. China-Japan Joint symposium on Recent Developments of Theory and Practice in Geotechnology,* Shanghai, China, pp. 1-8.

Chai, J.C., Miura, N., Sakajo, S., and Bergado, D., 1995, Behavior of vertical drain improved subsoil under embankment loading. *J. Soil and Foundations,* Japanese Geotechnical Society, Vol. 35, No. 4, pp. 49-61.

Christopher, B.R. and Holtz, R.D., 1985, *Geotextile Engineering Manual.* U.S. Federal Highway Administration, Report No. FHWA-TS-86/203, 1044 p.

Hansbo, S., 1960, Consolidation of clay with special reference to influence of vertical sand drains. *Swedish Geotechnical Institute, Proc.* No. 18, 160 p.

Hansbo, S., 1979, Consolidation of clay by band-shaped prefabricated drains. *Ground Eng.,* Vol. 12, No. 5, pp. 16-25.

Hansbo, S., 1981, Consolidation of fine-grained soils by prefabricated drains. *Proc. 10th Int. Conf. SMFE.,* Stockholm, Vol. 3, pp. 677-682.

Hansbo, S., 1987, Design aspects of vertical drains and lime column installation. *Proc. 9th Southeast Asian Geotechnical Conference,* Vol. 2, No. 8, pp. 1-12.

Hird, C.C., Pyrah, I.C., and Russell, D., 1992, Finite element modelling of vertical drains beneath embankments on soft ground. *Geotechnique,* Vol. 42, No. 3, pp. 499-511.

Hird, C.C., Pyrah, I.C., Russell, D., and Cinicioglu, F., 1995, Modelling the effect of vertical drains in two-dimensional finite element analyses of embankments on soft ground. *Can. Geotech. J.,* Vol. 32, pp. 795-807.

Holtz, R.D., Jamiolkowski, M., Lancellotta, R., and Pedroni, S., 1988, Behaviour of bent prefabricated vertical drains. *Proc. 12th ICSMFE,* Rio De Janeiro, Vol. 3, pp. 1657-1660.

Holtz, R.D., Jamiolkowski, M., Lancellotta, R., and Pedroni, S., 1991, Prefabricated vertical drains: design and performance, CIRIA ground engineering report: ground improvement. *Butterworth-Heinemann Ltd.,* UK, 131 p.

Indraratna, B., Balasubramaniam, A.S., and Balachandran, S., 1992, Performance of test embankment constructed to failure on soft marine clay. J. *Geotech. Eng., ASCE,* No. 118, pp. 12-33.

Indraratna, B., Balasubramaniam, A.S., and Ratnayake, P., 1994, Performance of embankment stabilized with vertical drains on soft clay. *J. Geotech. Eng., ASCE,* Vol. 120, No. 2, pp. 257-273.

Indraratna, B., Balasubramaniam, A.S., and Sivaneswaran, N., 1997, Analysis of settlement and lateral deformation of soft clay foundation beneath two full-scale embankments. *Int. J. for Numerical and Analytical Methods in Geomechanics,* Vol. 21, pp. 599-618.

Indraratna, B. and Redana, I.W., 1995, Large-scale, radial drainage consolidometer with central drain facility. *Australian Geomechanics,* Vol. 29, pp. 103-105.

Indraratna, B. and Redana, I.W., 1997, Plane strain modeling of smear effects associated with vertical drains. *J. Geotech. Eng.*, ASCE, Vol. 123, No. 5, pp. 474-478

Indraratna, B. and Redana, I.W., 1998, Laboratory determination of smear zone due to vertical drain installation. *J. Geotech. Eng.*, ASCE, Vol. 124, No. 2, pp. 180-184.

Indraratna, B. and Redana, I.W., 1999, Closure: Plane strain modelling of smear effects associated with vertical drains. *J. Geotechnical and Geoenvironmental Eng.* ASCE, Vol. 125, No. 1, pp. 96-99.

Jamiolkowski, M., and Lancellotta, R., 1981, Consolidation by vertical drains —uncertainties involved in prediction of settlement rates. *Panel Discussion, Proc. 10th Int. Conf. Soil Mech. and Found. Eng.*, Stockholm, Vol. 1, pp. 345-451.

Jamiolkowski, M., and Lancellotta, R., 1984, *Embankment on vertical drains-pore pressure during construction. Proc. Int. Conf. on Case Histories in Geotech. Enggr.*, St. Louis, Vol. 1, pp. 275-278.

Kjellman, 1948, Accelerating consolidation of fine grain soils by means of cardboard wicks. *Proc. 2nd ICSMFE,* Vol. 2, pp. 302-305.

Mesri, G., and Godlewski, P.M., 1977, Time and stress compressibility inter-relationship. *J. Geotech. Eng. Div.*, ASCE, Vol. 103, No. GT5, pp. 417-430.

Onoue, A., 1988b, Consolidation by vertical drains taking well resistance and smear into consideration. *J. Soils and Foundations,* Vol. 28, No. 4, pp. 165-174.

Onoue, A., Ting, N.H., Germaine, J.T. and Whitman, R.V., 1991, Permeability of disturbed zone around vertical drains. *Proc. ASCE Geotech. Enggr. Congress*, Colorado, pp. 879-890.

Pradhan, T.B.S., Imai, G., Murata, T., Kamon, M. and Suwa, S., 1993, Experiment study on the equivalent diameter of a prefabricated band-shaped drain. *Proc. 11th Southeast Asian Geotech. Conf.,* Vol. 1, pp. 391-396.

Ratnayake, A.M.P., 1991, Performance of test embankments with and without vertical drains at Muar Flats site, Malaysia. *MS thesis, Asian Institute of Technology,* Bangkok, Thailand.

Redana, I.W., 1999, Efficiency of vertical drains installed in soft clay with special reference to smear effect. *University of Wollongong,* Australia, PhD thesis.

Richart, F.E., Jr., 1959, A review of the theories for sand drains. *J. Soil Mech. and Foundation Enggr.* ASCE, Vol. 83, No. SM 3, pp. 1301(1-38).

Rixner, J.J., Kraemer, S.R., and Smith, A.D., 1986, Prefabricated Vertical Drains, Vol. I, II and III: Summary of Research Report-Final Report. Federal Highway Admin., *Report No. FHWA-RD-86/169,* Washington D.C., 433 p.

Roscoe, K.H., and Burland, J.B., 1968, *On the generalized stress strain behaviour of wet clay*. Engineering plasticity, Cambridge Univ. Press; Cambridge, U.K., pp. 535-609.

Schofield, A.N. and Wroth, C.P., 1968, *Critical State Soil Mechanics*. McGraw-Hill, London, 310 p.

Shogaki, T., Moro, H., Masaharu, M., Kaneko, M., Kogure, K. and Sudho, T., 1995, Effect of sample disturbance on consolidation parameters of anisotropic clays. *Proc. Int. Sym. on Compression and Consolidation of Clayey Soils*, Hiroshima, Japan, A. A. Balkema, Rotterdam, Vol. 1, pp. 561-566.

Takagi, S., 1957, Determination of the process of consolidation by sand drains under variable loading. *Proc. 4th Int. Conf. Soil Mech. and Found. Eng.*, London, Vol. 1, pp. 444-448.

Tavenas, F., Jean, P., and Leroueil, S., 1983b, The permeability of natural soft clays, Part 2: Permeability characteristics. *Can., Geotech.*, J., Vol. 20, pp. 645-660.

Terzaghi, K., 1925, *Erdbaumechanik auf bodenphysikalischer grundlage*. Franz Deuticke, Leipzig und Wein, 399 pp.

Ting, N.H., Onoue, A., Germaine, J.T., Whitman, R.V., and Ladd, C.C., 1990, *Effects of disturbance on soil consolidation with vertical drains*. Research Report R90-11, Dept. Civil Enggr., MIT, MA, USA, 245 p.

Yoshikuni, H., and Nakanodo, H., 1974, Consolidation of Fine-Grained Soils by Drain Wells with Finite Permeability. *Japan Soc. Soil Mech. and Found. Eng.* Vol. 14, No. 2, pp. 35-46.

Subject Index